ETHICS IN ENGINEERING

T0200700

ETHICS IN ENGINEERING

FIFTH EDITION

Qin Zhu

*Assistant Professor of Humanities,
Arts & Social Sciences
Colorado School of Mines*

Mike W. Martin

*Professor of Philosophy
Chapman University*

Roland Schinzinger

*Professor Emeritus of Electrical Engineering
University of California, Irvine*

ETHICS IN ENGINEERING

Published by McGraw Hill LLC, 1325 Avenue of the Americas, New York, NY 10121. Copyright © 2022 by McGraw Hill LLC. All rights reserved. Printed in the United States of America. No part of this publication may be reproduced or distributed in any form or by any means, or stored in a database or retrieval system, without the prior written consent of McGraw Hill LLC, including, but not limited to, in any network or other electronic storage or transmission, or broadcast for distance learning.

Some ancillaries, including electronic and print components, may not be available to customers outside the United States.

This book is printed on acid-free paper.

1 2 3 4 5 6 7 8 9 LCR 24 23 22 21

ISBN 978-1-265-25490-2
MHID 1-265-25490-7

Cover Image: *The Art Institute of Chicago, Mr. and Mrs. Martin A. Ryerson Collection*

ABOUT THE AUTHORS

Qin Zhu is Assistant Professor of Ethics and Engineering Education in the Department of Humanities, Arts & Social Sciences and an affiliate faculty member in the Department of Engineering, Design & Society and the Robotics Graduate Program at the Colorado School of Mines. Dr. Zhu is Editor for International Perspectives at the Online Ethics Center for Engineering and Science, Associate Editor for Engineering Studies, Program Chair of American Society for Engineering Education's Division of Engineering Ethics (2020–2021), Executive Committee Member of the Society for Ethics Across the Curriculum, and Treasurer of the Society for Philosophy and Technology. Dr. Zhu's research interests include the cultural foundations of engineering ethics, global engineering education, and ethics and policy of computing technologies and robotics.

Mike W. Martin and **Roland Schinzinger** participated as a philosopher-engineer team in the National Project on Philosophy and Engineering Ethics, 1978–1980. Since then they have coauthored articles, team-taught courses, and given presentations to audiences of engineers and philosophers. In 1992 they received the Award for Distinguished Literary Contributions Furthering Engineering Professionalism from The Institute of Electrical and Electronics Engineers, United States Activities Board.

Mike W. Martin received his B.S. (Phi Beta Kappa, Phi Kappa Phi) and M.A. from the University of Utah, and his Ph.D. from the University of California, Irvine, and he is currently professor of philosophy at Chapman University. In addition to publishing many articles, he is author, coauthor, or editor of eight books, including *Meaningful Work: Rethinking Professional Ethics* (2000) and *Everyday Morality* (3rd ed., 2001). He received the Arnold L. and Lois S. Graves Award for Teachers in the Humanities and two fellowships from the National Endowment for the Humanities.

Roland Schinzinger received his B.S., M.S., and Ph.D. in electrical engineering from the University of California at Berkeley. Born and raised in Japan where he had industrial experience with Bosch-Japan, Tsurumi Shipyard of Nippon Steel Tube Co., and Far Eastern Equipment Co., he worked in the United States as design/development engineer at Westinghouse Electric Corp. and taught at the University of Pittsburgh and at Robert College/Bosporus University (in Istanbul). He was a founding faculty member to the University of California at Irvine, from which he retired as Professor Emeritus of Electrical Engineering in 1993. He authored or co-authored *Conformal Mapping: Methods and Applications* (1991, 2003), *Emergencies in Water Delivery* (1979), and *Experiments in Electricity and Magnetism* (1961, accompanying a kit he designed for use in Turkey). His honors include the IEEE Centennial and Third Millennium medals, Fellow of IEEE, and Fellow of AAAS. He is a registered professional engineer.

FOR *ELLY B. ZHU,*
FOR *LILI GUAN*

QIN ZHU

FOR *SONIA* AND *NICOLE MARTIN,*
FOR *SHANNON SNOW MARTIN,*
AND IN MEMORY OF *THEODORE R. MARTIN*
AND *RUTH L. MARTIN.*

MIKE W. MARTIN

FOR *STEFAN, ANNELISE,* AND *BARBARA*
SCHINZINGER,
FOR *SHIRLEY BARROWS PRICE,*
AND IN MEMORY OF *MARY JANE HARRIS*
SCHINZINGER

ROLAND SCHINZINGER

BRIEF CONTENTS

CONTENTS

PREFACE

Technology has a pervasive and profound effect on the contemporary world, and engineers play a central role in all aspects of technological development. In order to hold paramount the safety, health, and welfare of the public, engineers must be morally committed and equipped to grapple with ethical dilemmas they confront.

Ethics in Engineering provides an introduction to the issues in engineering ethics. It places those issues within a philosophical framework, and it seeks to exhibit their social importance and intellectual challenge. The goal is to stimulate reasoning and to provide the conceptual tools necessary for responsible decision making.

In large measure we proceed by clarifying key concepts, sketching alternative views, and providing relevant case study material. Yet in places we argue for particular positions that in a subject like ethics can only be controversial. We do so because it better serves our goal of encouraging responsible reasoning than would a mere digest of others' views. We are confident that such reasoning is possible in ethics, and that, through engaged and tolerant dialogue, progress can be made in dealing with what at first seem irresolvable difficulties.

Sufficient material is provided for courses devoted to engineering ethics. Chapters of the book can also be used in modules within courses on engineering design, engineering law, engineering and society, safety, technology assessment, professional ethics, business management, and values and technology.

FIFTH EDITION

All chapters and appendixes in this edition have been updated with the most recent data, research findings, and teaching resources. Chapters 1, 3, 6, 8, 9, and 10 are either new or extensively reorganized and developed. This edition has extensively expanded the discussions on corporate social responsibility, research ethics in less traditional contexts (e.g., children, animals, cross-cultural, and online), environmental ethics in the Anthropocene, duty ethics, design ethics, life-cycle assessment, and the philosophy of technology. Particularly, one major strength added to this edition is the global and international dimension. Chapter 3 added one section on Confucian role ethics, which has not been well discussed in any other engineering ethics textbooks. Chapter 9 is completely new and it has incorporated a comprehensive review of four existing approaches to engineering ethics in the global context. Most recent studies in artificial intelligence and robotics have been added to Chapter 10. The pedagogical resources in Appendix A have been fully updated to 2021. Qin Zhu worked on revising this edition, with general approval from Mike W. Martin.

ACKNOWLEDGMENTS

Since the first edition of *Ethics in Engineering* appeared in 1983, many students, professors, and reviewers have provided helpful feedback or in other ways influenced our thinking. We wish to thank especially Robert J. Baum, Michael Davis, Dave Dorchester, Walter Elden, Charles B. Fleddermann, Albert Flores, Alastair S. Gunn, Charles E. (Ed) Harris, Joseph R. Herkert, Deborah G. Johnson, Ron Kline, Edwin T. Layton, Jerome Lederer, Heinz C. Luegenbiehl, Mark Maier, Nicole Marie Martin, Sonia Renée Martin, Carl Mitcham, Steve Nichols, Michael J. Rabins, Jimmy Smith, Michael S. Pritchard, Harold Sjursen, Carl M. Skooglund, John Stupar, Stephen H. Unger, Pennington Vann, P. Aarne Vesilind, Vivien Weil, Caroline Whitbeck, and Joseph Wujek.

And we thank the many authors and publishers who granted us permission to use copyrighted material as acknowledged in the notes, and also the professional societies who allowed us to print their codes of ethics in Appendix B.

Mike and Roland's deepest gratitude is to Shannon Snow Martin and to Shirley Barrows Price, whose love and insights have so deeply enriched our work and our lives. Qin's greatest gratitude is to Elly and Lili who have been unconditionally supportive while Qin was working on revising this edition. Qin also appreciates the longtime mentorship and encouragement from Carl Mitcham.

Qin Zhu
Mike W. Martin
Roland Schinzinger

CHAPTER

1

ETHICS
AND
PROFESSIONALISM

Engineers create products and processes to improve food production, shelter, energy, communication, transportation, health, and protection against natural calamities—and to enhance the convenience and beauty of our everyday lives. They make possible spectacular human triumphs once only dreamed of in myth and science fiction. Almost a century and a half ago in *From the Earth to the Moon,* Jules Verne imagined American space travelers being launched from Florida, circling the moon, and returning to splash down in the Pacific Ocean. In December 1968, three astronauts aboard an Apollo spacecraft did exactly that. Seven months later, on July 20, 1969, Neil Armstrong took the first human steps on the moon. This extraordinary event was shared with millions of earthbound people watching the live broadcast on television. Engineering had transformed our sense of connection with the cosmos and even fostered dreams of routine space travel for ordinary citizens.

Most technology, however, has double implications: As it creates benefits it raises new moral challenges. Just as exploration of the moon and planets stand as engineering triumphs, so the crashes of two new Boeing 737 Max series aircrafts (Lion Air Flight 610 in 2018 and Ethiopian Airlines Flight 302 in 2019) were tragedies that could have been prevented, had urgent warnings voiced by experienced engineers been heeded. We will examine these and other cases of human error, for in considering ethics and engineering alike we can learn from seeing how things go wrong. Technological risks, however, should not overshadow technological benefits, and ethics involves appreciating the many positive dimensions of engineering that so deeply enrich our lives.

This chapter introduces central themes, defines engineering ethics, and states the goals in studying it. Next, the importance of accepting and sharing moral responsibility is underscored. Finally, we attend to the corporate setting in which today most engineering takes place and the communal setting in which an increasing number of engineers are working, emphasizing the need for reflecting on the broader social and ethical implications of engineering work.

1.1 SCOPE OF ENGINEERING ETHICS

1.1.1 Overview of Themes

In this book we explore a wide variety of topics and issues, but seven themes recur. Taken together, the themes constitute a normative (value) perspective on engineering and on engineering ethics.

1. Engineering projects are social experiments that generate both new possibilities and risks, and engineers share responsibility for creating benefits, preventing harm, and pointing out dangers.
2. Moral values permeate all aspects of technological development, and hence ethics and excellence in engineering go together.
3. Personal meaning and commitments matter in engineering ethics, along with principles of responsibility that are stated in codes of ethics and are incumbent on all engineers.
4. Promoting responsible conduct and advocating good works is even more important than punishing wrongdoing.
5. Ethical dilemmas arise in engineering, as elsewhere, because moral values are myriad and can conflict.
6. Engineering ethics should explore both micro and macro issues, which are often connected and more ethical issues are arising from the global context of engineering.
7. Technological development especially in the age of artificial intelligence warrants cautious optimism—optimism, with caution.

Let us briefly introduce and illustrate each of these themes.

(1) ENGINEERING AS SOCIAL EXPERIMENTATION. When the space shuttle *Columbia* exploded on February 1, 2003, killing the seven astronauts on board, some people feared the cause was a terrorist attack, given the post–September 11 concerns about terrorism. The working hypothesis quickly emerged, however, that the cause was a piece of insulating foam from the external fuel tank that struck the left wing 82 seconds after launch. The panels on the leading edge of the wing were composed of reinforced carbon carbon, a remarkable material that protected it from 3000-degree temperatures caused by air friction upon reentry from space into the earth's atmosphere. Even a small gap allowed superheated gases to enter the wing, melt the wiring, and spray molten metal throughout the wing structure.

Investigators stated they were interested in far more than pinpointing the immediate cause of the disaster.[1] Several previous incidents involved insulating material breaking off from the fuel tank. Why were these occurrences not scrutinized more carefully? And why were so many additional hazards emerging, such as faulty "bolt catchers," which were chambers designed to capture bolts attaching the solid rocket boosters to the external fuel tank after their detonated-release? Had the safety culture at NASA eroded, contrary to assumptions that it had improved since the 1986 *Challenger* disaster, such that the independent judgment of engineers was not being heeded? Even during *Columbia*'s last trip, when crumbling shielding hit fragile tiles covering the craft's wings, some knowledgeable engineers were rebuffed when they requested that the impacts be simulated and observed without delay. Had the necessary time, money, personnel, and procedures for ensuring safety been shortchanged?

Very often technological development is double-edged, Janus-faced, morally ambiguous: As engineering projects create new possibilities they also generate new dangers. To emphasize the benefit-risk aspects in engineering, in chapter 4 we introduce a model of engineering as social experiments—experiments on a societal scale. This model underscores the need for engineers to accept and share responsibility for their work, exercise due care, imaginatively foresee hazards, conscientiously monitor their projects when possible, and alert others of dangers to permit them to give informed consent to risks. In highlighting risk, the model also accents the good made possible through engineering discoveries and achievements. And it underscores the need for preventive ethics: ethical reflection and action aimed at preventing moral harm and avoidable ethical dilemmas.

(2) ETHICS AND EXCELLENCE: MORAL VALUES ARE EMBEDDED IN ENGINEERING. Moral values are embedded in even the simplest engineering projects, not "tacked on" as external burdens. Consider the following assignment given to students in a freshman course at Harvey Mudd College:

> Design a chicken coop that would increase egg and chicken production, using materials that were readily available and maintainable by local workers [at a Mayan cooperative in Guatemala]. The end users were to be the women of a weaving cooperative who wanted to increase the protein in their children's diet in ways that are consistent with their traditional diet, while not appreciably distracting from their weaving.[2]

The task proved more complex than it at first appeared. The students had to identify plausible building materials, decide between cages or one open area, and design structures for strength and endurance. They had to create safe access for the villagers, including ample head and shoulder room at entrances and a safe floor for bare feet. They had to ensure humane conditions for the chickens, including adequate space and ventilation, comfort during climate changes, convenient delivery of food and water, and protection from local predators that could dig under fences. They also had to improve cleaning procedures to minimize damage to the environment while recycling chicken droppings as fertilizers. The primary goal, however, was to double current chicken and egg production. A number of design concepts were explored before a variation of a fenced-in concept proved preferable to a set of cages. In 1997 four students and their advisor, supported by a humanitarian aid group named Xela-Aid, traveled to San Martin Chiquito, Guatemala, and worked with villagers in building the chicken coop and additional structures such as a weaving building.

Moral values are embedded at several junctures in engineering projects, including: the basic standards of safety and efficiency, the social, cultural, and environmental contexts of the community, the character of engineers who spearhead technological progress, and the very idea of engineering as a profession that combines advanced skill with commitment to the public good. In engineering, as in other professions, excellence and ethics go together—for the most part and in the long run. In general, ethics involves much more than problems and punishment, duties and dilemmas.[3] Ethics involves the full range of moral values to which we aspire in guiding our endeavors and in structuring our relationships and communities. This emphasis on moral aspiration was identified by the ancient Greeks, whose word *arete* translates into English as either "excellence" or as "virtue."

(3) PERSONAL COMMITMENT AND MEANING. A team of engineers are redesigning an artificial lung marketed by their company. They are working in a highly competitive market, with long hours and high stress. The engineers have little or no contact with the firm's customers, and they are focused on technical problems, not people. It occurs to the project engineer to invite recipients of artificial lungs and their families to the plant to talk about how their lives were affected by the artificial lung. The change is immediate and striking: "When families began to bring in their children who for the first time could breathe freely, relax, learn, and enjoy life because of the firm's product, it came as a revelation. The workers were energized by concrete evidence that their efforts really did improve people's lives, and the morale of the workplace was given a great lift."[4]

Engineers' motives and commitments are as many and varied as those of all human beings. The desire for meaningful work, concern to make a living, care for other human beings, and the need to maintain self-respect all combine to motivate excellence in engineering. For the most part, they are mutually reinforcing in advancing a sense of personal responsibility for one's work. As we emphasize repeatedly, engineering is about people as well as products, and the people include engineers who stand in moral (as well as monetary) relationships with customers, colleagues, employers, and the general public.

All engineers are required to meet the responsibilities stated in their code of ethics. These requirements set a minimum, albeit a high standard of excellence. The personal commitments of individual engineers need to be aimed at and integrated with these shared responsibilities. Yet some responsibilities and sources of meaning are highly personal, and cannot be incumbent on every engineer. They include commitments concerning religion, the environment, military work, family, and personal ambitions. When we speak of "personal commitments" we have in mind both commitments to shared responsibilities and to these more individual commitments as they affect professional endeavors.

Engineers' motives and commitments are critical for them to actually devote themselves to ethical actions. Based on the findings in moral psychology, it is very likely that an engineer knows what the right action is but feels hesitant to do it as the engineer lacks motivation.[5] Engineering ethics education programs in the United States tend to teach students to separate their personal commitments and meaning from *professional ideals*. Arguably, the traditional approach to engineering ethics education often assumes that engineers are isolated, rational, and autonomous human beings and engineering as a profession needs to be *depersonalized*.[6] Therefore, personal traits such as emotion, virtues, and commitments are sometimes invisible in engineering education or are considered irrelevant.[7] Philosopher Michael Davis argues that emotion is quite normal and sometimes can be justified and necessary in the everyday practice of engineers. For instance, an engineer can feel angry when their company generates chemical pollutants to the community and the company leadership has kept overlooking this engineer's remonstration. The emotional state of this engineer in fact well demonstrates their commitment to the safety, health, and welfare of the public.

(4) PROMOTING RESPONSIBLE CONDUCT, PREVENTING WRONGDOING, AND ADVOCATING GOOD WORKS. Beginning in 2001, a wave of corporate scandals shook Americans' confidence in corporations.* In that year, Enron became the largest bankruptcy in U.S. history, erasing about $60 billion in shareholder value.[8] The following year the scandal-ridden WorldCom bankruptcy set another new record. Arthur Andersen, a large and respected accounting firm charged with checking the books of Enron and other corporations, was charged with complicity and was forced to dissolve. We return to these events later in this chapter.

Compliance issues are about making sure that individuals comply to professional standards and avoid wrongdoing. Procedures are needed in all corporations to deter fraud, theft, bribery, incompetence, and a host of other forms of outright immorality. Equally essential are reasonable laws and government regulation, including penalties for reckless and negligent conduct. We should examine the pressures that sometimes lead engineers to cooperate in wrongdoing, rather than reporting wrongdoing to proper authorities.[9]

Having said this, an important part of engineering ethics is preventing wrongdoing in the first place. There is a need for what we have referred to as

---

*The term "corporation" will be used freely to include companies that may not be incorporated. In its strict sense, a corporation is a legal construct that enables investors to pool their financial resources for carrying out large, costly, and often risky projects without the burden of individual responsibility for the outcome, physically and financially, beyond possible lack of return on investments. A corporation is treated as if it were an individual itself, taking the blame for the real individual investors. Such an arrangement, so common in our modern economy, raises many questions of accountability and responsibility, particularly shared responsibility.

"preventive ethics": ethical reflection and action aimed at preventing moral harm and unnecessary ethical problems. The main emphasis in ethics should be supporting responsible conduct. In fact, the vast majority of engineers are morally committed. So too are most corporations. Reinforcing the connection between ethics and excellence, individuals and corporations should primarily be "value-driven," rather than simply preoccupied with "compliance-based" procedures, to invoke terms used in management theory. More recently, Charles Harris and his colleagues have suggested that engineering ethics education needs to pay closer attention to the more positive aspects or the "aspirational ethics" of engineering.[10] Most articles in engineering codes of ethics often focus on preventative ethics and they do not provide much clear guidance on how engineering work can promote human well-being. Practicing aspirational ethics often requires engineers to go beyond what is obligatory for them. Nevertheless, we argue that advocating aspirational ethics is beneficial for building positive public images of engineering, cultivating ethical culture of the engineering profession, enhancing the mutual trust between engineers and the public, and generating positive impacts of technological change.

(5) MYRIAD MORAL REASONS GENERATE ETHICAL DILEMMAS. A chemical engineer working in the environmental division of a computer manufacturing firm learns that their company might be discharging unlawful amounts of lead and arsenic into the city sewer.[11] The city processes the sludge into a fertilizer used by local farmers. To ensure safety, it imposes restrictive laws on the discharge of lead and arsenic. Preliminary investigations convince the engineer that the company should implement stronger pollution controls, but their manager insists the cost of doing so is prohibitive and that technically the company is in compliance with the law. The engineer is responsible for doing what promotes the success of their company, but they also have responsibilities to the local community that might be harmed by the effluent. In addition, they have responsibilities to their family, and rights to pursue their career. What should they do?

Ethical dilemmas, or *moral dilemmas,* are situations in which moral reasons come into conflict, or in which the applications of moral values are problematic, and it is not immediately obvious what should be done. The moral reasons might be obligations, rights, goods, ideals, or other moral considerations. In engineering as elsewhere, moral values are myriad and they can come into conflict, requiring good judgment about how to reconcile and integrate them. Beginning in chapter 2 we discuss resources for understanding and resolving ethical dilemmas, including codes of ethics and ethical theories. We emphasize that ethical dilemmas need not be a sign that something has gone wrong; instead, they indicate the presence of moral complexity. That complexity would exist even if we could eliminate all preventable problems, such as the corporate scandals.

(6) MICRO AND MACRO ISSUES. *Micro issues* consider individuals and internal relations of the engineering profession. *Macro issues* concern much broader issues, such as the directions in technological development, the laws that should or should not be passed, and the collective responsibilities of groups such as engineering professional societies and consumer groups.[12] Both micro and macro issues are important in engineering ethics, and often they are interwoven.[13]

As an illustration, consider debates about sport utility vehicles (SUVs). Micro issues arose, for example, concerning the Ford Explorer and also Bridgestone/Firestone, who provided tires for the Explorer. During the late 1990s, reports began to multiply about the tread on Explorer tires separating from the rest of the tire, leading to blowouts and rollovers. By 2002, estimates were that 300 people had died and another thousand were injured and more recent estimates place the numbers much higher since then.[14] Ford and Bridgestone/Firestone blamed each other for the problem, leading to the breakup of a century-old business partnership. As it turned out, the hazard had multiple sources. Bridgestone/Firestone used a flawed tire design and poor quality control at a major manufacturing facility. Ford chose tires with a poor safety margin, relied on drivers to maintain proper inflation within a very narrow range, and then dragged its feet in admitting the problem and recalling dangerous tires.

In contrast, macro issues center on charges that SUVs are among the most harmful vehicles on the road, even *the* most harmful, given their numbers. The problems are many: instability because of their height that leads to rollovers, far greater "kill rate" of other drivers during accidents, reducing the vision of drivers in shorter cars behind them on freeways, blinding other drivers' vision because of high-set lights, gas-guzzling, and excessively polluting. Keith Bradsher estimates that SUVs are causing about 3,000 deaths in excess of what cars would have caused: "Roughly 1,000 extra deaths occur each year in SUVs that roll over, compared to the expected rollover death rate if these motorists had been driving cars. About 1,000 more people die each year in cars hit by SUVs than would occur if the cars had been hit by other cars. And up to 1,000 additional people succumb each year to respiratory problems because of the extra smog caused by SUVs."[15] Bradsher believes these numbers will continue to increase as more SUVs are added to the road each year and as older vehicles are resold to younger and more dangerous drivers.

Should "the SUV issue" be examined within engineering as a whole, or at least by representative professional and technical societies? If so, what should be done? Or, in a democratic and capitalistic society, should engineers play a role only as individuals, but not as organized groups? Should engineers remain uninvolved, leaving the issue entirely to consumer groups and lawmakers? Even larger macro issues surround public transportation issues, in relation to all automobiles and SUVs, as we look to the future with a dramatically increasing population and a shrinking of our traditional resources.

(7) CAUTIOUS OPTIMISM ABOUT TECHNOLOGY. The most general macro issues pertain to technology in its entirety, including its overall promise and perils, an issue taken up in chapter 10. Pessimists view advanced technology as ominous and often out of our control. They point to pollution, depletion of natural resources, fears of biological and chemical weapons, and the lingering threat of robotics taking human jobs. Optimists highlight how technology profoundly improves all our lives. Each of us benefits in some ways from the top 20 engineering achievements of the twentieth century, as identified by the National Academy of Engineering: electrification, automobiles, airplanes, water supply and distribution, electronics, radio and television, agricultural mechanization, computers, telephones, air-conditioning and refrigeration, highways, spacecrafts, Internet, imaging technologies in medicine and elsewhere, household appliances, health technologies, petrochemical technologies, laser and fiber optics, nuclear technologies, and high-performance materials.[16]

As authors, we are cautiously optimistic about technology. Nothing is more central to human progress than sound technology, and no aspect of creative human achievement is less appreciated by the public than engineers' ingenuity. At the same time, consistent with the social experimentation model, the exuberant confidence and hope—so essential to technological progress—needs to be accompanied by sober realism about dangers.

Such a cautiously optimistic attitude is even more critical in the age of AI. Given the huge potential of AI-enabled technologies in improving human well-being and production efficiency, it is unlikely that humans will completely terminate or abandon the development of these technologies. As philosopher Peter-Paul Verbeek has suggested, we as humans need to learn how to morally accompany technology. We are required to thoroughly engage with designers and engineers and "look for points of application for moral reflection and anticipate the social impact of technology-in-design."[17]

1.1.2 What Is Engineering Ethics?

With this overview of themes and sampling of issues in mind, we can now define engineering ethics. The word *ethics* has several meanings. In the sense used in the title of this book, ethics is synonymous with morality. It refers to moral values that are sound, actions that are morally required (right) or morally permissible (all right), policies and laws that are desirable. Accordingly, *engineering ethics consists of the responsibilities and rights that ought to be endorsed by those engaged in engineering, and also of desirable ideals and personal commitments in engineering.*

In a second sense, ethics is the study of morality; it is an inquiry into ethics in the first sense. It studies which actions, goals, principles, policies, and laws are morally justified. Using this meaning, which also names the field of study of this book, *engineering ethics is the study of the decisions, policies, and values that are morally desirable in engineering practice and research.*

These two senses are *normative:* They refer to justified values and choices, to things that are desirable (not merely desired). Normative senses differ from *descriptive* senses of ethics. In one descriptive sense, we speak of Henry Ford's ethics or the ethics of American engineers, referring thereby to what specific individuals or groups believe and how they act, without implying that their beliefs and actions are justified. In another descriptive sense, social scientists study ethics when they describe and explain what people believe and how they act; they conduct opinion polls, observe behavior, examine documents written by professional societies, and uncover the social forces shaping engineering ethics.

As it turns out, morality is not easy to define. Of course, we can all give examples of moral values, but the moment we try to provide a comprehensive definition of morality we are drawn into at least rudimentary ethical theory—a normative theory about morality. For example, if we say that morality consists in promoting the most good, we are invoking an ethical theory called utilitarianism. If we say that morality is about human rights, we invoke rights ethics. And if we say that morality is essentially about good character, we might be invoking virtue ethics.

These and other ethical theories are discussed in chapter 3. For now, let us simply say that morality concerns respect for persons, both others and ourselves. It involves being fair and just, meeting obligations and respecting rights, and not causing unnecessary harm by dishonesty and cruelty or by hubris. In addition, it involves ideals of character, such as integrity, gratitude, and willingness to help people in severe distress.[18] And it implies minimizing suffering to animals and damage to the environment.

1.1.3 Why Study Engineering Ethics?

Engineering ethics should be studied because it is *important,* both in contributing to safe and useful technological products and in giving meaning to engineers' endeavors. It is also *complex,* in ways that call for serious reflection throughout a career, beginning with earning a degree. But beyond these general observations, what specific aims should guide the study of engineering ethics?

In our view, the direct aim is to increase one's ability to deal effectively with moral complexity in engineering. Accordingly, the study of engineering ethics strengthens one's ability to reason clearly and carefully about moral questions. To invoke a term widely used in ethics, the unifying goal is to increase moral autonomy.

Autonomy means "self-determining" or "independent." But not just any kind of independent reflection about ethics amounts to moral autonomy. Moral autonomy can be viewed as the skill and habit of thinking rationally about ethical issues on the basis of moral concern. This foundation of moral concern, or general responsiveness to moral values, derives primarily from the training we receive as children in being sensitive to the needs and rights of others, as well as of ourselves. When such training is absent, as it often is with seriously abused children, the tragic result can be an adult sociopath who lacks any sense of moral right and wrong.[19] Sociopaths (or psychopaths) are not morally autonomous, regardless of how "independent" their intellectual reasoning about ethics might be.

Improving the ability to reflect carefully on moral issues can be accomplished by improving various practical skills that will help produce autonomous thought about moral issues. As related to engineering ethics, these skills include the following:

1. Moral awareness: Proficiency in recognizing moral problems and issues in engineering.
2. Cogent moral reasoning: Comprehending, clarifying, and assessing arguments on opposing sides of moral issues.
3. Moral coherence: Forming consistent and comprehensive viewpoints based upon a consideration of relevant facts.
4. Moral imagination: Discerning alternative responses to moral issues and receptivity to creative solutions for practical difficulties.
5. Moral communication: Precision in the use of a common ethical language, a skill needed to express and support one's moral views adequately to others.

These are the direct goals in college courses. They center on cognitive skills—skills of the intellect in thinking clearly and cogently. But it is possible to have these skills and yet not act in morally responsible ways. Should we therefore add to our list of goals the following goals that specify aspects of moral commitment and responsible conduct?

6. Moral reasonableness: The willingness and ability to be morally reasonable.
7. Respect for persons: Genuine concern for the well-being of others as well as oneself.
8. Tolerance of diversity: Within a broad range, respect for ethnic and religious differences, and acceptance of reasonable differences in moral perspectives.
9. Moral hope: Enriched appreciation of the possibilities of using rational dialogue in resolving moral conflicts.
10. Integrity: Maintaining moral integrity, and integrating one's professional life and personal convictions.
11. Moral emotions: Social emotions (feelings or intuitions) that "are linked to the interests or welfare either of society as a whole or at least of persons other than the judge or agent."[20]

In our view we should add these goals to the study of engineering ethics, for there would be little moral point to studying ethics without the expectation that doing so contributes to the goals. At the same time, these goals are often best pursued implicitly and indirectly, more in how material is studied and taught than in preaching and testing. A foundation of moral concern must be presupposed, as well as evoked and expanded, in studying ethics at the college level.

DISCUSSION QUESTIONS

1. Identify the moral values, issues, and dilemmas, if any, involved in the following cases, and explain why you consider them *moral* values and dilemmas.

a. An engineer notified his firm that for a relatively minor cost, a flashlight could be made to last several years longer by using a more reliable bulb. The firm decides that it would be in its interests not to use the new bulb, both to keep costs lower and to have the added advantage of "built-in obsolescence" so that consumers would need to purchase new flashlights more often.

b. A linear electron accelerator for therapeutic use was built as a dual-mode system that could either produce X-rays or electron beams. It had been in successful use for some time, but every now and then some patients received high overdoses, resulting in painful aftereffects and several deaths. One patient on a repeat visit experienced great pain, but the remotely located operator was unaware of any problem because of lack of communication between them: the intercom was broken and the video monitor had been unplugged. There also was no way for the patient to exit the examination chamber without help from the outside, and hence the hospital was partly at fault. Upon cursory examination of the machine, the manufacturer insisted that the computerized and automatic control system could not possibly have malfunctioned and that no one should spread unproven and potentially libelous information about the design. It was the painstaking, day-and-night effort of the hospital's physicist that finally traced the problem to a software error introduced by the manufacturer's efforts to make the machine more user-friendly.[21]

2. Regarding the artificial lung example, comment on why you think simple human contact made such a large difference. What does it say about what motivated the engineers, both before and after the encounter? Is the case too unique to permit generalizations to other engineering products?

3. Should SUV problems at the macro level be of concern to engineers as a group and their professional societies? And should individual automotive engineers, in their daily work, be concerned about the general social and environmental impacts of SUVs?

4. It is not easy to define *morality* in a simple way, but it does not follow that morality is a hopelessly vague notion. For a long time, philosophers thought that an adequate definition of any idea would specify a set of logically necessary and sufficient conditions for applying the idea. For example, each of the following features is logically necessary for a triangle, and together they are sufficient: a plane figure, having three straight lines, closed to form three angles. The philosopher Ludwig Wittgenstein (1889–1951), however, argued that most ordinary (nontechnical) ideas cannot be neatly defined in this way. Instead, there are often only "family resemblances" among the things to which words are applied, analogous to the partly overlapping similarities among members of a family—similar eye color, shape of nose, body build, temperament, and so forth.[22] Thus, a book might be hardback, paperback, or electronic; printed or handwritten; in English or German; etc. Can you specify necessary and sufficient conditions for the following ideas: chairs, buildings, energy, safety, morality?

5. Unfortunately, the mention of ethics sometimes evokes groans, rather than engagement, because it brings to mind onerous constraints and unpleasant disagreements. Worse, it evokes images of self-righteousness, hypocrisy, and excessively punitive attitudes of blame and punishment—attitudes that are themselves subject to moral critique. Think of a recent event that led to a public outcry. With regard to the event, discuss the difference between being morally reasonable and being "moralistic" in a pejorative sense. In doing so, consider such things as breadth of vision, tolerance, sensitivity to context, and commitment.

1.2 ACCEPTING AND SHARING RESPONSIBILITY

Before he became president of the United States, Herbert Hoover was a mining engineer. In his memoirs he reflects on engineering in general:

> It is a great profession. There is the fascination of watching a figment of the imagination emerge through the aid of science to a plan on paper. Then it moves to realization in stone or metal or energy. Then it brings jobs and homes to men. Then it elevates the standards of living and adds to the comforts of life. That is the engineer's high privilege.
>
> The great liability of the engineer compared to men of other professions is that his works are out in the open where all can see them. His acts, step by step, are in hard substance. He cannot bury his mistakes in the grave like the doctors. He cannot argue them into thin air or blame the judge like the lawyers. He cannot, like the architects, cover his failures with trees and vines. He cannot, like the politicians, screen his shortcomings by blaming his opponents and hope that the people will forget. The engineer simply cannot deny that he did it. If his works do not work, he is damned.[23]

Hoover is reflecting on an era when engineering was dominated, at least in outlook, by the independent consultant, rather than by the corporate engineer. In his day, it was easier for individual engineers to work with a sense of personal responsibility for an entire project. When a bridge fell or a ship sank, the engineers responsible could be more easily identified. This made it easier to endorse Hoover's vision of individualism in regard both to creativity and personal accountability within engineering.

Today, the products of engineering are "out in the open" as much as they were in Hoover's time. In fact, mass communication ensures that major mistakes receive even closer public scrutiny. And there are more engineers than ever. Yet despite their greater numbers, engineers of today are less visible to the public than were those of Hoover's era. Technological progress is taken for granted as being the norm, and technological failure is blamed on corporations, if not government. And in the public's eye, the representative of any corporation is its top manager, who is often far removed from the daily creative endeavors of the company's engineers. This "invisibility" can make it difficult for engineers to retain a sense of mutual understanding with and accountability to the public. Nevertheless, individuals who accept responsibility for their work can make an enormous difference, as the following case illustrates.

1.2.1 Saving Citicorp Tower

Structural engineer Bill LeMessurier (pronounced "LeMeasure") and architect Hugh Stubbins faced a challenge when they worked on the plans for New York's fifth highest skyscraper. St. Peter's Lutheran Church owned and occupied a corner of the lot designated in its entirety as the site for the new structure. An agreement was reached: The bank tower would rise from nine-story-high stilts positioned at the center of each side of the tower, and the church would be offered a brand new St. Peter's standing freely underneath one of the cantilevered corners. Completed in 1977, the Citicorp Center appears as shown in figure 1-1. The new church building is seen below the lower left corner of the raised tower.

LeMessurier's structure departed from the usual in that the massive stilts are not situated at the corners of the building, and half of its gravity load as well all of its wind load is brought down an imaginatively designed trussed frame, which incorporates wind braces, on the outside of the tower.[24] In addition, LeMessurier installed a tuned mass damper, the first of its kind in a tall building, to keep the building from swaying in the wind.

Questions asked by an engineering student a year after the tower's completion prompted LeMessurier to review certain structural aspects of the tower and pose some questions of his own.[25] For instance, could the structure withstand certain loads due to strong quartering winds? In such cases, two sides of the building receive the oblique force of the wind, and the resultant force is 40 percent larger than when the wind hits only one face of the structure straight on. The only requirement stated in the building code specified adequacy to withstand certain perpendicular wind loads, and that was the basis for the design of the wind braces. But there was no need to worry since the braces as designed could handle such an excess load without difficulty, provided the welds were of the expected high quality.

Nevertheless, the student's questions prompted LeMessurier to place a call from his Cambridge, Massachusetts, office to his New York office, to ask Stanley Goldstein, his engineer in charge of the tower erection, how the welded joints of the bracing structure had worked out. How difficult was the job? How good was

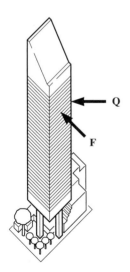

FIGURE 1-1

Axonometric view of Citicorp tower with the church in the lower left-hand corner. Wind loads: F, frontal and Q, Quartering. (Adaptation of an axonometric drawing by Henry Dong, Anspach Grossman Portugal, Inc., in Buildings Type Study 492, *Architectural Record,* Mid-August Special Issue, 1976, p. 66.)

the workmanship? To his dismay, Goldstein answered, "Oh, didn't you know? [The joints] were never welded at all because Bethlehem Steel came to us and said they didn't think we needed to do it." The New York office, as it was allowed to do, had approved the proposal that the joints be bolted instead. But again the diagonal winds had not been taken into account.

At first, LeMessurier was not too concerned; after all, the tuned mass damper would still take care of the sway. So he turned to his consultant on the behavior of high buildings in wind, Alan Davenport at the University of Western Ontario. On reviewing the results of his earlier wind tunnel tests on a scaled-down Citicorp Center, Davenport reported that a diagonal wind load would exceed the perpendicular wind load by much more than the 40 percent increase in stress predicted by an idealized mathematical model. Winds sufficient to cause failure of certain critical bolted joints—and therefore of the building—could occur in New York every 16 years. Fortunately, those braces that required strengthening were accessible, but the job would be disruptive and expensive, exceeding the insurance LeMessurier carried.

LeMessurier faced an ethical dilemma involving a conflict between his responsibilities to ensure the safety of his building for the sake of people who use it, his responsibilities to various financial constituencies, and his self-interest, which might be served by remaining silent. What to do? He retreated to his summerhouse on an island on Sebago Lake in Maine. There, in the quiet, he worked once more through all the design and wind tunnel numbers. Suddenly he was struck with "an almost giddy sense of power," as he realized that only he could prevent an eventual disaster by taking the initiative.

Having made a decision, he acted quickly. He and Stubbins met with their insurers, lawyers, the bank management, and the city building department to describe the problem. A retrofit plan was agreed upon: The wind braces would be strengthened at critical locations "by welding two-inch-thick steel plates over each of more than 200 bolted joints." Journalists, at first curious about the many lawyers converging on the various offices, disappeared when New York's major newspapers were shut down by a strike. The lawyers sought the advice of Leslie Robertson, a structural engineer with experience in disaster management. He alerted the mayor's Office of Emergency Management and the Red Cross so the surroundings of the building could be evacuated in case of a high wind alert. He also arranged for a network of strain gages to be attached to the structure at strategic points. This instrumentation allowed actual strains experienced by the steel to be monitored at a remote location. LeMessurier insisted on the installation of an emergency generator to assure uninterrupted availability of the damper.

When hurricane Ella appeared off the coast, there was some cause for worry, but work on the critical joints had almost been completed. Eventually the hurricane veered off and evacuation was not required. Even so, the retrofit and the tuned mass damper had been readied to withstand as much as a 200-year storm.

The parties were able to settle out of court, with Stubbins held blameless; LeMessurier and his joint-venture partners were charged the $2 million his insurance agreed to pay. The total repair bill had amounted to over $12.5 million. Not only did LeMessurier save lives and preserve his integrity, but his reputation was enhanced rather than tarnished by the episode.

1.2.2 Meanings of "Responsibility"

If we say that LeMessurier was responsible, as a person and as an engineer, we might mean several things: he met his responsibilities (obligations); he was responsible (accountable) for doing so; he acted responsibly (conscientiously); and he is admirable (praiseworthy). Let us clarify these and related senses of "responsibility," beginning with obligations—the core idea around which all the other senses revolve.[26]

1. *Obligations.* Responsibilities are *obligations*—types of actions that are morally mandatory. Some obligations are incumbent on each of us, such as to be honest, fair, and decent. Other obligations are *role responsibilities,* acquired when we take on special roles such as parents, employees, or professionals. Thus, a safety engineer might have responsibilities for making regular inspections at a building site, or an operations engineer might have responsibilities for identifying potential benefits and risks of one system as compared to another.

2. *Accountable.* Being responsible means being accountable. This means having the general capacities for moral agency, including the capacity to understand and act on moral reasons. It also means being answerable for meeting particular obligations, that is, liable to be held to account by other people in general or by specific individuals in positions of authority. We can be called upon to explain why we acted as we did, perhaps providing a justification or perhaps offering reasonable excuses. We also hold ourselves accountable for meeting our obligations, sometimes responding with emotions of self-respect and pride, other times responding with guilt for harming others and shame for falling short of our ideals.

    Wrongdoing takes two primary forms: voluntary wrongdoing and negligence. Voluntary actions occur when we knew what we were doing was wrong and we were not coerced. Some voluntary wrongdoing is recklessness, that is, flagrant disregard of known risks and responsibilities. Other voluntary wrongdoing is due to weakness of will, whereby we give in to temptation or fail to try hard enough. In contrast, negligence occurs when we unintentionally fail to exercise due care in meeting responsibilities. We might not have known what we were doing, but we should have. Shoddy engineering, due to sheer incompetence, usually falls into this category.

3. *Conscientious.* Morally admirable engineers like LeMessurier accept their obligations and are conscientious in meeting them. They diligently try to do the right thing, and they largely succeed in doing so, even under difficult circumstances. Of course, no one is perfect, and it is possible to be conscientious in some areas of life, such as one's work, and less conscientious in other areas, such as raising a child.

4. Blameworthy/Praiseworthy. In contexts where it is clear that accountability for wrongdoing is at issue, "responsible" becomes a synonym for *blameworthy.* In contexts where it is clear that right conduct is at issue, "responsible" is a synonym for *praiseworthy.* Thus, the question "Who is responsible for designing the antenna tower?" might be used to ask who is blameworthy for its collapse or who deserves credit for its success in withstanding a severe storm.

The preceding meanings all concerned *moral* responsibility. Moral responsibility overlaps with, but is distinguishable from, causal, job, and legal responsibility. *Causal responsibility* consists simply in being a cause of some event. (A young child playing with matches causes a house to burn down, but the adult who left the child with the matches is morally responsible.) *Job responsibility* consists of one's assigned tasks at the place of employment. And *legal responsibility* is whatever the law requires—including legal obligations and accountability for meeting them. Within large domains, the causal, job, and legal responsibilities of engineers overlap with their moral responsibilities, but not completely. Indeed, it makes sense to say that a particular law is morally unjustified. Moreover, professional responsibilities transcend narrow job assignments. For example, LeMessurier recognized and accepted a responsibility to protect the public even though his particular job description left it unclear exactly what was required of him.

1.2.3 Dimensions of Engineering

Let us now gain a more detailed understanding of the complexity of sharing responsibility within corporations. Doing so will also reveal to us a wider range of moral issues that arise in engineering, as well as a richer appreciation of how moral values are embedded in all aspects of engineering.

Ethical issues arise as a product develops from a mental concept to physical completion. Engineers encounter both moral and technical problems concerning variability in the materials available to them, the quality of work by coworkers at all levels, pressures imposed by time and the whims of the marketplace, and relationships of authority within corporations. Figure 1-2 charts the sequence of tasks that leads from the concept of a product to its design, manufacture, sale, use, and ultimate disposal.

For convenience, several terms are used in broad, generic senses. *Products* can be mass-produced household appliances, an entire communication system, or an oil refinery complex. *Manufacturing* can occur on a factory floor or at a construction site. *Engineers* might be employees of large or small corporations, entrepreneurs, or consultants. *Organizations* might be for-profit organizations, consulting firms, the public works department of a city, or non-for-profit organizations devoted to community development. *Tasks* include creating the concept of a new product, improving an existing product, detailed design of part of an engine, or manufacture of a product according to complete drawings and specifications submitted by another party.

The idea of a new product is first captured in a conceptual design, which will lead to establishing performance specifications and conducting a preliminary

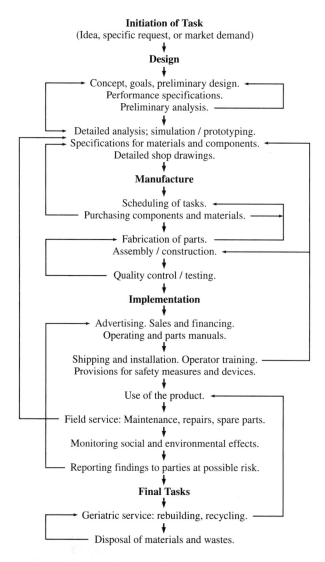

FIGURE 1-2
Progression of engineering tasks (→ ideal progression, — typical iterations)

analysis based on the functional relationships among design variables. These activities lead to a more detailed analysis, possibly assisted by computer simulations and physical models or prototypes. The end product of the design task will be detailed specifications and shop drawings for all components.

Manufacturing is the next major task. It involves scheduling and carrying out the tasks of purchasing materials and components, fabricating parts and subassemblies, and finally assembling and performance-testing the product.

Selling comes next, or delivery if the product is the result of a prior contract. Thereafter, either the manufacturer's or the customer's engineers perform installation, personnel training, maintenance, repair, and ultimately recycling or disposal.

Seldom is the process carried out in such a smooth, continuous fashion as indicated by the arrows progressing down the middle of figure 1-2. Instead of this uninterrupted sequence, intermediate results during or at the end of each stage often require backtracking to make modifications in the design developed thus far. Errors need to be detected and corrected. Changes may be needed to improve product performance or to meet cost and time constraints. An altogether different, alternative design might have to be considered. In the words of Herbert Simon, "Design is usually the kind of problem solving we call ill-structured . . . you don't start off with a well-defined goal. Nor do you start off with a clear set of alternatives, or perhaps any alternatives at all. Goals and alternatives have to emerge through the design process itself: one of its first tasks is to clarify goals and to begin to generate alternatives."[27]

This results in an iterative process, with some of the possible recursive steps indicated by the thin lines and arrows on either side of figure 1-2. As shown, engineers are usually forced to stop during an initial attempt at a solution when they hit a snag or think of a better approach. They will then return to an earlier stage with changes in mind. Such reconsiderations of earlier tasks do not necessarily start and end at the same respective stages during subsequent passes through design, manufacture, and implementation. That is because the retracing is governed by the latest findings from current experiments, tempered by the outcome of earlier iterations and experience with similar product designs.

Changes made during one stage will not only affect subsequent stages but may also require an assessment of prior decisions. Requests for design changes while manufacture or construction is in progress must be handled with particular care, or else tragic consequences such as the Hyatt-Regency walkway failure illustrated in figure 1-3 may result. Dealing with this complexity requires close cooperation among the engineers of many different departments and disciplines such as chemical, civil, electrical, industrial, and mechanical engineering. It is not uncommon for engineering organizations to suffer from "silo mentality," which makes engineers disregard or denigrate the work carried out by groups other than their own. It can be difficult to improve a design or even to rectify mistakes under such circumstances. Engineers do well to establish contact with colleagues across such artificial boundaries so that information can be exchanged more freely. Such contacts become especially important when there is a need to tackle morally complex problems.

To repeat, engineering generally does not consist of completing designs or processes one after another in a straightforward progression of isolated tasks. Instead, it involves a trial-and-error process with backtracking based on decisions made after examining results obtained along the way. The design iterations resemble feedback loops,[28] and like any well-functioning feedback control system, engineering takes into account natural and social environments that affect the product and people using it. Let us therefore revisit the engineering tasks, this time as listed in table 1-1, along with examples of problems that might arise.

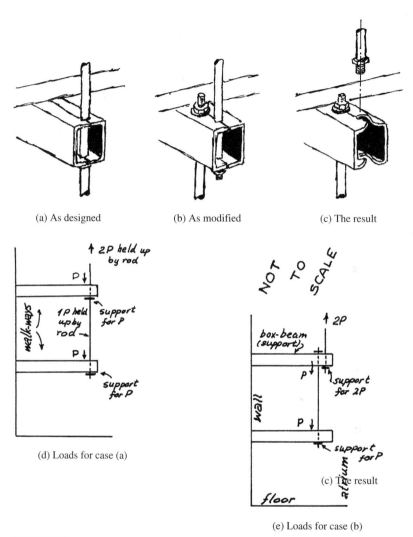

(a) As designed   (b) As modified   (c) The result

(d) Loads for case (a)

(e) Loads for case (b)

FIGURE 1-3

The Kansas City Hyatt-Regency walkway collapse. Two walkways—one above the other—along one wall of a large atrium are to be supported by welded box-beams, which in turn are held up along the atrium side by long rods extending from the ceiling. Because of perceived difficulties in implementing design (a), the modification (b) using two shorter rods to replace each long rod was proposed and approved. What is the result? Let the expected load on each box-beam at its atrium end be P (the same on each floor). Then, in design (a) an upper-floor beam would have to support P pounds as shown in sketch (d), but the design change raised that to 2P as shown in (e). This overload caused the box-beam/rod/nut supports on the upper floor to fail as shown in (c). In turn, the upper and lower walkways collapsed, causing a final death toll of 114 with 200 injured. Later it was found that the design change had been stamped "approved" but not checked. [(For more, see M. Levy and M. Salvadori, *Why Buildings Fall Down* [Norton & Co., 1992].)]

TABLE 1-1
Engineering tasks and possible problems

Tasks	A selection of possible problems
Conceptual design	Blind to new concepts. Violation of patents or trade secrets. Product to be used illegally.
Goals; performance specifications	Unrealistic assumptions. Design depends on unavailable or untested materials.
Preliminary analysis	Uneven: Overly detailed in designer's area of expertise, marginal elsewhere.
Detailed analysis	Uncritical use of handbook data and computer programs based on unidentified methodologies.
Simulation, prototyping	Testing of prototype done only under most favorable conditions or not completed.
Design specifications	Too tight for adjustments during manufacture and use. Design changes not carefully checked.
Scheduling of tasks	Promise of unrealistic completion date based on insufficient allowance for unexpected events.
Purchasing	Specifications written to favor one vendor. Bribes, kickbacks. Inadequate testing of purchased parts.
Fabrication of parts	Variable quality of materials and workmanship. Bogus materials and components not detected.
Assembly/construction	Workplace safety. Disregard of repetitive-motion stress on workers. Poor control of toxic wastes.
Quality control/testing	Not independent, but controlled by production manager. Hence, tests rushed or results falsified.
Advertising and sales	False advertising (availability, quality). Product oversold beyond client's needs or means.
Shipping, installation, training	Product too large to ship by land. Installation and training subcontracted out, inadequately supervised.
Safety measures and devices	Reliance on overly complex, failure-prone safety devices. Lack of a simple "safety exit."
Use	Used inappropriately or for illegal applications. Overloaded. Operations manuals not ready.
Maintenance, parts, repairs	Inadequate supply of spare parts. Hesitation to recall the product when found to be faulty.
Monitoring effects of product	No formal procedure for following life cycle of product, its effects on society and environment.
Recycling/disposal	Lack of attention to ultimate dismantling, disposal of product, public notification of hazards.

The grab-bag of problems in table 1-1 may arise from shortcomings on the part of engineers, their supervisors, vendors, or the operators of the product. The underlying causes can have different forms:

a. *Lack of vision,* which in the form of tunnel vision biased toward traditional pursuits overlooks suitable alternatives, and in the form of *groupthink* (a term coined by Irving Janis) promotes acceptance at the expense of critical thinking.

b. Incompetence among engineers carrying out technical tasks.

c. Lack of time or *lack of proper materials,* both ascribable to poor management.

d. A *silo mentality* that keeps information compartmentalized rather than shared across different departments.

e. The notion that there are safety engineers *somewhere down the line* to catch potential problems.

f. Improper use or disposal of the product by an unwary owner or user.

g. Dishonesty in any activity shown in figure 1-2, and pressure by management to take shortcuts.

h. Inattention to how the product is performing after it is sold and when in use.

Although this list is not complete, it hints at the range of problems that can generate moral challenges for engineers. It also suggests why engineers need foresight and caution, especially in imagining who might be affected indirectly by their products and by their decisions, in good or harmful ways.

DISCUSSION QUESTIONS

1. Upon identifying the structural danger in the Citicorp building, should LeMessurier have immediately notified the workers in the building, surrounding neighbors, and the general public who might do business in the building? Or was it enough that he made sure evacuation plans were in place and that he was prepared to provide warning to people affected in the event of a major storm?

2. Laws play an enormously important role in engineering, but sometimes they overshadow and even threaten morally responsible conduct. Thus, attorneys often advise individuals not to admit responsibility. Bring to mind some occasions where that is good advice. Then discuss whether it would have been sound advice to LeMessurier in the Citicorp Tower case.

3. Herbert Hoover assumes that engineers are accountable for whether the products they make actually work according to expectations. But suppose, as is typical, that an engineer works on only a small part of a building or computer. Is Hoover mistaken in saying that the engineer shares responsibility for the product in its entirety? Does what he says apply only to the project engineer responsible for overseeing an entire project? Distinguish the applicable senses of "responsibility."

1.3 RESPONSIBLE PROFESSIONALS AND ETHICAL CORPORATIONS

From its inception as a profession, as distinct from a craft, much engineering has been embedded in corporations. That is due to the nature of engineering, both in its goal of producing economical and safe products for the marketplace and in its usual complexity of large projects that requires that many individuals work together.

Engineer and historian Edwin T. Layton, Jr., identifies two main stages in the development of engineering as a profession during the nineteenth century. First, the growth of public resources during the first half of the century made possible the extensive building of railroads, canals, and other large projects that only large technological organizations could undertake. Second, from 1880 to 1920 the demand for engineers exploded, increasing their ranks 20 times over. Along with this increase came a demand for science- and mathematics-based training, as engineering schools began to multiply. About the same time, the dominance of independent consulting engineers began to fade, as engineering became increasingly tied to corporations.

Layton also suggests that corporate control underlies the primary ethical dilemmas confronted by engineers: "The engineer's problem has centered on a conflict between professional independence and bureaucratic loyalty," and "the role of the engineer represents a patchwork of compromises between professional ideals and business demands."[29]

We will encounter ethical dilemmas that provide some support for Layton's generalization. But we emphasize that corporate influence is by no means unique to engineering. Today, all professions are interwoven with corporations, including medicine, law, journalism, and science. Professional ethics and business ethics should be connected from the outset, although by no means equated. Let us begin with a brief characterization of professional ethics and then turn to business ethics.

1.3.1   What Are Professions?

In a broad sense, a profession is any occupation that provides a means by which to earn a living. In the sense intended here, however, professions are those forms of work involving advanced expertise, self-regulation, and concerted service to the public good.[30]

1. *Advanced expertise.* Professions require sophisticated skills ("knowing-how") and theoretical knowledge ("knowing-that") in exercising judgment that is not entirely routine or susceptible to mechanization. Preparation to engage in the work typically requires extensive formal education, including technical studies in one or more areas of systematic knowledge as well as broader studies in the liberal arts (humanities, sciences, arts). Generally, continuing education and updating knowledge are also required.

2. *Self-regulation.* Well-established societies of professionals are allowed by the public to play a major role in setting standards for admission to the profession, drafting codes of ethics, enforcing standards of conduct, and representing the profession before the public and the government. Often this is referred to as the "autonomy of the profession," which forms the basis for individual professionals to exercise autonomous professional judgment in their work.

3. *Public good.* The occupation serves some important public good, or aspect of the public good, and it does so by making a concerted effort to maintain high ethical standards throughout the profession. For example, medicine is directed toward promoting health, law toward protecting the public's legal rights, and engineering toward technological solutions to problems concerning the public's well-being, safety, and health. The aims and guidelines in serving the public good are detailed in professional codes of ethics, which, in order to ensure the public good is served, need to be taken seriously throughout the profession.

Drawing attention to the positive, honorific connotations of "profession," some critics argue that the attempt to distinguish professions from other forms of work is an elitist attempt to elevate the prestige and income of certain groups of workers. Innumerable forms of work contribute to the public good, even though they do not require advanced expertise: for example, hair cutting, selling real estate, garbage collection, and professional sports.

In reply, we agree that these are valuable forms of work and that professionalism should not be primarily about social status. Nevertheless, we believe that concerted efforts to maintain high ethical standards, together with a sophisticated level of required skill and the requisite autonomy to do so, warrants the recognition traditionally associated with the word *profession*. We readily acknowledge, however, that in taking seriously this traditional idea of professions, we are tacitly asserting a value perspective. In this way, how one defines professions expresses one's values, a point to which we return in the Discussion Questions.

More recently, along with other scholars in engineering education, Roel Snieder and Qin Zhu have been advocating a value-based approach to professional education. Such an approach encourages engineering students and practicing engineers to articulate, critically examine, and cultivate the values that guide through their decision-making processes. Such approach to professional education encourages students to explore their "self-knowledge" and ask questions fundamental to their own personal and professional goals.[31] These questions are well articulated by Darshan Karwat. Karwat suggests that engineering educators should teach students to step back from the nuances of their work and be able to ask some big questions: Why am I an engineer? For whose benefit do I work? What is the full measure of my moral and social responsibility?[32] Due to the personal nature of values, it is often difficult for engineering students and practicing engineers to clearly articulate the values that drive through their decision-making. Snieder and Zhu have explored multiple ways to help students cultivate capabilities to reflect on their own personal and professional values. These methods include moral exemplars exercise, personal ethics statement, and ethics autobiography. In the Discussion Questions section, we will include the moral exemplars exercise. More details about how personal ethics statement and ethics autobiography pedagogies can help students develop self-reflective competency can be found in a most recent paper by Zhu and Sandy Woodson.[33]

1.3.2   Morally Committed Corporations

Return for a moment to the wave of corporate scandals that, beginning in 2001, shook the confidence of Americans. In that year, Enron became the largest bankruptcy in U.S. history, erasing about $60 billion in shareholder value.[34] Created in 1985, Enron grew rapidly, selling natural gas and wholesale electricity in a new era of government deregulation. In the 1990s it began using fraudulent accounting practices, partly indulged by auditors from Arthur Andersen, a major accounting firm that collapsed in the aftermath of the Enron scandal. Enron created

"Special Purpose Entities," nicknamed "Raptors" after the dinosaurs portrayed in the movie *Jurassic Park*—off–balance sheet partnerships designed to conceal hundreds of millions of dollars in debt and to inflate reported profits. Other unethical practices included price manipulation in sales of electricity to California resulting in massive financial losses to the state. For a time, the game of smoke and mirrors worked, keeping Enron's credit rating buoyant so that it could continue to borrow and invest heavily in ever-expanding markets, often where it lacked expertise. Indeed, for five consecutive years, between 1996 and 2000, Enron was voted in a *Fortune* magazine poll to be the most innovative corporation in the United States.

Fortunately, most corporations are not like Enron. Many, indeed most, companies place a high priority on concern for worthwhile products and ethical procedures. We will encounter many examples as we proceed, and here one illustration will suffice. Quickie Designs, which manufactures wheelchairs.[35] The company was founded in 1980 by Marilyn Hamilton, a schoolteacher and athlete who two years earlier was paralyzed in a hang-gliding accident. Her desire to return to an active life was frustrated by the unwieldiness of the heavy wheelchairs then available. At her request, two of her friends designed a highly mobile, lightweight, and versatile wheelchair made from aluminum tubing originally developed in the aerospace industry and now used to make hang gliders. The friends created a company that rapidly expanded to make a variety of *innovatively* engineered products for people with disabilities. The company went on to create and support Winners on Wheels, a not-for-profit organization that sponsors sports events for young people in wheelchairs.

Quickie Designs is both relatively small and exceptionally committed to what has been called "caring capitalism." Larger corporations characterized by more intense competition and profit-making pressures face a greater challenge in maintaining an ethical climate. But many of them are finding ways to deal with these pressures.

1.3.3 Social Responsibility Movement

Since the 1960s, a "social responsibility movement" has raised attention to product quality, the well-being of workers, the wider community, and the environment. The movement is reflected in what is called "stakeholder theory": corporations have responsibilities to all groups that have a vital stake in the corporation, including employees, customers, dealers, suppliers, local communities, and the general public.[36]

Thus, beyond being concerned with employee relations and other internal organizational matters, responsible corporations also strive to be good neighbors by supporting local schools, cultural activities, civic groups, and charities. But often the wider question of how a corporation's product is ultimately used, and by whom, is conveniently put aside because the effects often do not appear nearby or early on, and important questions are therefore not raised. For instance, what happens to used dry-cell batteries? In the United States nearly three billion of them, along with their noxious ingredients, end up in the municipal waste stream annually. Worldwide, 15 billion are produced per year.

While many corporations are genuinely concerned about what happens to a product once it leaves the factory, others have ready excuses that contain at most partial truths: "We cannot control who buys the product, how it is used, how it is discarded!" Obviously the task is not easy and usually requires industry-wide and government efforts, but socially responsible corporations participate in finding solutions, a task that satisfies even shareholders when common action throughout a particular industry is in the offing, or when the corporation can shine as an industry pioneer. A good example of a corporation's efforts to be in touch with its customers is Cardiac Pacemakers Inc. of St. Paul, Minnesota. Heart patients are invited to the plant so they may share and perhaps alleviate their concerns while employees working on the pacemakers develop a heightened awareness of their responsibilities to turn out high-quality product.

The social responsibility movement in business is not without its critics who contend that corporations should concentrate solely on maximizing profits for stockholders and that there are no additional responsibilities to society, customers, and employees. In a famous essay, "The Social Responsibility of Business is to Increase Its Profits," Nobel Laureate economist Milton Friedman attacked the social responsibility movement. He argued that the paramount, indeed the sole, responsibility of management is to satisfy the desires of stockholders who entrust corporations with their money in order to maximize return on their investment. Management acts irresponsibly and violates stockholders' trust when it adopts further social goals, such as protecting the environment, training disadvantaged workers, using affirmative action hiring practices, or making philanthropic donations to local communities or the arts. The responsibility of managers is "to conduct the business in accordance with their [stockholders'] desires, which generally will be to make as much money as possible while conforming to the basic rules of the society, both those embodied in law and those embodied in ethical custom."[37]

Ironically, Friedman's allusion to heeding "ethical custom" invites recognition of the wider corporate responsibilities he inveighs against. In our society, the public expects corporations to contribute to the wider community good and to protect the environment, and that becomes a moral custom (as indeed it largely has). It seems clear, however, that by "ethical custom" Friedman means only refraining from fraud and deception, and he opposes everything but the most minimum regulation of business needed to protect contracts.[38]

In its extreme form, Friedman's view is ultimately self-defeating. As quickly as the public learns that corporations are indifferent to anything but profit, it will pass restrictive laws that make profit-making difficult. Conversely, when the public perceives corporations as having wider social commitments, it is more willing to cooperate with them to assure reasonable regulations and to selectively purchase products from such socially responsible corporations. Even many investors will be more likely to stay with companies whose ethical commitments promise long-lasting success in business. For these reasons, it would be difficult to find a CEO today who would publicly say that maximum profits are all his or her company is devoted to, that greed is good, and that the environment, workers, and customer safety are mere means to profit.

Sound ethics and good business go together, for the most part and in the long run. Hence at a fundamental level, the moral roles of engineers and their corporations are symbiotic, despite occasional tensions between engineers and managers. As a result of their different experience, education, and roles, higher management tends to emphasize corporate efficiency and productivity—the bottom line. Engineers and other professionals tend to emphasize excellence in creating useful, safe, and quality products.[39] But these differences should be a matter of emphasis rather than opposition.

In order to ensure the confluence of good engineering, good business, and good ethics, it is essential for engineering and corporations, in their major dimensions, to be "morally aligned." As Howard Gardner and his coauthors of *Good Work* contend, professions make possible "good work—work that is both excellent in quality and socially responsible" when the aims of professionals, their corporations, clients, and the general public are congruent, if not identical.[40] Professions go through periods in which these aims become misaligned. For example, Gardner cites journalism as a profession that is currently in upheaval as marketplace pressures subvert standards of objectivity and newsworthiness. Genetic science, by contrast, is a profession in which professional standards of integrity and joy in research are for the most part in line with the expectations of corporations and the public. There are "storm clouds" developing, however, as biotechnology companies begin to patent life forms and pursue controversial but profitable procedures in attempts to corner the market.

Like journalism and genetic science, engineering is periodically subjected to extreme marketplace forces that threaten professional standards. Most corporations respond to those forces responsibly, but some do not. Even under normal economic conditions, some corporations are more committed to quality, safety, and ethics than others. Scholars in engineering education such as Jessica Smith and Juan Lucena have recently advocated for the integration of corporate social responsibility (CSR) into engineering ethics and engineering education.[41] According to Smith's research, when over 70 engineers were interviewed about their decision-making, none of them referred to their professional codes of ethics. Nor did they remember any case studies previously learned. These engineers often attributed to their employer's CSR policies that allowed them to generate internal support for their work devoted to social responsibility. Therefore, these engineers and their employers framed the community development projects they worked on as efforts to support their company's own policies rather than externally exposed obligations such as codes of ethics. In this way, corporate responsibility is well aligned with engineers' professional responsibility and their everyday decision-making.

1.3.4 Senses of Corporate Responsibility

We have been talking about corporate responsibility, but the word *responsibility* is ambiguous, as noted earlier. All the senses we distinguished in connection with individuals also apply to corporations.

1. Just as individuals have *responsibilities* (obligations), so do corporations. To be sure, corporations are communities of individuals, structured within legal frameworks. Yet corporations have internal structures consisting of policy manuals and flowcharts assigning responsibilities to individuals.[42] When those individuals act (or should act) in accordance with their assigned responsibilities, the corporation as a unity can be said to act. Thus, when we say that Intel created a new subsidiary, we understand that individuals with the authority took certain steps.

2. Just as individuals are *accountable* for meeting their obligations, so corporations are accountable to the general public, to their employees and customers, and to their stockholders. Corporations, too, have the capacity for morally responsible agency because it is intelligible to speak of the corporation as acting. The actions of the corporation are performed by individuals and subgroups within the corporation, according to how the flowchart and policy manual specifies areas of authority.

3. Just as individuals manifest the *virtue* of responsibility when they regularly meet their obligations, so too corporations manifest the virtue of responsibility when they routinely meet their obligations. In general, it makes sense to ascribe virtues such as honesty, fairness, and public spiritedness to certain corporations and not to others.[43]

4. In contexts where it is clear that accountability for wrongdoing is at issue, "responsible" becomes a synonym for *blameworthy,* and in contexts where it is clear that right conduct is at issue, "responsible" is a synonym for *praiseworthy.* This is as true for corporations as it is for individuals.

All these moral meanings are distinct from *causal responsibility,* which consists simply in being a cause of some event. The meanings are also distinct from *legal responsibility,* which is simply what the law requires. Engineering firms can be held legally responsible for harm that was so unlikely and unforeseeable that little or no moral responsibility is involved. One famous court case involved a farmer who lost an eye when a metal chip flew off the hammer he was using.[44] He had used the hammer without problems for 11 months before the accident. It was constructed from metals satisfying all the relevant safety regulations, and no specific defect was found in it. The manufacturer was held legally responsible and required to pay damages. The basis for the ruling was the doctrine of *strict legal liability,* which does not require proof of defect or negligence in design. Yet surely the manufacturing firm was not morally guilty or blameworthy for the harm done. It is morally responsible only insofar as it has an obligation (based on the special relationship between it and the farmer created by the accident) to help repair, undue, or compensate for the harm caused by the defective hammer.

DISCUSSION QUESTIONS

1. Michael Davis defines professions as follows: "A profession is a number of individuals in the same occupation voluntarily organized to earn a living by openly serving a certain moral ideal in a morally permissible way beyond what law, market, and [everyday] morality would otherwise require."[45] He argues that carpenters, barbers, porters, and other groups who organize their work around a shared code of ethics should be recognized as professionals. Do you agree or disagree, and why? Can this issue be settled by reference to a dictionary?

2. Do the following definitions, or partial definitions, of professionalism express something important, or do they express unwarranted views?

   a. "Professionalism implies a certain set of attitudes. A professional analyzes problems from a base of knowledge in a specific area, in a manner which is objective and independent of self-interest and directed toward the best interests of his client. In fact, the professional's task is to know what is best for his client even if his client does not know himself."[46]

   b. "So long as the individual is looked upon as an employee rather than as a free artisan, to that extent there is no professional status."[47]

   c. "A truly professional man will go beyond the call to duty. He will assume his just share of the responsibility to use his special knowledge to make his community, his state, and his nation a better place in which to live. He will give freely of his time, his energy, and his worldly goods to assist his fellow man and promote the welfare of his community. He will assume his full share of civic responsibility."[48]

3. Disputes arise over how a person becomes or should become a member of an accepted profession. Such disputes often occur in engineering. Each of the following has been proposed as a criterion for being a "professional engineer" in the United States. Assess these definitions to determine which, if any, captures what you think should be part of the meaning of "engineers."

   a. Earning a bachelor's degree in engineering at a school approved by the Accreditation Board for Engineering and Technology. (If applied in retrospect, this would rule out Leonardo da Vinci, Thomas Edison, and Nikola Tesla.)

   b. Performing work commonly recognized as what engineers do. (This rules out many engineers who have become full-time managers, but embraces some people who do not hold engineering degrees.)

   c. In the United States, being officially registered and licensed as a Professional Engineer (PE). Becoming registered typically includes (1) passing the Engineer-in-Training Examination or Professional Engineer Associate Examination shortly before or after graduation from an engineering school, (2) working four to five years at responsible engineering, (3) passing a professional examination, and (4) paying the requisite registration fees. (Only those engineers whose work directly affects public safety and who sign official documents such as drawings for buildings are required to be registered as PEs. Engineers who practice in manufacturing or teach at engineering schools are exempt. Nevertheless, many acquire their PE licenses out of respect for the profession or for prestige.)

   d. Acting in morally responsible ways while practicing engineering. The standards for responsible conduct might be those specified in engineering codes of ethics or an even fuller set of valid standards. (This rules out scoundrels, no matter how creative they may be in the practice of engineering.)

4. Milton Friedman argues that the sole responsibility of managers is to stockholders, to maximize their profits within the bounds of law and without committing fraud. An alternative view is stakeholder theory: managers have responsibilities to all individuals and organizations that make contracts with a corporation or otherwise are directly affected by them.[49] Clarify what you see as the implications of these alternative views as they apply to decisions about relocating a manufacturing facility in order to lower costs for workers' salaries. Then, present and defend your view as to which of these positions is the more defensible morally.

5. Enron CEO Kenneth Lay betrayed his employees by strongly encouraging them to purchase Enron stock, even after he knew the stock was in trouble—indeed, because he knew it was in trouble—and had begun to sell large amounts of his own shares. In addition, when the stock meltdown began, a company policy prevented employees from selling their stock until it became worthless, thereby causing huge losses in employee retirement programs. Friedman and stakeholder theory would join in condemning such practices. What might each say, however, about Enron's "rank and yank" program? According to one account, every six months all employees were ranked on a 1-to-5 scale, with managers forced to place 15 percent of employees in the lowest category.[50] Those ranked lowest were given six months to improve, although usually they were given severance packages, especially because at the next six-month ranking the 15 percent rule still applied. What are the pros and cons of such employee policies for sustaining both an ethical climate and excellence?

6. Although many engineering corporations are run by professionals who move up the corporate ladder, in some corporations higher management is dominated by individuals with a very different background, as we noted. Certainly that was true of Enron, which employed engineers and scientists but was run by managers whose education was in business, economics, or accounting. For example, CEO Kenneth Lay's bachelor's, master's, and doctorate degrees were all in economics; subsequent CEO Jeffrey Skilling held a bachelor's in applied science and an M.B.A.; and chief financial officer Andrew Fastow had a bachelor's degree in economics and an M.B.A. It is obviously unfair to stereotype persons by their degrees, but could a person who accepts the responsibilities in an engineering code of ethics do what they did? (Review an engineering code of ethics for its general implications.)

7. Moral exemplars exercise: write down the names of seven people (i.e., exemplars) you consider as positive role models. These can be scientists or engineers (e.g., Albert Einstein); famous public figures (e.g., Martin Luther King); friends (e.g., postdoctoral researcher Tom in our lab); family members (e.g., my grandfather); or they can be fictional characters (e.g., Frodo Baggins). Next, list three character traits for each of the exemplars. Finally, pick the character traits that are most frequently selected across the role models you chose or the traits that appeal to you in particular.[51]

To see the moral exemplars exercise worksheet ready to be used in class, visit Dr. Roel Snieder's website: https://inside.mines.edu/~rsnieder/

KEY CONCEPTS

—**Central themes in this book:** (1) Engineering projects are social experiments that generate both new possibilities and risks, and engineers share the responsibility for creating benefits, preventing harm, and informing of dangers. (2) Moral values permeate all aspects of technological development, and hence ethics and excellence in engineering go together. (3) Personal meaning and commitments matter in engineering ethics, along with principles of responsibility that are stated in codes of ethics and incumbent on all engineers. (4) Promoting responsible conduct is even more important than punishing wrongdoing. (5) Ethical dilemmas arise in engineering, as elsewhere, because moral values are myriad and can conflict. (6) Engineering ethics should explore both micro and macro issues, which are often connected. (7) Technological development warrants cautious optimism.

—**Preventive ethics:** ethical reflection and action aimed at preventing moral harm and avoidable ethical dilemmas.

—**Engineering ethics** has two normative (value-laden) meanings. As a set of values, engineering ethics consists of the responsibilities and rights that ought to be endorsed by those engaged in engineering, and also of desirable ideals and personal commitments in engineering. As an area of inquiry, engineering ethics is the study of the decisions, policies, and values that are morally desirable in engineering practice and research.

—**Ethical dilemmas,** or **moral dilemmas:** situations in which moral reasons come into conflict, or in which the application of moral values is problematic, and it is not immediately obvious what should be done.

—**Micro ethical issues** in engineering concern the decisions made by individuals and companies. **Macro ethical issues** concern the general direction of technological development and collective responsibilities of engineers, engineering professional societies, and industrial associations.

—**Morality** concerns obligations and rights, ideals of character, and minimizing harm to humans, animals, and the environment. Ethical theories provide more detailed characterizations.

—**Goals in studying engineering ethics:** improving skills in moral awareness, moral reasoning, moral coherence, moral imagination, moral communication; and, perhaps more indirectly, strengthening moral reasonableness, respect for persons, tolerance of diversity, and confidence in resolving moral conflicts, and preserving moral integrity.

—**Moral responsibility** (of individuals or corporations) has several meanings: obligations, moral accountability, the virtue of being conscientious, and praiseworthiness (for desirable actions) and blameworthiness (for wrongdoing).

—**Silo mentality:** keeping information and discussion compartmentalized rather than shared across different departments within an organization.

—**Professions:** those forms of work involving advanced expertise, independent judgment, self-regulation, and concerted service to the public good as usually formulated in a code of ethics.

—**Stakeholder theory:** corporations have responsibilities to all groups that have a vital stake in the corporation, including employees, customers, dealers, suppliers, local communities, and the general public.

REFERENCES

1. John Schwartz, "Investigators Link Shuttle's Breakup to Damage to Wing," *New York Times* (May 7, 2003), pp. A1, A25; Nick Anderson, "NASA Is Chided for Faults in Safety Teams," *Los Angeles Times* (May 15, 2003), p. A15; John Schwartz, "Shuttle Inquiry Finds New Risks," *New York Times* (June 14, 2003), pp. A1 and A13.

2. Clive L. Dym and Patrick Little, *Engineering Design: A Project-Based Introduction* (New York: John Wiley & Sons, 2000), pp. 73–74.

3. Edmund L. Pincoffs, *Quandaries and Virtues* (Lawrence, KS: University Press of Kansas, 1986), p. 15.

4. Mihaly Csikszentmihalyi, *Good Business* (New York: Viking, 2003), p. 206.

5. Roel Snieder and Qin Zhu, "Connecting to the Heart: Teaching Value-based Professional Ethics," *Science and Engineering Ethics* 26, no. 4 (2020): 2243.

6. Ibid, p. 2238.

7. Ibid.

8. Peter C. Fusaro and Ross M. Miller, *What Went Wrong at Enron?* (New York: John Wiley & Sons, 2002); Loren Fox, *Enron: The Rise and Fall* (New York: John Wiley & Sons, 2003).

9. Banks McDowell draws attention to the importance of compliance issues in *Ethics and Excuses: The Crisis in Professional Responsibility* (Westport, CT: Quorum Books, 2000).

10. Charles E. Harris, Michael S. Pritchard, Ray W. James, Elaine E. Englehardt and Michael J. Rabins, *Engineering Ethics: Concepts and Cases* (6th ed.) (Boston, MA: Cengage, 2019).

11. A similar case is dramatized in the video "Gilbane Gold," produced by the National Society of Professional Engineers. (See Appendix A.)

12. Joseph R. Herkert, "Ways of Thinking about and Teaching Ethical Problem-solving: Microethics and Macroethics in Engineering," *Science and Engineering Ethics* 11, no. 3 (2005): p. 373.

13. John Ladd drew this distinction in "The Quest for a Code of Professional Ethics: An Intellectual and Moral Confusion," in Rosemary Chalk, Mark S. Frankel, and Sallie B. Chafer, eds., *AAAS Professional Ethics Project: Professional Ethics Activities in the Scientific and Engineering Societies* (Washington, D.C.: AAAS, 1980), reprinted in Deborah G. Johnson, ed., *Ethical Issues in Engineering* (Englewood Cliffs, NJ: Prentice Hall, 1991), pp. 130–36. The distinction is explored insightfully by Joseph R. Herkert in "Future Directions in Engineering Ethics Research: Microethics, Macroethics and the Role of Professional Societies," *Science and Engineering Ethics* 7 (2001): 403–14.

14. Keith Bradsher, *High and Mighty* (New York: PublicAffairs, 2002), p. 305.

15. Ibid., pp. xvii–xviii.

16. National Academy of Engineering, www.greatachievements.org.

17. Peter-Paul Verbeek, "Accompanying Technology: Philosophy of Technology after the Ethical Turn," *Techné* 14, no. 1 (2010): 52.

18. Edmund L. Pincoffs, *Quandaries and Virtues* (Lawrence, KS: University Press of Kansas, 1986), p. 15.

19. In clinical terms, sociopaths (or psychopaths) have "antisocial personality disorder." Only a minority of sociopaths are violent, which is fortunate given that by some estimates 3 percent of American males and 1 percent of American females have antisocial personality disorder. American Psychiatric Association, *Diagnostic and Statistical Manual of Mental Disorders,* Fourth Edition, Text Revision (Washington, DC: American Psychiatric Association, 2000), p. 704.

20. Jonathan Haidt, "The Moral Emotions," in Richard J. Davidson, Klaus R. Scherer and H. Hill Goldsmith (eds.), *Handbook of Affective Sciences* (Oxford, UK: Oxford University Press, 2003), p. 852.

21. N. B. Leveson and C. Turner, "An Investigation of the Therac-25 Accidents," *Computer* (IEEE, July 1993): 18–41; Roland Schinzinger, "Ethics on the Feedback Loop," *Control Engineering Practice 6* (1998): 239–45.

22. Ludwig Wittgenstein, *Philosophical Investigations,* 3rd ed., trans. G. E. M. Anscombe (New York: Macmillan, 1958), p. 32.

23. Herbert Hoover, "The Profession of Engineering," in *The Memoirs of Herbert Hoover,* vol. 1 (New York: Macmillan, 1961), pp. 131–34. Quotation in text used with permission of the Hoover Foundation.

24. Buildings Type Study 492, Engineering for Architecture: "Citicorp Center and St. Peter's Lutheran Church," *Architectural Record* (Mid-August Special Issue, 1976): 61–71; Charles Thornton, "Conversation with William LeMessurier," in C. H. Thornton et al., *Exposed Structures in Building Design* (New York: McGraw-Hill, 1993).

25. Joe Morgenstern, "The Fifty-Nine Story Crisis," *The New Yorker* (May 29, 1995): 45–53. Check also the Online Ethics Center for Engineering and Science (http://www.onlineethics.org/).

26. H. L. A. Hart, *Punishment and Responsibility* (Oxford, England: Clarendon, 1973), pp. 211–30; Graham Haydon, "On Being Responsible," *The Philosophical Quarterly* 28 (1978): 46–57; and

R. Jay Wallace, *Responsibility and the Moral Sentiments* (Cambridge, MA: Harvard University Press, 1996).

27. Herbert A. Simon, "What We Know about Learning," *Journal of Engineering Education* (American Society of Engineering Education) 87 (October 1998): 343–48.

28. Roland Schinzinger, *"Ethics* on the Feedback Loop," *Control Engineering Practice* 6 (1998): 239–45. See also Harris, Pritchard, and Rabins, *Engineering Ethics;* and Carolyn Whitbeck, *Ethics in Engineering Practice and Research,* for use of "feedback" in resolving ethical problems.

29. Edwin T. Layton, Jr., *The Revolt of the Engineers: Social Responsibility and the American Engineering Profession* (Baltimore: Johns Hopkins University Press, 1986), pp. 1, 5.

30. Michael Bayles, *Professional Ethics,* 2nd ed. (Belmont, CA: Wadsworth, 1989). Two wide-ranging anthologies on professional ethics are: Joan C. Callahan (ed.), *Ethical Issues in Professional Life* (New York: Oxford University Press, 1988); and John Rowan and Samuel Zinaich, Jr. (eds.), *Ethics for the Professions* (Belmont, CA: Wadsworth/Thomson Learning, 2003).

31. Roel Snieder and Qin Zhu, "Connecting to the Heart: Teaching Value-based Professional Ethics," p. 2244.

32. Darshan M. A. Karwat, "Self-reflection for Activist Engineering," *Science and Engineering Ethics* 26, no. 3 (2020): 1329–1352.

33. Qin Zhu and Sandy Woodson, "Educating Self-Reflective Engineers Ethics Autobiography as a Tool for Moral Pedagogy in Engineering," *Teaching Ethics* (2021). Available at: https://doi.org/10.5840/tej20214190

34. Peter C. Fusaro and Ross M. Miller, *What Went Wrong at Enron?* (New York: John Wiley & Sons, 2002); Loren Fox, *Enron: The Rise and Fall* (New York: John Wiley & Sons, 2003).

35. Mary Scott and Howard Rothman, *Companies with a Conscience* (New York: Carol Publishing Group, 1992), pp. 103–17.

36. R. Edward Freeman, "The Politics of Stakeholder Theory," *Business Ethics Quarterly* 4 (1994): 409–21. See also James J. Brummer, *Corporate Responsibility and Legitimacy* (New York: Greenwood Press, 1991), pp. 144–64; Ronald M. Green, *The Ethical Manager* (New York: Macmillan, 1994), pp. 25–42.

37. Milton Friedman, "The Social Responsibility of Business Is to Increase Its Profits," *The New York Times Magazine,* September 13, 1970.

38. Milton Friedman, *Capitalism and Freedom* (Chicago: University of Chicago Press, 1963), p. 133. See Thomas Carson, "Friedman's Theory of Corporate Social Responsibility," *Business and Professional Ethics Journal* 12 (Spring 1993): 3–32. For critiques of Friedman's view, see Peter Drucker, *An Introductory View of Management* (New York: Harper & Row, 1974), p. 61; Robert C. Solomon, *Ethics and Excellence: Cooperation and Integrity in Business* (New York: Oxford University Press, 1992), p. 120.

39. Joseph A. Raelin, *The Clash of Cultures: Managers Managing Professionals* (Boston: Harvard Business School Press, 1991).

40. Howard Gardner, Mihaly Csikszentmihalyi, and William Damon, *Good Work: When Excellence and Ethics Meet* (New York: Basic Books, 2001), p. xi.

41. Jessica M. Smith and Juan C. Lucena, "Socially Responsible Engineering," in Diane P. Michelfelder and Neelke Doorn (eds.), *The Routledge Handbook of the Philosophy of Engineering* (London, UK: Routledge, 2021), p. 670.

42. Peter A. French, *Corporate Ethics* (New York: Harcourt Brace, 1995).

43. Michael D. Smith, "The Virtuous Organization," *Journal of Medicine and Philosophy* 7, no. 1 (February 1982): 35–42.

44. Richard C. Vaughn, *Legal Aspects of Engineering,* 3rd ed. (Dubuque, IA: Kendall/Hunt, 1977), pp. 41–47.

45. Michael Davis, *Profession, Code, and Ethics* (Burlington, VT: Ashgate Publishing Company, 2002), p. 3.

46. Lawrence Storch, "Attracting Young Engineers to the Professional Society," *Professional Engineer* 41 (May 1971): 3.

47. Robert L. Whitelaw, "The Professional Status of the American Engineer: A Bill of Rights," *Professional Engineer* 45 (August 1975): 37–38.
48. Harry C. Simrall, "The Civic Responsibility of the Professional Engineer," *The American Engineer* (May 1963): 39.
49. James J. Brummer, *Corporate Responsibility and Legitimacy* (New York: Greenwood Press, 1991), pp. 144–64; Ronald M. Green, *The Ethical Manager* (New York: Macmillan, 1994), pp. 25–42.
50. Peter C. Fusaro and Ross M. Miller, *What Went Wrong at Enron?*, p. 51.
51. Roel Snieder and Qin Zhu, "Connecting to the Heart: Teaching Value-based Professional Ethics," p. 2249.

CHAPTER
2

MORAL REASONING AND CODES OF ETHICS

Ethical (or moral) dilemmas are situations in which moral reasons come into conflict, or in which the applications of moral values are unclear, and it is not immediately obvious what should be done. Ethical dilemmas arise in engineering, as elsewhere, because moral values are many and varied. In this chapter we discuss the steps in confronting and resolving moral dilemmas. We also emphasize that even routine moral decision making in engineering requires weighing and balancing conflicting moral values. In doing so, we explore the importance of professional codes of ethics in guiding ethical conduct.

2.1 RESOLVING ETHICAL DILEMMAS

2.1.1 Steps in Resolving Ethical Dilemmas

Reasonable solutions to ethical dilemmas are clear, informed, and well-reasoned. *Clear* refers to moral clarity—clarity about which moral values are at stake and how they pertain to the situation. It also refers to conceptual clarity—precision in using the key concepts (ideas) applicable in the situation. *Informed* means knowing and appreciating the implications of the available facts that are morally relevant, that is, relevant in light of the applicable moral values. In addition, it means being aware of alternative courses of action and what they entail. *Well-reasoned* means that good judgment is exercised in integrating the relevant moral values and facts to arrive at a morally desirable solution.

These criteria for reasonable solutions also enter as steps in resolving ethical dilemmas. By "steps" we do not mean single-file movements, but instead activities that are carried out jointly and in iterative patterns. Thus, a preliminary survey of the applicable moral values and relevant facts might be followed by conceptual clarification and additional fact gathering, which in turn evince a more nuanced understanding of the applicable values and the implications of the relevant facts. Let us illustrate these steps by considering an example from chapter 1.

A chemical engineer working in the environmental division of a computer manufacturing firm learns that the company might be discharging unlawful amounts of lead and arsenic into the city sewer.[1] The city processes the sludge into a fertilizer used by local farmers. To ensure the safety of both the discharge and the fertilizer, the city imposes restrictive laws on the discharge of lead and arsenic. Preliminary investigations convince the engineer that the company should implement stronger pollution controls, but their supervisor tells them the cost of doing so is prohibitive and that technically the company is in compliance with the law. The engineer is also scheduled to appear before town officials to testify in the matter. What should they do?

(1) MORAL CLARITY: IDENTIFY THE RELEVANT MORAL VALUES. The most basic step in confronting ethical dilemmas is to become aware of them! This means identifying the moral values and reasons applicable in the situation, and bearing them in mind as further investigations are made. These values and reasons might be obligations, duties, rights, goods, ideals, or other moral considerations. It matters which kinds of considerations we are considering. Are we dealing with morally mandatory minimums, in the form of strict duties? Or are we dealing with ideals that are desirable to pursue where possible, but not strictly mandatory?

Exactly how we articulate the relevant values reflects our moral outlook. Hence, the moral frameworks discussed in chapter 3 are relevant even in stating what the ethical dilemma is. Another resource is talking with colleagues, who can help sharpen our thinking about what is at stake in the situation. But the most useful resource in identifying ethical dilemmas in engineering are professional codes of ethics, as interpreted in light of one's ongoing professional experience.

Like most codes of ethics, the code of ethics of the American Institute of Chemical Engineers (AIChE) indicates the engineer has at least three responsibilities in the situation. One responsibility is to be honest: "Issue statements or present information only in an objective and truthful manner." A second responsibility is to the employer: "Act in professional matters for each employer or client as faithful agents or trustees, avoiding conflicts of interest and never breaching confidentiality." A third responsibility is to the public, and also to protect the environment: "Hold paramount the safety, health and welfare of the public and protect the environment in performance of their professional duties." In the case at hand, the members of the public most directly affected are the local farmers, but the dangerous chemicals could affect more persons as lead and arsenic are drawn into the food chain. Additional moral considerations, not cited in the code, include duties to maintain personal and professional integrity, and rights to pursue one's career.

(2) CONCEPTUAL CLARITY: CLARIFY KEY CONCEPTS. Professionalism requires being a faithful agent of one's employer, but does that mean doing what one's supervisor directs or doing what is good for the corporation in the long run? These might be different things, in particular when one's supervisor is adopting a short-term view that could harm the long-term interests of the corporation. Again, what does it mean to "hold paramount the safety, health and welfare of the public" in the case at hand? Does it pertain to all threats to public health, or just serious threats, and what is a "serious" threat? Again, does being "objective and truthful" simply mean never lying (intentionally stating a falsehood), or does it mean revealing all pertinent facts (withholding nothing important) and doing so in a way that gives no preference to the interests of one's employer over the needs of the public to be informed of hazards?

(3) INFORMED ABOUT THE FACTS: OBTAIN RELEVANT INFORMATION. This means gathering information that is relevant in light of the applicable moral values (as identified in step 1). Sometimes the primary difficulty in resolving moral dilemmas is uncertainty about the facts, rather than conflicting values per se. For instance, in the United States, the dispute over abortion between pro-life and pro-choice views is not that they value life differently but they hold different views on some facts (e.g., when a life actually starts).[2] Certainly in the case at hand, the chemical engineer needs to check and recheck her findings, perhaps asking colleagues for their perspectives. Her corporation might be violating the law, but is it actually doing so? We, like the engineer, need to know more about the possible harm caused by the minute quantities of lead and arsenic over time. How serious is it, and how likely to cause harm?

(4) INFORMED ABOUT THE OPTIONS: CONSIDER ALL OPTIONS. Initially, ethical dilemmas seem to force us into a two-way choice: Do this or do that. Either bow to one's supervisor's orders or blow the whistle to the town authorities. A closer look often reveals additional options. (Sometimes writing down the main options and suboptions as a matrix or decision tree ensures that all options are considered.) The chemical engineer might be able to suggest a new course of research that will improve the removal of lead and arsenic. Or they might discover that the city's laws are needlessly restrictive and should be slightly revised. Perhaps they can think of a way to convince their supervisor to be more open-minded about the situation, especially given the possible damage to the corporation's image if it should later be found in violation of the law. Unless an emergency develops, these and other steps should be attempted before informing authorities outside the corporation—a desperate last resort, especially given the likely penalties for whistleblowing. (See chapter 6.)

(5) WELL-REASONED: MAKE A REASONABLE DECISION. Arrive at a carefully reasoned judgment by weighing all the relevant moral reasons and facts. This is not a mechanical process, something that a computer or simple algorithm might do for us. Instead, it is a deliberation aimed at taking into account all the relevant reasons, facts, and values—and doing so in a morally reasonable manner. If there is no ideal solution, we seek at least a satisfactory one, what Herbert Simon dubbed "satisficing."

Often a code of ethics provides a straightforward solution to dilemmas, but not always. Codes are not recipe books that contain a comprehensive list of absolute (exceptionless) rules together with precise hierarchies of relative stringency among the rules. What about the case at hand? The code does assert one very important hierarchy: Hold paramount the public safety, health, and welfare. Nevertheless, sometimes it is quite challenging to clearly determine what "the public" means in specific cases. Does the public include future generations? If so, how much value should we assign to them?[3] The AIChE code also requires engineers to "formally advise their employers or clients (and consider further disclosure, if warranted) if they perceive that a consequence of their duties will adversely affect the present or future health or safety of their colleagues or the public." This statement, combined with the statement of the paramount responsibility, makes it clear that the responsibility to be a faithful agent of the employer does not override professional judgment in important matters of public safety.

At the same time, the recommendation to "consider further disclosure, if warranted" seems somewhat lukewarm, both because it is placed parenthetically and because it only says "consider." It suggests something to think about, rather than a firm statement of duty. As such, it is weaker than statements in the NSPE and other codes that require notification of appropriate authorities when one's judgment is overridden in matters where public safety is endangered. Which of these codes takes precedence?

Furthermore, exactly what does the paramount statement entail in the case at hand? If the engineer is convinced the company produces valuable computers, might they reasonably conclude that the public good is held paramount by coming "close enough" to obeying the law? As for the requirement to be "objective and truthful," that certainly implies not lying to the town officials, but might the engineer reasonably conclude they are being objective by not divulging information their supervisor says is confidential? Obviously, such conclusions might be products of rationalization (biased reasoning), rather than sound moral reasoning. We mention them only to suggest that codes are no substitute for morally good judgment—honest, fair, responsible moral judgment. Indeed, as we have just seen, good judgment is needed even in interpreting the code of ethics.[4] The development of good moral judgment is part and parcel of developing experience in engineering. It is also a primary goal in studying ethics.

Michael Davis's eight moral tests can provide some intuitive assessment of the plausibility of moral judgment:

- Harm test: Does this option do less harm than any alternative?
- Publicity test: Would I want my choice of this option published in the newspaper?
- Reversibility test: Would I still think the choice of this option good if I were one of those adversely affected by it?

- Rights test: Would I be violating someone's human rights?
- Virtue test: What would I become if I choose this option often?
- Professional test: What might my profession's ethics committee say about this option?
- Colleague test: What do my colleagues say when I describe my problem and suggest this option as my solution?
- Organization test: What does the organization's ethics officer or legal counsel say about this?[5]

2.1.2   Right-Wrong or Better-Worse?

We might divide ethical dilemmas into two broad categories. On the one hand, many, perhaps most, dilemmas have solutions that are either right or wrong. "Right" means that one course of action is obligatory, and failing to do that action is unethical (immoral). In most instances a code of ethics specifies what is clearly required: obey the law and heed engineering standards, do not offer or accept bribes, speak and write truthfully, maintain confidentiality, and so forth. On the other hand, some dilemmas have two or more solutions, no one of which is mandatory but one of which should be chosen. These solutions might be better or worse than others in some respects, but not necessarily in all respects.

In illustrating the two types of dilemmas, we will continue discussing the requirement to hold paramount the safety, health, and welfare of the public. We will also draw upon examples from the National Society of Professional Engineers' (NSPE) Board of Ethical Review (BER). This board, which currently consists of seven members, provides the valuable service of applying the NSPE code to cases that are fictionalized but based on actual events. These cases including their detailed analyses can be found at NSPE's website (https://www.nspe.org/).

Consider BER Case 93-7:

> Engineer A, an environmental engineer, is retained by a major industrial owner to examine certain lands adjacent to an abandoned industrial facility formerly owned and operated by the owner. Owner's attorney, Attorney X, requests that as a condition of the retention agreement that Engineer A sign a secrecy provision whereby Engineer A would agree not to disclose any data, findings, conclusions or other information relating to his examination of the owner's land to any other party unless ordered by a court. Engineer A signs the secrecy provision.[6]

What is the ethical problem? Although the NSPE code does not explicitly forbid signing the secrecy provision, it does in fact require engineers to hold paramount the public safety and, if their judgment should be overruled in matters of public safety, to notify proper authorities. This implies that Engineer A should not sign a secrecy provision that precludes acting according to the code. As the Board of Ethical Review states, "We do not believe an engineer should ever agree, either by contract or other means, to relinquish his right to exercise professional judgment in such matters." The board also cites the provisions in the code requiring confidentiality about clients, not only proprietary (legally protected) information, but all information obtained in the course of providing professional services. Nevertheless, the paramount clause requires that the public safety, health, and welfare be an overriding consideration. The spirit, if not the letter, of the code indicates that it is unethical for Engineer A to sign the secrecy provision.

As it stands, the decision about whether to sign the secrecy agreement was a dilemma involving lack of clarity about how two moral values applied in the situation: confidentiality and the paramount responsibility to protect the public safety, health, and welfare. (Similar dilemmas arise concerning restrictive confidentiality agreements between salaried engineers and their corporations, although engineers and their corporations are usually granted much wider leeway in reaching confidentiality agreements.) According to NSPE, the solution to this dilemma involves one mandatory action: Refrain from signing the agreement.

But Engineer A does sign the secrecy agreement, and so what happens at that point? The board does not address itself to this question, but clearly another ethical dilemma arises: A commitment and perhaps an obligation to keep the agreement is created, but the paramount responsibility still applies. Hence, if dangers to the public are discovered and if the client refuses to remedy them, the engineer would be obligated to notify proper authorities. But should Engineer A go back to the client and ask to have the secrecy provision revoked? And if the client refuses, should Engineer A break the contract, a step that might have legal repercussions? Or should Engineer A simply hope that no problems will arise and continue with his or her contracted work, postponing any hard decisions until later? As these questions indicate, dilemmas can generate further dilemmas! In this instance, possibly more than one option is reasonable—if not ideal, at least permissible.

To underscore the possibility of several solutions, no one of which is ideal in every regard, consider another case, BER Case 96-4.

> Engineer A is employed by a software company and is involved in the design of specialized software in connection with the operations of facilities affecting the public health and safety (i.e., nuclear, air quality control, water quality control). As part of the design of a particular software system, Engineer A conducts extensive testing, and although the tests demonstrate that the software is safe to use under existing standards, Engineer A is aware of new draft standards that are about to be released by a standard setting organization—standards which the newly designed software may not meet. Testing is extremely costly and the company's clients are eager to begin to move forward. The software company is eager to satisfy its clients, protect the software company's finances, and protect existing jobs; but at the same time, the management of the software company wants to be sure that the software is safe to use. A series of tests proposed by Engineer A will likely result in a decision whether to move forward with the use of the software. The tests are costly and will delay the use of the software at least six months, which will put the company at a competitive disadvantage and cost the company a significant amount of money. Also, delaying implementation will mean the state public service commission utility rates will rise significantly during this time. The company requests Engineer A's recommendation concerning the need for additional software testing.[7]

Here the answer seems obvious enough. In tune with our theme that good engineering and ethics go together, Engineer A should write an honest report. Indeed, it might seem that there is no dilemma for Engineer A at all because what should be done is so obvious. To be sure, the software company faces an ethical dilemma: Is it all right to proceed without the additional testing? But that is a dilemma for the managers, it would seem, not the engineer. The engineer should focus solely on safety issues and fully inform management about the risks, the new draft standards, and the proposed tests. That is what the Board of Ethical Review concludes: "Engineer A has a professional obligation under the Code of Ethics to explain why additional testing is required and to recommend to his company that it be undertaken. By so doing, the company can make an informed decision about the need for additional testing and its effects on the public health, safety, and welfare."

In reaching this conclusion, the board suggests the engineer should focus solely on safety, leaving consideration of other nontechnical matters (such as financial impacts) to management. Yet the board also concludes that the recommendation should be for further testing. As authors, we do not find that conclusion altogether obvious from the facts presented. Much depends on exactly what the risks and circumstances are, and here we need further information.

The case also can be used to suggest that there are sometimes better or worse decisions, perhaps both of which are permissible in the situation. By "better or worse" we mean that two (or more) options are morally permissible—"all right"—but that one is likely to bring about more good than the other. Because multiple moral values are involved, one decision might be better in some respects, and the other decision better in other respects.

Perhaps the public health and safety might well be served by having the company do the further tests even at the risk of severe economic hardship or even bankruptcy. It would be better, however, for employees and customers that this not occur. The paramountcy clause apparently requires bankruptcy rather than imposing unacceptable and severe risks on the public, but it is unclear that such risks are posed in this case. Hence, there might be two morally permissible courses of action: do the tests; do not do the tests. Each option might have further options under it. For example: do the tests, but interrupt them if economic conditions worsen; or do the tests, but devise a quicker version of them; or do the tests, but go ahead with the present sale, being willing to make modifications if the tests raise concerns. In a moment, we return to the possibility of more than one permissible option in resolving dilemmas.

DISCUSSION QUESTIONS

With regard to each of the following cases, answer several questions. First, what is the moral dilemma (or dilemmas)? In stating the dilemma, make explicit the competing moral reasons involved—for example, rights, responsibilities, duties, good consequences, or admirable features of character (virtues). Second, are there any concepts (ideas) involved in dealing with the moral issues that it would be useful to clarify? Third, what factual inquiries do you think might be needed in making a reliable judgment about the case? Fourth, what are the options you see available for solving the dilemma? Fifth, which of these options is required (obligatory, all things considered) or permissible (all right)?

Case 1. An inspector discovers faulty construction equipment and applies a violation tag, preventing its continued use. The inspector's supervisor, a construction manager, views the case as a minor infraction of safety regulations and orders the tag removed so the project will not be delayed. What should the inspector do?

Case 2. A software engineer discovers that a colleague has been downloading restricted files that contain trade secrets about a new product that the colleague is not personally involved with. The software engineer knows the colleague has been having financial problems, and fears the colleague is planning to sell the secrets or perhaps leave the company and use them in starting up a new company. Company policy requires the software engineer to inform their supervisor, but the colleague is a close friend. Should they first talk with the friend about what the friend is doing, or should they immediately inform their supervisor?

Case 3. An aerospace engineer is volunteering as a mentor for a high school team competing in a national contest to build a robot that straightens boxes. The plan was to help the students on weekends for at most 8 to 10 hours. As the national competition nears, the robot's motor overheats and the engine burns out. The aerospace engineer wants to help the dispirited students, and believes their mentoring commitment requires they do more. But doing so would involve additional evening work that could potentially harm his work, if not his family.

Case 4. During an investigation of a bridge collapse, Engineer A investigates another similar bridge, and finds it to be only marginally safe. He contacts the governmental agency responsible for the bridge and informs them of his concern for the safety of the structure. He is told that the agency is aware of this situation, and has planned to provide in next year's budget for its repair. Until then, the bridge must remain open to traffic. Without this bridge, emergency vehicles such as police and fire apparatus would have to use an alternate route which would increase their response time about twenty minutes. Engineer A is thanked for his concern and asked to say nothing about the condition of the bridge. The agency is confident that the bridge will be safe.[8]

2.2 MAKING MORAL CHOICES

Moral dilemmas comprise the most difficult occasions for moral reasoning. Nevertheless, they constitute a relatively small percentage of *moral choices,* that is, decisions involving moral values. Most moral choices are routine and straightforward.[9] The following example illustrates how choices involving moral values enter into routine decisions during technological development, punctuated by periodic moral dilemmas.

2.2.1 Designing Aluminum Cans

Henry Petroski chronicles the development of aluminum beverage cans with stay-on tab openers.[10] Aluminum cans are now ubiquitous—about 180 billion are produced in the United States each year. The first aluminum can was designed in 1958 by Kaiser Aluminum, in the attempt to improve upon heavier and more expensive tin cans. Aluminum proved ideal as a lightweight, flexible material that allowed manufacturing of the bottom and sides of the can from a single sheet, leaving the top to be added after the can was filled. The trick was to make the can strong enough to keep the pressurized liquid inside, while being thin enough to be cost-effective. The can also had to fit conveniently in the hand and reliably satisfy customers' needs. Design calculations solved the problem of suitable thickness of material, but improvements came gradually in shaping of the inward-dished bottom in order to improve stability when the can is set down, as well as to provide some leeway for expansion of the can.

The first aluminum cans, like the tin cans before them, were opened with a separate opener, which required additional manufacturing costs to make them readily available to consumers. The need for separate openers also caused inconvenience, as Ermal Fraze discovered when, forgetting an opener while on a picnic in 1959, he had to resort to using a car bumper. Fraze, who owned Dayton Reliable Tool and Manufacturing Company and was hence familiar with metal, envisioned a design for a small lever that was attached to the can but which was removed as the can opened. The idea proved workable and was quickly embraced by manufacturers. Gradual improvements were made over subsequent years to ensure easy opening and prevention of lip and nose injuries from the jagged edges of the opening.

Within a decade an unanticipated crisis arose, however, creating an ethical dilemma. Fraze had not thought through the implications of billions of discarded pull tabs causing pollution, foot injuries, and injuries to fish and infants who ingested them. The dilemma was what to do in order to balance usefulness to consumers with protection of the environment. A technological innovation solved the dilemma in a manner that integrated all the relevant values. In 1976 Daniel F. Cudzik invented a simple, stay-attached opener of the sort familiar today. Once again, minor design improvements came as problems were identified. Indeed, the search for improvements continues today to create a product fully accessible to whoever may use it. All the while, of course, the broader problem of pollution from cans themselves prompted recycling programs that now recycle more than 6 out of 10 cans (leaving room for further improvement here as well).

Petroski recounts these developments in order to illustrate how engineering progresses by learning from design failures—that is, designs that cause unacceptable risks or other problems. At each stage of the design process, engineers are preoccupied with what might go wrong. The hope is to anticipate and prevent failures, drawing on knowledge about past failures. Here, however, our interest is in how moral values were embedded in the design process at all stages, in addition to surfacing in explicit ethical dilemmas concerning the environment.

If we understand moral choices broadly, as decisions involving moral values, then the development of aluminum cans can be understood as a series of routine moral choices interspersed with occasional moral dilemmas. Moral values entered implicitly into the decision-making process of engineers and their managers—decisions that probably appeared to be purely technical or purely economic. This appearance is misleading, for the technical and economic decisions had moral dimensions in four general directions: safety, environmental protection, consumer usefulness, and economic benefits.

First, human safety is obviously a moral value, rooted directly in the moral worth of human beings. Some aspects of safety seem minor—slight cuts to lips and noses from poorly designed openers and minor injuries to feet in recreation areas like beaches. But minor injuries might cause infections, and even by themselves they have some moral significance. Again, various kinds of poisoning might occur unless all materials were tested under a range of conditions, and there are potential industrial accidents during the manufacturing process. Finally, extensive testing was needed to ensure that exploding cans, while not inherently dangerous, did not cause automobile accidents when drivers opened cans.

A second set of moral values concern the environment. Many of them overlap with the first set, safety. Billions of detached can openers raised the level of hazards to people walking with bare feet. Injuries to fish and other wildlife posed additional concerns. As we discuss in a later chapter, the damage to wildlife can be understood in different ways. Depending on one's environmental ethic, they might be understood either as indirect moral harms due to further impacts on human beings or as direct moral harms to creatures recognized as having inherent worth. The broader problem of environmental pollution from aluminum cans and their openers required both corporate action in paying for recycled materials and community action in developing the technologies for recycling, not to mention changes in public policy and social attitudes about recycling.

Third, some moral values are masked under terms like "useful" and "convenient" products. We tend to think of such matters as nonmoral, especially with regard to trivial things like sipping a carbonated beverage with a pleasing taste. But there are moral connections, however indirect or minor. After all, drinking liquids is a basic need, and convenient access to pleasant-tasting liquids contributes to human well-being. However slightly, these pleasures bear on human happiness and well-being, especially when considered on the scale of mass-produced products. In addition, the aesthetic values pertaining to the shape and appearance of cans also have some relevance to satisfying human needs.

Finally, the economic benefits to stakeholders in the corporation have moral implications. Money matters, and it matters morally. Jobs provide the livelihood for workers and their families that make possible the material goods that contribute to happiness—and survival. The corporation's success contributes as well to the livelihood of suppliers and retailers, as well as to stockholders.

All these values—safety, environmental protection, usefulness, and monetary—were relevant throughout the development of aluminum cans, not merely when they explicitly entered into moral dilemmas. Hence, the case illustrates how moral values permeate engineering practice.

2.2.2 Design Analogy: Whitbeck

We have been discussing engineering design as a domain where moral choices are made. Turning things around, some thinkers suggest that engineering design provides an illuminating model for thinking about all moral decision making, not just decisions within engineering.

More recently Caroline Whitbeck suggests that engineering design is in many respects a model for "designing" courses of action in many moral situations, in engineering and elsewhere.[11] She emphasizes that in devising courses of action we are engaged *participants* who discover or "design" good choices, rather than detached *spectators* who merely criticize choices already made by others. Moral judgments and criticisms are involved in making moral choices, of course, but as part of multifaceted courses of action, rather than as simple "right versus wrong" verdicts about another person's choice.

As an illustration, Whitbeck cites a class assignment in which she supervised several mechanical engineering students. The assignment was to design a child seat that fits on top of standard suitcases with wheels. She specified several constraints. Some pertained to size: The child seat must be easily removable and storable under seats and in overhead storage bins. Others pertained to use: The seat must have multiple uses, including the possibility of strapping it into a seat on an airplane. Still others set safety limits: conformity to applicable safety laws plus avoiding unnecessary dangers. Yet there were many areas of uncertainty and ambiguity surrounding how to maximize safety (for example, when carrying the infant in the seat) and how many convenience features to include, such as storage spaces for baby bottles and diapers.

The students arrived at strikingly different designs, varying in size and shape as well as in the basic structure of the crossbar that held the infant in place. Several were reasonable solutions to the design problem. Yet no design was ideal in every regard, and each had strengths and weaknesses. For example, one was larger and would accommodate older infants, but the added size increased the cost of manufacturing. Again, the bar securing the infant was more convenient in some directions of motion and less convenient in other directions. As for the dynamic feature, in the real world the design of the child seat would go through many iterations, as feedback was received from testing and use of the child seat.

Whitbeck identifies several aspects of engineering decisions that highlight important aspects of moral decisions in general. First, usually there are alternative solutions to design problems, more than one of which is satisfactory or "satisfices." Moral issues, too, frequently have more than one satisfactory solution. We tend to overlook the possibility of several good options because we are preoccupied with moral dilemmas that focus our choice between two mutually exclusive options, leaving only one "right" choice.

Second, multiple moral factors are involved, and among the satisfactory solutions for design problems, one solution is typically better in some respects and less satisfactory in other respects when compared with alternative solutions.

That is, even if two options are equally satisfactory overall, there might be genuine strengths and weaknesses among the many specific features. Analogously, when multiple moral values are applicable, some of them might be more fully satisfied by particular solutions, but with the trade-off of lessened satisfaction of others.

Third, some design solutions are clearly unacceptable. Designs of the child seat that violate the applicable laws or impose unnecessary hazards on infants are ruled out. In general, there are many "background constraints" that limit the range of reasonable options. The same is true in typical moral choices: Some solutions are ruled out from the outset, for example, by minimum standards of justice and decency.

Fourth, engineering design often involves uncertainties and ambiguities, not only about what is possible and how to achieve it, but also about the specific problems that will arise as solutions are developed. Obviously this aspect of engineering highlights a familiar feature of moral decisions in general.

Finally, design problems are dynamic. Usually there is not just one problem to be solved, but instead a cluster of problems that evolve over time. Finding one part of the overall solution often generates new problems, or even a revised understanding of problems, means, and goals. Moral choices, too, are often dynamic and involve ongoing series of choices, rather than one final choice.

Whitbeck argues that the analogies between engineering design and ethical decision making apply to moral dilemmas as well as to routine decision making. What should I do, she asks, if I am an engineer and my supervisor tells me to dump a toxic substance down a drain? There are ambiguities in the situation. Is the substance regulated by law, what does the law say, and how serious is the hazard in dumping the substance? Once such questions are answered, it seems the dilemma is clearly structured: Either obey or disobey the supervisor. Much more is involved, however. I must design a course of action, and do so in an appropriate way within a situation that is dynamic.

> I need to figure out what to do about the supervisor's order. Shall I ignore it? Refuse it? Report it to someone? To someone in the company? To the Environmental Protection Agency? Should I do something else altogether? Is there any place I can go for advice about my options in a situation like this? What are the likely consequences of using those channels (if they exist)? Where could I find out those consequences? Also, what do I do with that toxic waste, at least for the present?[12]

Realistic options will take full account of the circumstances, especially the business in which I am working—its resources and recommended procedures—and my relationship with the supervisor. But odds are there will be more than one possible approach. Sometimes a rhetorical question will do: "Is it all right if I dispose of it in another (legal) way?" Sometimes polite persuasion works, as one gently reminds the supervisor that the chemical is a regulated substance. Moreover, the situation might be as fluid as the toxic waste. Perhaps a firm refusal will settle the matter, but it might also provoke the supervisor to grab the waste in anger and pour it down the drain. What do I do then? In general, good judgment in engineering needs to be both ethically justifiable and practically plausible (e.g., enriching the learning experience of the decision-maker, cultivating healthy working relationships, improving existing organizational cultures).[13]

DISCUSSION QUESTIONS

1. Consider Caroline Whitbeck's example of an engineer who is told by a supervisor to dump a toxic substance down a drain. Is she correct in saying that there might be more than one reasonable choice here, or is there a required course of action? Review the AIChE code: Does it indicate one required course of conduct?
2. Consider Whitbeck's example of designing the car seat. Given the constraints specified in the course, there was more than one satisfactory solution to the design assignment. But within an actual corporation, is it more likely that there would be one best solution in light of what is attractive to most potential buyers (within the constraints of the law), in order to maximize sales?
3. A cafeteria in an office building has comfortable tables and chairs, indeed too comfortable: They invite people to linger longer than the management desires.[14] You are asked to design uncomfortable ones, to discourage such lingering. (a) Is there a moral dilemma here? (b) Are there moral choices involved in whether and how to design the new furniture?
4. Moral skeptics challenge whether sound moral reasoning is possible. An extreme form of moral skepticism is called *ethical subjectivism:* moral judgments merely express feelings and attitudes, not beliefs that can be justified or unjustified by appeal to moral reasons. The most famous version of ethical subjectivism is called emotivism: moral statements are merely used to express emotions—to emote—and to try to influence other people's behavior, but they are not supportable by valid moral reasons.[15] What might be said in reply to the ethical subjectivist? Using Whitbeck's two examples, discuss how moral reasons and values can be objective (justified) even though they sometimes allow room for different applications to particular situations.

2.3 CODES OF ETHICS

2.3.1 Importance of Codes

Codes of ethics state the moral responsibilities of engineers as seen by the profession, and as represented by a professional society. Because they express the profession's collective commitment to ethics, codes are enormously important, not only in stressing engineers' responsibilities but also the freedom to exercise them.

Codes of ethics play at least eight essential roles: serving and protecting the public, providing guidance, offering inspiration, establishing shared standards, supporting responsible professionals, contributing to education, deterring wrongdoing, and strengthening a profession's image.

(1) SERVING AND PROTECTING THE PUBLIC. Engineering involves both advanced expertise that professionals, but not the general public, have, and considerable risks to a vulnerable public. Professionals stand in a fiduciary relationship with the public: trust and trustworthiness are essential. A code of ethics functions as a commitment by the profession as a whole that engineers will serve the public health, safety, and welfare.

(2) GUIDANCE. Codes provide helpful guidance concerning the main obligations of engineers. Since codes should be brief to be effective, they offer mostly general guidance. Nonetheless, when well written, they identify primary responsibilities. More specific directions may be given in supplementary statements or guidelines, which tell how to apply the code. Further specificity may also be attained by the interpretation of codes.

(3) INSPIRATION. Because codes express a profession's collective commitment to ethics, they provide a positive stimulus (motivation) for ethical conduct. In a powerful way, they voice what it means to be a member of a profession committed to responsible conduct in promoting the safety, health, and welfare of the public. Although this paramount ideal is somewhat vague, it, together with more focused guidelines, constitutes a collective commitment to the public good that inspires individuals to have similar aspirations.

(4) SHARED STANDARDS. The diversity of moral viewpoints among individual engineers makes it essential that professions establish explicit standards, in particular minimum standards. In this way, the public is assured of a minimum standard of excellence on which it can depend, and professionals are provided a fair playing field in competing for clients.

(5) SUPPORT FOR RESPONSIBLE PROFESSIONALS. Codes give positive support to professionals seeking to act ethically. A publicly proclaimed code allows an engineer, under pressure to act unethically, to say: "I am bound by the code of ethics of my profession, which states that. . . ." This by itself gives engineers some group backing in taking stands on moral issues. Moreover, codes can potentially serve as legal support for engineers criticized for living up to work-related professional obligations.

(6) EDUCATION AND MUTUAL UNDERSTANDING. Codes can be used by professional societies and in the classroom to prompt discussion and reflection on moral issues. NSPE's Board of Ethical Review actively promotes moral discussion by applying the NSPE code to cases for educational purposes. Widely circulated and officially approved by professional societies, codes encourage a shared understanding among professionals, the public, and government organizations about the moral responsibilities of engineers. In practice, engineers develop relationships with other stakeholders such as contractors, inspectors, and financiers. These social groups also learn from codes of ethics about what they expect from engineers. To a large extent, codes of ethics are written for diverse audiences more than just engineers.[16]

(7) DETERRENCE AND DISCIPLINE. Codes can also serve as the formal basis for investigating unethical conduct. Where such investigation is possible, a deterrent for immoral behavior is thereby provided. Such an investigation generally requires paralegal proceedings designed to get at the truth about a given charge without violating the personal rights of those being investigated. Unlike the American Bar Association and some other professional groups, engineering societies cannot by themselves revoke the right to practice engineering in the United States. Yet some professional societies (e.g., American Society of Civil Engineers) do suspend or expel members whose professional conduct has been proven unethical, and this alone can be a powerful sanction when combined with the loss of respect from colleagues and the local community that such action is bound to produce.

(8) CONTRIBUTING TO THE PROFESSION'S IMAGE. Codes can present a positive image to the public of an ethically committed profession. Where the image is warranted, it can help engineers more effectively serve the public. It can also win greater powers of self-regulation for the profession itself, while lessening the demand for more government regulation. The reputation of a profession, like the reputation of an individual professional or a corporation, is essential in sustaining the trust of the public.

2.3.2 Abuse of Codes

When codes are not taken seriously within a profession, they amount to a kind of window dressing that ultimately increases public cynicism about the profession. Worse, codes occasionally stifle dissent within the profession and are abused in other ways.

Probably the worst abuse of engineering codes is to restrict honest moral effort on the part of individual engineers in the attempt to preserve the profession's public image and protect the status quo. Preoccupation with keeping a shiny public image may silence healthy dialogue and criticism. And an excessive interest in protecting the status quo may lead to a distrust of the engineering profession on the part of both government and the public. The best way to increase trust is by encouraging and helping engineers to speak freely and responsibly about public safety and well-being. This includes a tolerance for criticisms of the codes themselves, rather than allowing codes to become sacred documents that have to be accepted uncritically.

On rare occasions, abuses have discouraged moral conduct and caused serious harm to those seeking to serve the public. In 1932, for example, two engineers were expelled from ASCE for violating a section of its code forbidding public remarks critical of other engineers. Yet the actions of those engineers were essential in uncovering a major bribery scandal related to the construction of a dam for Los Angeles County.[17]

Moreover, codes have sometimes placed unwarranted "restraints of commerce" on business dealings to benefit those within the profession. Obviously there is disagreement about which, if any, entries function in these ways. Consider the following entry in the pre-1979 versions of the NSPE code: The engineer "shall not solicit or submit engineering proposals on the basis of competitive bidding." This prohibition was felt by the NSPE to best protect the public safety by discouraging cheap engineering proposals that might slight safety costs in order to win a contract. The Supreme Court ruled, however, that it mostly served the self-interest of established engineering firms and actually hurt the public by preventing the lower prices that might result from greater competition. (*The National Society of Professional Engineers v. the United States* (1978).)

2.3.3   Limitations of Codes

Codes are no substitute for individual responsibility in grappling with concrete dilemmas. For instance, most codes are restricted to general wording, and hence inevitably contain substantial areas of vagueness. Thus, they may not be able to straightforwardly address all situations. At the same time, vague wording may be the only way new technical developments and shifting social and organizational structures can be accommodated.

Other uncertainties can arise when different entries in codes or different components of one entry come into conflict with each other. Usually codes provide little guidance as to which entry or which component of an entry should have priority in those cases. For example, this is not a very good example as philosophers such as Michael Davis have argued for the central/dominant role of the responsibility to the public. Also, we don't need to include two examples here. One is sufficient. Duties to speak honestly—not just to avoid deception, but also to reveal morally relevant truths—are sometimes in tension with duties to maintain confidentiality.

A further limitation of codes results from their proliferation. Andrew Oldenquist (a philosopher) and Edward Slowter (an engineer and former NSPE president) point out how the existence of separate codes for different professional engineering societies can give members the feeling that ethical conduct is more relative and variable than it actually is.[18] But Oldenquist and Slowter have also demonstrated the substantial agreement to be found among the various engineering codes, and they call for the adoption of a unified code. Indeed, attempts are now being undertaken in that direction by umbrella organizations of engineering, such as the Accreditation Board for Engineering and Technology (ABET). The National Society of Professional Engineers (NSPE) provides a unifying code for individuals who are registered professional engineers. The World Federation of Engineering Organizations (WFEO) has created a model code of ethics which will be employed to support member institutions to create their own codes of ethics.

Most important, despite their authority in guiding professional conduct— akin to the authority of law in structuring societies—codes are not always the complete and final word.[19] Practicing engineers can often be more familiar with their own corporate codes of ethics than professional codes of ethics. The journal *Chemical Engineering* conducted a survey among members of the

AIChE and found that AIChE members fully ignored their code of ethics in ethical decision making in the workplace.[20] Codes can be flawed, both by omission and commission. An example of omission in many codes—although this is now changing—is the absence of explicit mention of responsibilities concerning the environment. We also note that codes invariably emphasize responsibilities but say nothing about the rights of professionals (or employees) to pursue their endeavors responsibly. An example of commission is the former ban in engineering codes on competitive bidding. Codes never be treated as sacred canon in silencing healthy moral debate, including debate about how to improve them.

This limitation of codes connects with a wider issue about whether professional groups or entire societies can create sets of standards for themselves that are both morally authoritative and not open to criticism, or whether group standards are always open to moral scrutiny in light of wider values familiar in everyday life. This is the issue of ethical relativism.[21]

2.3.4 Ethical Relativism and Justification of Codes

Does a profession's code of ethics create the obligations that are incumbent on members of the profession, so that engineers' obligations are relative to their code of ethics? Or does it simply record the obligations that already exist? And once the code is established, does it, like the law, impose requirements as a kind of self-certifying document—or rather, as certified by the professional society?

One view is that codes try to put into words obligations that already exist, whether or not the code is written. That seems to be the view of Stephen Unger, who has been active in discussions about the IEEE Code of Ethics. Unger writes that codes "recognize" obligations that already exist: "A code of professional ethics may be thought of as a collective recognition of the responsibilities of the individual practitioners."[22] Unger adds that codes cannot be "used in cookbook fashion to resolve complex problems," but instead they are "valuable in outlining the factors to be considered."[23] To be sure, Unger takes codes very seriously as a profession's shared voice in articulating the responsibilities of its practitioners. A good code provides valuable focus and direction in thinking about engineers' responsibilities. But it does not itself generate obligations so much as articulate the obligations that already exist.

Michael Davis gives a different emphasis regarding professional codes of ethics. In his view, codes are conventions established within professions in order to promote the public good. As such, they are morally authoritative. The code itself generates obligations: "a code of ethics is, as such, not merely good advice or a statement of aspiration. It is a standard of conduct which, if generally realized in the practice of a profession, imposes a *moral* obligation on each member of the profession to act accordingly."[24] Notice the word "imposes," as distinct from "recognizing" an obligation that already exists. To violate the code is inherently wrong, and it also creates an unfair advantage in competing with other professionals in the marketplace.

Davis has been accused of endorsing *ethical relativism,* also called ethical conventionalism, which says that moral values are entirely relative to and reducible to customs—to the conventions, laws, and norms of the group to which one belongs.[25] What is right is simply what conforms to custom, and it is right solely *because* it conforms to customs. We can never say an act is objectively right or obligatory without qualification, but only that it is right for members of a given group because it is required by their customs. In particular, professional ethics is simply the set of conventions embraced by members of a profession, as expressed in their code.

There are problems with ethical relativism, whether we are talking about the conventions of a profession like engineering or the conventions of a society in its entirety. By viewing customs as self-certifying, ethical relativism rules out the possibility of critiquing the customs from a wider moral framework. For example, it leaves us without a basis for criticizing genocide, the oppression of women and minorities, child abuse, and torture, when these things are the customs of another culture. Regarding professional ethics, ethical relativism implies that we cannot morally critique a given code of ethics, giving reasons for why it is justified in certain ways and perhaps open to improvement in other ways.

Ethical relativism also seems to allow any group of individuals to form its own society with its own conventions, perhaps ones that common sense tells us are immoral.

In our view, then, Unger and Davis are both partly correct. Unger is correct in holding that many of the entries in codes of ethics state responsibilities that would exist regardless of the code—for example, to protect the safety, health, and welfare of the public. Davis is correct that some parts of codes are conventions arrived at by mutual agreement within the profession, and they create moral responsibilities because of the mutual commitments within the profession to abide by them.

If codes of ethics do not merely state conventions, as ethical relativists hold, what does justify those responsibilities that are not mere creations of convention? A code, we might say, specifies the (officially endorsed) "customs" of the professional "society" that writes and promulgates it as incumbent on all members of a profession (or at least members of a professional society). When these values are specified as responsibilities, they constitute *role responsibilities*—that is, obligations connected with a particular social role as a professional. These responsibilities are not self-certifying, any more than other customs are.

A sound professional code will stand up to three tests: (1) It will be clear and coherent; (2) it will organize basic moral values applicable to the profession in a systematic and comprehensive way, highlighting what is most important; and (3) it will provide helpful guidance that is compatible with our most carefully considered moral convictions (judgments, intuitions) about concrete situations. (In chapter 3 we will also apply these criteria to ethical theories.) In addition, it will be widely accepted within the profession. But how can we determine whether the code meets these criteria? One way is to test the code against ethical theories of the sort discussed in chapter 3—theories that attempt to articulate wider moral principles. Obviously, testing the code in light of an ethical theory will need to take close account of both the morally relevant features of engineering and the kinds of public goods engineering seeks to provide for the community.

To conclude, any set of conventions, whether codes of ethics or actual conduct, should be open to scrutiny in light of wider values. At the same time, professional codes should be taken very seriously. They express the good judgment of many morally concerned individuals, the collective wisdom of a profession at a given time. Certainly codes are a proper starting place for an inquiry into professional ethics; they establish a framework for dialogue about moral issues; and more often than not they cast powerful light on the dilemmas confronting engineers.

DISCUSSION QUESTIONS

1. From appendix A, or from the website of an engineering professional society, select a code of ethics of interest to you, given your career plans; for example, the American Society of Civil Engineers, the American Society of Mechanical Engineers, or the Institute of Electrical and Electronics Engineers. Compare and contrast the code with the NSPE code (in appendix B of this book), selecting three or four specific points to discuss. Do they state the same requirements with the same emphasis?

2. With regard to the same two codes you used in question 1, list three examples of responsibilities that you believe would be incumbent on engineers even if the written code did not exist, and explain why. Also list two examples, if any, of responsibilities created (entirely or in part) because the code was written as a consensus document within the profession.

3. Is the following argument for ethical relativism a good argument? That is, is its premise true and does the premise provide good reason for believing the conclusion?

   (1) People's beliefs and attitudes in moral matters differ considerably from society to society. (Call this statement "descriptive relativism," because it simply describes the way the world is.)

   (2) Therefore, the dominant conventional beliefs and attitudes in the society are morally justified and binding (ethical relativism).

4. Reflection on the Holocaust led many anthropologists and other social scientists to reconsider ethical relativism as defined in question 3. The Holocaust also reminds us of the power of custom, law, and social authority to shape conduct. Nazi Germany relied on the expertise of engineers, in carrying out genocide, as well as its war efforts. (a) Do you agree that the Holocaust is a clear instance of where a cross-cultural judgment about moral wrong and right can be made? Can we expect every morally reasonable person to condemn the actions of the Nazis? (b) Judging actions to be immoral is one thing; blaming persons for wrongdoing is another (where blame is a morally negative attitude toward a person). Present and defend your view about whether the Nazi engineers are blameworthy for their contributions to the Holocaust. Or is cross-cultural blame, at least in this extreme instance, an important way of asserting values that we cherish?

KEY CONCEPTS

—**Ethical (or moral) dilemmas** are situations in which moral reasons come into conflict, or in which the applications of moral values are problematic, and it is not immediately obvious what should be done.

—**Steps in resolving ethical dilemmas:** (1) Moral clarity: Identify the relevant moral values. (2) Conceptual clarity: Clarify key concepts. (3) Informed about the facts: Obtain relevant information. (4) Informed about the options: Consider all genuine options. (5) Well-reasoned: Make a reasonable decision.

—**Right-wrong, better-worse:** Some ethical dilemmas have solutions that are either right (obligatory) or wrong (morally forbidden); other dilemmas have more than one permissible solution, some of which are better or worse than others either in some respects or overall.

—**Design analogy:** Engineering design as a metaphor or model for thinking about moral decision making—in general, not just within engineering. Like design, moral choice often involves alternative permissible solutions to dilemmas, integrating multiple values, some clearly unacceptable solutions, uncertainties and ambiguities, and dynamic processes involving series of problems.

—**Importance of codes of ethics:** serving and protecting the public, providing guidance, offering inspiration, establishing shared standards, contributing to education, deterring wrongdoing, and strengthening a profession's image.

—**Abuse of codes:** window-dressing, stifling dissent.

—**Limitations of codes:** Codes contain areas of vagueness, possible internal conflict among entries, possible conflicts among different codes in engineering.

—**Ethical subjectivism:** the view that moral judgments merely express feelings and attitudes, not beliefs that can be justified or unjustified by appeal to moral reasons. The most famous version of ethical subjectivism is called **emotivism:** moral statements are merely used to express emotions ("emote") and to try to influence other people's behavior, but they are not supportable by valid moral reasons.

—**Ethical relativism,** or ethical conventionalism: the view that actions are morally right within a particular society when, and only because, they are approved by law, custom, or other conventions of that society.

REFERENCES

1. This example is a variation of the case in "Gilbane Gold," a video made by the National Society of Professional Engineers. The "Gilbane Gold" case is discussed by Charles E. Harris, Jr., Michael S. Pritchard, and Michael J. Rabins in *Engineering Ethics,* 2nd ed. (Belmont, CA: Wadsworth, 2000), pp. 69–72.

2. Rachels, J., & Rachels, S. (2018). *The elements of moral philosophy* (9th ed.). New York, NY: McGraw-Hill Education.

3. Starrett, S. K., Lara, A. L., & Bertha, C. (2017). *Engineering ethics: Real world case studies.* Reston, VA: ASCE Press.

4. On interpreting codes intelligently, see Michael Davis, "Professional Responsibility as Just Following the Rules," in Michael Davis, *Profession, Code, and Ethics* (Burlington, VT: Ashgate Publishing Company, 2002), pp. 83–98.

5. Davis, M. (2011). The usefulness of moral theory in teaching practical ethics: A reply to Gert and Harris. *Teaching Ethics,* 12(1), 51–60.

6. National Society of Professional Engineers, *Opinions of the Board of Ethical Review,* Vol. VII (Alexandria, VA: National Society of Professional Engineers, 1994), Case No. 93-7, p. 101.

7. National Society of Professional Engineers, Opinions of the Board of Ethical Review, Case 96-4. http://www.niee.org/cases/case96-4.

8. Unpublished case study written by and used with the permission of L. R. Smith and Sheri Smith.

9. James D. Wallace, *Moral Relevance and Moral Conflict* (Ithaca, NY: Cornell University Press, 1988), p. 63.

10. Henry Petroski, *Invention By Design: How Engineers Get From Thought to Thing* (Cambridge, MA: Harvard University Press, 1996), pp. 89–103.

11. Caroline Whitbeck, *Ethics in Engineering Practice and Research* (New York: Cambridge University Press, 1998), pp. 53–68.

12. Ibid., p. 54.

13. Zhu, Q., & Jesiek, B. (2017). A Pragmatic approach to ethical decision-making in engineering practice: Characteristics, evaluation criteria, and implications for instruction and assessment. *Science and Engineering Ethics*, 23(3), 663–679.

14. The case is a variation on one described by Donald A. Norman, *The Design of Everyday Things* (New York: Doubleday, 1988), p. 154. Also published as *The Psychology of Everyday Things* (New York: Basic Books, 1988).

15. Charles Stevenson, *Ethics and Language* (New Haven, CT: Yale University Press, 1944).

16. Johnson, D. (2020). *Engineering ethics: Contemporary and enduring debates*. New Heaven, CT: Yale University Press.

17. Edwin T. Layton, "Engineering Ethics and the Public Interest: A Historical View," in *Ethical Problems in Engineering,* vol. 1, ed. Albert Flores (Troy, NY: Rensselaer Polytechnic Institute, 1980), pp. 26–29.

18. Andrew G. Oldenquist and Edward E. Slowter, "Proposed: A Single Code of Ethics for All Engineers," *Professional Engineer* 49 (May 1979): 8–11.

19. John Ladd, "The Quest for a Code of Professional Ethics," in Rosemary Chalk, Mark Frankel, and Sallie B. Chafer (eds.), *AAAS Professional Ethics Project* (Washington, DC: American Association for the Advancement of Science, 1980), pp. 154–59. For a defense of codes as even more strongly authoritative, see Michael Davis, *Thinking Like an Engineer* (New York: Oxford University Press, 1998).

20. Luegenbiehl, H. C. (1983). Codes of ethics and the moral education of engineers. *Business and Professional Ethics*, 2(4), 41–61.

21. See John Ladd, "The Quest for a Code of Professional Ethics: An Intellectual and Moral Confusion," in Rosemary Chalk, Mark S. Frankel, and Sallie B. Chafer (eds.), *AAAS Professional Ethics Project: Professional Ethics Activities in the Scientific and Engineering Societies* (Washington, DC: American Association for the Advancement of Science, 1980), pp. 154–59; and Heinz C. Luegenbiehl, "Codes of Ethics and the Moral Education of Engineers," *Business and Professional Ethics Journal* 2, no. 4 (1983): 41–61. Both are reprinted in Deborah G. Johnson (ed.) *Ethical Issues in Engineering* (Englewood Cliffs, NJ: Prentice Hall, 1991), pp. 130–36 and pp. 137–54, respectively.

22. Stephen H. Unger, *Controlling Technology,* 2nd ed. (New York: John Wiley & Sons, 1994), p. 106.

23. Ibid.

24. Michael Davis, *Thinking Like an Engineer* (New York: Oxford University Press, 1998), p. 111.

25. In replying to this criticism, Davis confusingly switches from his defense of actual codes to ideal codes: "When an immoral provision appears in an actual code, it is, strictly speaking, not part of it (not, that is, what 'Obey your profession's code' commands obedience to)." Michael Davis, *Profession, Code, and Ethics,* p. 32.

CHAPTER
3

MORAL FRAMEWORKS:
A GLOBAL SURVEY

An ethical theory is a comprehensive perspective on morality that clarifies, organizes, and guides moral reflection. If successful, it provides a framework for making moral choices and resolving moral dilemmas—not a simple formula, but rather a comprehensive way to identify, structure, and integrate moral reasons. Ethical theories also ground the requirements in engineering codes of ethics by reference to broader moral principles.

We discuss five types of ethical theories that have been especially influential: utilitarianism, rights ethics and duty ethics (discussed together), virtue ethics, and self-realization ethics. *Utilitarianism* says that we ought to maximize the overall good, taking into equal account all those affected by our actions. *Rights ethics* says we ought to respect human rights, and *duty ethics* says we ought to respect individuals' autonomy. *Virtue ethics* says that good character is central to morality. *Self-realization ethics* emphasizes the moral significance of self-fulfillment. None of these theories has won a consensus, and each has different versions. Nevertheless, suitably modified, the theories complement and enrich each other to the extent that they usually agree with respect to the right action in particular situations. Taken individually and together, they provide illuminating perspectives on engineering ethics.

As indicated in the title of this chapter, this chapter adopts a *global* approach to surveying the ethical theories. The meaning of the term "global" has two facets. First, most of the ethical theories introduced historically were derived from the Western cultures; however, we also include philosophical and religions thoughts (e.g., Confucian role ethics) from Eastern traditions. Second, this chapter covers a much wider range of ethical theories than other engineering ethics books. In addition to the ethical theories prominent in the West, such as utilitarianism, rights ethics, duty ethics, and virtue ethics, it also includes theories less prominent (e.g., self-realization ethics) but are critical for understanding the professional and ethical identities of engineers.

3.1 UTILITARIANISM

3.1.1 Utilitarianism versus Cost-Benefit Analysis

Utilitarianism is the view that we ought always to produce the most good for the most people, giving equal consideration to everyone affected. The standard of right conduct is maximization of good consequences. "Utility" is sometimes used to refer to the effects brought by a particular action.

At first glance, the utilitarian standard seems simple and plausible. Surely morality involves producing good consequences—especially in engineering! Utilitarianism even seems a straightforward way to interpret the central principle in most engineering codes: "Engineers shall hold paramount the safety, health and welfare of the public in the performance of their professional duties." After all, "welfare" is a rough synonym for "overall good" (utility), and safety and health might be viewed as especially important aspects of that good. Although utilitarianism considers giving equal consideration to all people affected by a particular action, sometimes it is unclear what the term "public" refers to in most engineering codes of ethics. Does public mainly refer to users, community members, or the general public? Could it also include corporations, engineers themselves, or even foreigners? Imagine those engineers who were involved in the boarder wall project of the Trump administration, whose welfare did they promote? American people's welfare or Mexicans'? Furthermore, what exactly is the good to be maximized? And should we maximize the good effects of individual actions or the good effects of general rules (policies, laws, principles in codes of ethics)? Depending on how these questions are answered, utilitarianism takes different forms.

Before discussing these different forms, let us compare and contrast utilitarianism with cost-benefit analyses familiar in engineering.[1] A typical cost-benefit analysis identifies the good and bad consequences of some action or policy, usually in terms of dollars. It weighs the total goods against the total bads, and then compares the results to similar tallies of the consequences of alternative actions or rules. This sounds just like utilitarianism, but often it is not. To see this, we need to look closely at whose good and bad is considered and promoted, as well as how good and bad are measured. Usually the answers center around the good of a corporation, rather than the good of everyone affected, considered impartially.

Consider the cost-benefit analysis performed by Ford Corporation in developing its Pinto automobile, which for years was the largest-selling subcompact car in America. During the early stages of its development, crashworthiness tests revealed that the Pinto could not sustain a front-end collision without the windshield breaking. A quick-fix solution was adopted: The drive train was moved backward. As a result, the differential was moved very close to the gas tank.[2] Thus many gas tanks collapsed and exploded upon rear-end collisions at low speeds.

In 1977, Mark Dowie published an article in *Mother Jones* magazine that divulged the cost-benefit analysis developed by Ford Motor Company in 1971 to decide whether to add an $11 part per car that would greatly reduce injuries by protecting the vulnerable fuel tank—a tank that exploded in rear-end collisions under 5 miles per hour.[3] The $11 seems an insignificant expense, even adjusting to current dollars, but in fact it would make it far more difficult to market a car that was to be sold for no more than $2000. Moreover, the costs of installing the part on 11 million cars and another 1.5 million light trucks added up. The cost of not installing the part, and instead paying out costs for death and injuries from accidents, was projected using a cost-benefit analysis. The analysis estimated the worth of a human life at about $200,000, a figure borrowed from the National Highway Traffic Safety Administration. The cost per non-death injury was $67,000. These figures were arrived at by adding together such costs as a typical worker's future earnings, hospital and mortuary costs, and legal fees, although these figures did not fully consider some other less visible costs such as harms to the worker's family and community and the negative impact on the company's reputation. In addition, it was estimated that about 180 burn deaths and another 180 serious burn injuries would occur each year. Multiplying these numbers together, the annual costs for death and injury was $49.5 million, far less than the estimated $137 million for adding the part, let alone the lost revenue from trying to advertise a car for the uninviting figure of $2,011, or else reducing profit margins.

Ford's cost-benefit analysis is usually understood to be a utilitarian calculation, and certainly it was much like one. It appealed solely to the sum of good and bad consequences, and it sought to maximize the good over the bad. To be sure, its calculations were seriously flawed. The deaths and injuries turned out to be more than were estimated—Dowie estimated 3000 per year. Also, juries awarded larger damage verdicts once Dowie's article appeared, and the negative publicity Ford received greatly damaged its reputation and adversely affected all of its sales for a decade. Even if it had been accurate, however, the cost-benefit analysis was not strictly a utilitarian calculation. It implicitly focused on the costs and benefits to Ford Motor Company. In particular, it omitted the bad consequences of not informing consumers of known dangers. It also focused on costs that could be quantified in dollars, rather than taking into account additional good consequences such as human happiness, and it calculated costs in the short run, for each year, rather than in the long run.

In contrast, utilitarian analyses consider the costs and benefits to everyone affected by a project or proposal. They weigh the interests of each person affected equally, giving no preference to members of a corporation. They adopt a long-term view, and they usually do not reduce good and bad to dollars. With these observations in mind, let us turn to the main versions of utilitarianism.

3.1.2 Act-Utilitarianism versus Rule-Utilitarianism

Act-utilitarianism focuses on each situation and the alternative actions possible in the situation. A particular action is right if it is likely to produce the most good for the most people in a given situation, compared to alternative choices that might be made. The standard can be applied at any moment, and according to act-utilitarians it should be. Right now, should you continue reading this chapter? You might instead take a break, go to sleep, see a movie, or pursue any number of other options. Each option would have both immediate and long-term consequences that can be estimated. The right action is the one that produces the most overall good, taking into account everyone affected.

Of course, even the time spent in making such calculations needs to be considered, and usually we operate according to rules of thumb, such as "complete assignments on time." Such rules, however, provide only rough guidance based on past experience. According to John Stuart Mill (1806–1873), the same is true of everyday moral rules such as "do not deceive" and "keep your promises." These are rules of thumb that summarize past human experience about the types of actions that usually maximize utility.[4] The rules should be broken whenever doing so will produce the most good in a specific situation. The same is true regarding rules stated in engineering codes of ethics.

An alternative version of utilitarianism says we should take rules, rather than isolated actions, much more seriously. Justified rules are morally authoritative, rather than loose guidelines. According to this view, called *rule-utilitarianism,* right actions are those required by rules that produce the most good for the most people. Because rules interact with each other, we need to consider a set of rules. Thus, Richard Brandt (1910–1997), who introduced the term *rule-utilitarianism,* argued that individual actions are morally justified when they are required by an *optimal moral code*—that set of rules which maximizes the public good more than alternative codes would (or at least as much as alternatives).[5] Brandt had in mind society-wide standards, but the same idea applies to engineering codes of ethics. In particular, an engineering code of ethics is justified in terms of its overall good consequences (compared to alternative codes), and so engineers should abide by it even when an exception might happen to be beneficial. For example, if codified rules forbidding bribes and deception are justified, then even if a particular bribe or deception is beneficial in some situations (e.g., bringing significant benefits to the company), one should still refrain from them.

There are philosophical debates over precisely how much rule-utilitarianism and act-utilitarianism differ from each other, but at least sometimes they seem to point in different directions. Indeed, rule-utilitarianism was developed during the twentieth century primarily as a way of correcting several problems with act-utilitarianism.[6]

One problem, just noted, is that act-utilitarianism apparently permits some actions that we know (on other grounds) are patently immoral. Suppose that stealing a computer from my employer, an old one scheduled for replacement anyway, benefits me significantly and causes only miniscule harm to the employer and others. We know that the theft is unethical, and hence act-utilitarianism seems to justify wrongdoing. Rule-utilitarians express this moral knowledge by demonstrating the overall good is promoted when engineers heed the principle, "Act as faithful agents or trustees of employers."

A special problem concerns justice. Act-utilitarianism seems to permit injustice by promoting social good at the expense of individuals. Suppose that in a particular situation more good is promoted by keeping the public ignorant about serious dangers, for example, by not informing them about a hidden fault in a car or building. Or suppose it will improve company morale if several disliked engineers are fired after being blamed for mistakes they did not make. Doing so is unfair, but the overall good is promoted. Rule-utilitarians avoid this result by emphasizing the general good in heeding rules like "corporations should inform the public of dangers," "discipline or punish only the guilty." A similar concern can be found in employing utilitarianism to guide through the career decision-making among engineering students. For instance, is it ever okay to take a harmful job in order to do more good?[7] Act-utilitarianism seems to provide an affirmative response to this question. For instance, act-utilitarianism may think it is okay for an engineer to work at a lab developing dangerous biotech so that they can blow the whistle if they see something particularly dangerous happening.

Yet another problem, ironically, is that act-utilitarianism seems to be too morally demanding. Right now, each of us could promote the overall good by foregoing luxuries and redirecting our careers in order to give to worthy causes, such as alleviating world hunger. Our own well-being might be adversely affected, but surely saving people from starvation produces more good than missing a few movies and driving a less expensive car. But, using iterative reasoning, it follows that we should abandon virtually all luxuries and give in a degree that only saints could consider mandatory. To avoid this result, rule-utilitarians agree that relatively wealthy people should increase their philanthropic giving, but they also think the general good is promoted by allowing individuals to act in accord with a rule such as "Give to help others, while keeping sufficient resources for the security and reasonable luxuries for oneself and one's family."

3.1.3 Theories of Good

There is another area of disagreement among utilitarians. Justified actions or rules should maximize good consequences, but what is the standard for "good" consequences? In particular, what is *intrinsic good*—that is, good considered just by itself (apart from its consequences)? All other good things are *instrumental goods* in that they provide means (instruments) for gaining happiness.

Some utilitarians consider pleasure to be the only intrinsic good. But that seems counterintuitive—there is nothing good about the pleasures of rapists and sadistic torturers! More plausibly, Mill believes that happiness is the only intrinsic good, and hence he understands utilitarianism as the requirement to produce the greatest amount of happiness. But what is happiness? Mill thinks of it as (a) a life rich in pleasures, mixed with some inevitable pains, plus (b) a pattern of activities and relationships that one can affirm as valuable overall, as the way one wants one's life to be.

Especially in his book *On Liberty,* Mill emphasized the importance of individual choices in charting a path to happiness. Nevertheless, he also believed that the happiest life is rich in *higher pleasures,* those that are preferable in kind or quality. For example, Mill contended that the pleasures derived from love, friendship, intellectual inquiry, creative accomplishment, and appreciation of beauty are inherently better than the bodily pleasures derived from eating, sex, and exercise. That contention is questionable, however. How, after all, do we determine which pleasures are better than others, apart from their subjective "feel"? Mill suggested that one pleasure is higher than another if it is favored by the majority of people who have experienced both, but why should the majority view matter here? (Mill's Victorian peers supported his view that physical pleasures have less worth than mental ones, but probably most people today would question such a general ranking.) If we rank pleasures, it is probably because we are actually ranking the types of activities and relationships that generate them, thereby shifting to a new theory of good as a list of especially valuable activities and relationships.[8]

In contrast, Brandt argues that things like love, creativity, and other perfectionist goals that contribute to an excellent human life such as gaining knowledge are good because they satisfy rational desires. *Rational desires* are those that we can affirm after fully examining them in light of all relevant information about the world and our own deepest needs. Some self-destructive desires, such as the desire to use dangerous drugs, are not rational since if we saw their full implications we would not approve of them. Desires (and pleasures) such as those of rapists and sadists are also not rational.

Mill and Brandt both try to use an objective standard on what counts as good. Other utilitarians, especially economists, adopt a "preference theory": What is good is what individuals prefer, as manifested in their choices in the marketplace. Economists base their cost-benefit analyses on the preferences that people express through their buying habits. In this version, utilitarianism becomes the view that right actions produce the greatest satisfaction of the preferences of people affected.

DISCUSSION QUESTIONS

1. Apply act-utilitarianism and rule-utilitarianism in resolving the following moral problems. Do the two versions of utilitarianism lead to the same or different answers to the problems?

   a. George had a bad reaction to an illegal drug he accepted from friends at a party. He calls in sick the day after, and when he returns to work the following day he looks ill. His supervisor asks him why he is not feeling well. Is it morally permissible for George to lie by telling his supervisor that he had a bad reaction to some medicine his doctor prescribed for him?

b. Jillian was aware of a recent company memo reminding employees that office supplies were for use at work only. Yet she knew that most of the other engineers in her division thought nothing about occasionally taking home notepads, pens, computer disks, and other office "incidentals." Her eight-year-old daughter had asked her for a company-inscribed ledger like the one she saw her carrying. The ledger costs less than $20, and Jillian recalls that she has probably used that much from her personal stationery supplies during the past year for work purposes. Is it all right for her to take home a ledger for her daughter without asking her supervisor for permission?

2. Can utilitarianism provide a moral justification for engineers who work for tobacco companies, for example, in designing cigarette-making machinery? In your answer take account of the following facts (and others you may be aware of).[9] Cigarettes kill more than 400,000 Americans each year, which is more than the combined deaths caused by alcohol and drug abuse, car accidents, homicide, suicide, and AIDS. Cigarette companies do much good by providing jobs (Philip Morris employs more than 150,000 people worldwide), through taxes (over $4 billion paid by Philip Morris in a typical year), and through philanthropy. Most new users of cigarettes in the United States are teenagers (under 18). There is disagreement over just how addictive cigarettes are, but adults have some choice in deciding whether to continue using cigarettes, and they may choose to continue using for reasons beyond the addictive potential of nicotine.

3. Some cost-benefit analyses place a price tag on the loss of life. Is doing so inherently offensive, or can it be a reasonable procedure for limited purposes? In the Pinto case, even if Ford was justified in making the cost-benefit analysis, were there additional moral considerations that they should have used in deciding whether to improve the safety of the car?

4. Make a list of the things (activities, relationships, etc.) that are intrinsically good. Do you believe that every intelligent person would agree with your list? How much of your list is either culture bound or applicable only to individuals who share your interests? Is there any reason why engineers should adopt a particular theory of (intrinsic) good as either pleasure, a list of desirable activities and relationships, happiness, satisfaction of rational goods, or preference satisfaction?

3.2   RIGHTS ETHICS AND DUTY ETHICS

Rights ethics regards human rights as fundamental, and duty ethics regards duties of respect for autonomy as fundamental. Historically, the theories developed as distinct moral traditions, but their similarities are far more pronounced than their differences. Both theories emphasize respect for individuals' dignity and worth, in contrast with utilitarians' emphasis on the general good. Furthermore, rights ethics and duty ethics are largely mirror images of each other: Because you have a right to life, I have a duty not to kill you; and if I have a duty not to deceive you then you have a right not to be deceived.

3.2.1   Human Rights

Rights enter into engineering in many ways. Holding paramount the safety, health, and welfare of the public can be interpreted as having respect for the public's rights to life (by producing safe products), rights to privacy, rights not to be injured (by dangerous products), and rights to receive benefits through fair and honest exchanges in a free marketplace. In addition, the basic right to liberty implies a right to give informed consent to the risks accompanying technological products, an idea developed in chapter 4. Again, employers have rights to faithful service from employees, and employees have rights to reciprocal fair and respectful treatment from employers, as discussed in chapter 6. And rights to life imply a right to a livable environment, an idea explored in chapter 8.

Nearly all ethical theories leave room for rights. Rights ethics makes human rights the ultimate appeal—the moral bottom line. At its core, morality is about respecting the inherent dignity and worth of individuals as they exercise their liberty. Human rights constitute a moral authority to make legitimate moral demands on others to respect our choices, recognizing that others can make similar claims on us. As such, rights ethics provides a powerful foundation for the special ethical requirements in engineering and other professions.[10]

Rights ethics should sound familiar, for it provides the moral foundation of the political and legal system of the United States. Thus, in the Declaration of Independence Thomas Jefferson wrote: "We hold these truths to be self-evident; that all men are created equal; that they are endowed by their Creator with certain unalienable Rights, that among these are Life, Liberty, and the pursuit of Happiness." Unalienable—or inalienable, natural, human—rights cannot be taken away (alienated) from us, although of course they are sometimes violated. Human rights have been appealed to in all the major social movements of the twentieth century, including the women's movement, the civil rights movement, the farm workers' movement, and the gay rights movement. Human rights have been used as the basis for critiquing the violation of rights in other countries, such as the former Soviet Union and current dictatorships. They are also embodied in the Universal Declaration of Human Rights proclaimed by the General Assembly of the United Nations in 1948. Indeed, the idea of human rights is the single most powerful moral concept in making cross-cultural moral judgments about customs and laws.

3.2.2   Varieties of Rights Ethics

Rights ethics gets more complex as we ask which rights exist. Thus, human rights might come in two forms: liberty rights and welfare rights. *Liberty rights* are rights to exercise one's liberty, and they place duties on other people not to interfere with one's freedom. (The "not" explains why they are also called negative rights.) *Welfare rights* are rights to benefits needed for a decent human life, when one cannot earn those benefits (perhaps because one is severely handicapped) and when the community has them available. (As a contrast to negative rights, they are sometimes called positive rights.)

The extent of welfare rights is controversial, especially when they enter into the law. But most rights ethicists affirm that both liberty and welfare human rights exist. Indeed, they contend that liberty rights imply at least some basic welfare rights. What, after all, is the point of saying that we have rights to liberty if we are utterly incapable of exercising liberty because, for example, we are unable to obtain the basic necessities, such as jobs, worker compensation for serious injuries, and health care? Shifting to legal rights, most Americans also support selective welfare rights, including a guaranteed public education of kindergarten through twelfth grade, Medicare and Medicaid, Social Security, and reasonable accommodations for persons with disabilities.

This first version of rights ethics conceives of human rights as intimately related to communities of people. A. I. Melden, for example, argues that having moral rights presupposes the capacity to show concern for others and to be accountable within a moral community.[11] Melden's account, like that of most rights ethicists, allows for more "positive" welfare rights to community benefits needed for living a minimally decent human life (when one cannot earn those benefits on one's own and when the community has them available). Thus it lays the moral groundwork for recognizing the limited welfare system in the United States. The extent of welfare rights, just like that of liberty rights, always has to be determined contextually—for example, by what the community has available by way of resources and the severity of the obstacles to freedom confronted by various individuals.

A second version of rights ethics denies there are welfare human rights. *Libertarians* believe that only liberty rights exist; there are no welfare rights. John Locke (1632–1704), who was the first philosopher to carefully articulate a rights ethics, is often interpreted as a libertarian.[12] He believed that the three most basic human rights are to life, liberty, and property. His views provided the moral foundation of contemporary American society. Indeed, Jefferson simply modified Locke's triad of basic rights, changing property to the pursuit of happiness.

The individualistic aspect of Locke's thought is reflected in the contemporary political scene in the Libertarian political party and outlook, with its emphasis on the protection of private property and the condemnation of welfare systems. Libertarians take a harsh view of taxes and government involvement beyond the bare minimum necessary for national defense and the preservation of free enterprise. Locke's followers tend to insist that property is sacrosanct and that governments continually intrude on property rights, particularly in the form of excessive taxation and regulation. They also oppose extensive government regulation of business and the professions. Thus, Milton Friedman (discussed in chapter 1) is a leading libertarian thinker who argues against both government regulation and requiring corporations to accept responsibilities beyond maximizing profit (within the bounds of minimum laws, such as forbidding fraud).

We have been speaking of human rights, but there are also *special moral rights*—rights held by particular individuals rather than by every human being. For example, engineers and their employers have special moral rights that arise from their respective roles and the contracts they make with each other. Special rights are grounded in human rights, however indirectly. Thus, contracts and other types of promises create special rights because people have human rights to liberty that are violated when the understandings and commitments specified in contracts and promises are violated. And when the public purchases products, there is an implicit contract, based on an implicit understanding, that the products will be safe and useful.

Finally, few rights are absolute, in the sense of being unlimited and having no justifiable exceptions. Libertarians and other rights ethicists agree that members of the public do not have an absolute right not to be harmed by technological products. If people purchase hang gliders and then injure themselves by flying them carelessly or under bad weather conditions, their rights have not been violated—assuming that advertisements about the joys of hang gliding did not contain misleading information. But human rights to pursue one's legitimate interests do imply rights not to be poisoned, maimed, or killed by technological products whose dangers are not obvious or are deliberately hidden. These rights also imply a right to informed consent when purchasing or using products or services that might be dangerous, for example, buying an airline ticket. We might think of this as a right to make an "informed purchase."

In the political philosophy literature, there have been various approaches to categorizing human rights. Therefore, it is common that these approaches sometimes can overlap with each other. For instance, global ethicists often talk about "three generations of human rights": (1) first-generation rights are civil and political rights; (2) second-generation rights are economic, social, and cultural rights; and (3) third-generation rights are rights of peoples (e.g., indigenous persons and group rights).[13] It is worth noting that first-generation rights here mostly belong to liberty rights; second-generation rights are mainly welfare rights; and third-generation rights are some special rights held by particular individuals. In the global context, different cultures prioritize these different human rights differently in their social governance and technological development. Compared to Western cultures, Chinese philosophers argue that Confucian cultures tend to prioritize second-generation rights (e.g., education and economic security) over first-generation rights.[14] However, it does not necessarily mean that Confucian cultures do not care about first-generation, civil, and political rights.

3.2.3 Duty Ethics

Duty ethics says that right actions are those required by duties to respect the liberty or autonomy (self-determination) of individuals. One writer suggests the following list of important duties: "1. Don't kill. 2. Don't cause pain. 3. Don't disable. 4. Don't deprive of freedom. 5. Don't deprive of pleasure. 6. Don't deceive. 7. Keep your promise. 8. Don't cheat. 9. Obey the law. 10. Do your duty [referring to work, family, and other special responsibilities]."[15]

How do we know that these are our duties? Immanuel Kant (1724–1804), the most famous duty ethicist, argued that all such specific duties derive from one fundamental duty to respect persons. Persons deserve respect because they are moral agents—capable of recognizing and voluntarily responding to moral duty (or, like children, they potentially have such capacities). *Autonomy*—moral self-determination or self-governance—means having the capacity to govern one's life in accordance with moral duties. Hence, respect for persons amounts to respect for their moral autonomy.

Immorality occurs when we "merely use" others, reducing them to *mere* means to our ends, treating them as mere objects to gratify our needs. Violent acts such as murder, rape, and torture are obvious ways of treating people as mere objects serving our own purposes. We also fail to respect persons if we fail to provide support for them when they are in desperate need and we can help them at little inconvenience to ourselves. Of course we need to "use" one another as means all the time: business partners, managers and engineers, and faculty and students use each other to obtain their personal and professional ends. Immorality involves treating persons as "mere" means to our goals, rather than as autonomous agents who have their own goals.

We also have duties to ourselves, for we, too, are rational and autonomous beings. As examples, Kant says we have a duty not to commit suicide, which would bring an end to a valuable life; we have duties to develop our talents, as part of unfolding our rational natures; and we should avoid harmful drugs that undermine our ability to exercise our rationality. Obviously, Kant's repeated appeal to the idea of rationality makes a number of assumptions about morally worthy aims. After beginning with the minimal idea of rationality as the capacity to obey moral principles, he builds in a host of specific goals as part of what it means to be rational.

In a famous sentence, Kant stated the fundamental duty of respect for persons as rational and autonomous beings: "Act so that you treat humanity, whether in your own person or in that of another, always as an end and never as a means only."[16] As a moral "end," each person (including ourselves) places moral limits on our conduct. These limits are itemized by all valid moral rules stating obligations to others. Some obligations are to refrain from interfering with a person's liberty, and some express requirements to help them when they are in need, thereby paralleling the distinction between liberty and positive rights.

Kant also emphasized that duties are universal: They apply equally to all rational beings. He stated this idea in another famous sentence: "Act only according to that maxim [that is, rule of action] by which you can at the same time will that it should become a universal law."[17] Here again, the idea is that valid principles of duty apply to all rationally autonomous beings, and hence valid duties will be such that we can envision everyone acting on them. This idea of universal principles is often compared to the Golden Rule: Do unto others as you would have them do unto you; or, in its negative version, Do not do unto others what you would not want them to do to you.[18]

Finally, Kant insisted that moral duties are "categorical imperatives." As imperatives, they are injunctions or commands that we impose on ourselves as well as other rational beings. As categorical, they require us to do what is right *because* it is right, unconditionally and without special incentives attached. For example, we should be honest because honesty is required by duty; it is required by our basic duty to respect the autonomy of others, rather than to deceive and exploit them for our own selfish purposes. "Be honest!" says morality—not because doing so benefits us, but because honesty is our duty. Morality is not an "iffy" matter that concerns hypothetical (conditional) imperatives, such as "If you want to prosper, be honest." A businessperson who is honest solely because honesty pays—in terms of profits from customers who return and recommend their services, as well as from avoiding jail for dishonesty—fails to fully meet the requirements of morality. In this way, morality involves attention to motives and intentions, an idea also important in virtue ethics.

3.2.4 Prima Facie Duties

Kant thought that everyday principles of duty, such as "Do not lie" and "Keep your promises," are *absolute* in the sense of never having justifiable exceptions. In doing so, he conflated three ideas: (1) universality—moral rules apply to all rational agents; (2) categorical imperatives—moral rules command what is right because it is right; and (3) absolutism—moral rules have no exceptions. Nearly all ethicists reject Kant's absolutism, even ethicists who embrace his ideas of universality and categorical imperatives.

The problem with absolutism should be obvious. As we have emphasized, moral reasons are many and varied, including those expressed by principles of duty. Given the complexity of human life, they invariably come into conflict with each other, thereby creating moral dilemmas. Contemporary duty ethicists recognize that many moral dilemmas are resolvable only by recognizing some valid exceptions to simple principles of duty. Thus, engineers have a duty to maintain confidentiality about information owned by their corporations, but that duty can be overridden by the paramount duty to protect the safety, health, and welfare of the public.

To emphasize that most duties have some justified exceptions, the philosopher David Ross (1877–1971) introduced the expression *prima facie duties*. In this technical sense, prima facie simply means "might have justified exceptions" (rather than "at first glance"). Most duties are prima facie ones—they sometimes have permissible or obligatory exceptions. Indeed, the same is true of most rights and other moral principles, and hence today the term *prima facie* is often applied to rights and rules.

Ross believed that prima facie duties are intuitively. He emphasized, however, that it is not always obvious how best to balance conflicting duties, so as to arrive at our actual duty—our duty in a situation, all things considered. How, then, do we tell which duties should override others when they come into conflict? Ross noted that some principles, such as "Do not kill" and "Protect innocent life," clearly involve more pressing kinds of respect for persons than other principles, such as "Don't lie." Usually, however, general priorities cannot be established. Instead, he argued, we must simply reflect carefully on particular situations, weighing all relevant duties in light of all the facts, and trying to arrive at a sound judgment or intuition.

Ross points out that six categories of duties are relevant to our moral decision-making: (1) fidelity and reparation; (2) gratitude; (3) justice; (4) beneficence; (5) self-improvement; and (6) nonmaleficence. Philosopher Glen Miller argues that it is common that some of Ross's six duties can conflict with each other in engineering ethics cases. Engineers need to exercise their moral intuition to compare and prioritize these conflicting duties.[19] Miller invites us to imagine a typical engineering ethics scenario: an engineer feels concerned that their company may release a pollutant that may generate some minor harm to a few members of the local community. Some of Ross's duties may apply in this case. The engineer has the obligation to be loyal to their employer that such an obligation is associated with the duties of fidelity and gratitude. The duty of beneficence would encourage the engineer to take actions that benefit the majority. The duty of nonmaleficence requires the engineer to not generate harm to all people. Finally, the duty of justice allows the engineer to address the issue that the suffering is distributed disproportionally in the community. A major task of the engineer is to carefully reflect on this particular situation, consider all relevant duties and facts, and make an appropriate judgment that would appeal to the intuition.

In emphasizing the need to reflect contextually, as well as acknowledging human fallibility in doing that reasoning, Ross greatly improves on Kant's version of duty ethics. Nevertheless, Ross relies heavily on intuition, and persons sometimes differ in their moral intuitions. Hence, Ross is often criticized for not providing sufficiently detailed moral guidance. Most contemporary duty ethicists seek ways to minimize the need for intuitions (immediate judgments) in morality—for example, by underscoring the need for rational dialogue with others and periodic reflection in connecting general rules with specific applications.

DISCUSSION QUESTIONS

1. In the Pinto case, did Ford Motor Company have a duty to inform the public of the hazard with its gas tank, and did the public have rights to be so informed? If so, how might such information have been made available to the public?
2. Revisit the Citicorp tower case in chapter 1. Identify the rights of the various stakeholders involved in the case. How might a rights ethicist proceed in resolving what should have been done?
3. Suppose that you or your family owns, free and clear (without debt) a piece of land. Does a right to property permit you, morally speaking, to do anything you please with it, and are laws that say otherwise immoral? Some philosophers argue that what it means to say it is your property is largely a matter of what the law says you can and cannot do with it. What would Locke say, and do you agree with him? In your response, which will clarify your conception of what property is, consider right-of-way laws that allow the government to purchase your land at market value in order to construct a road or railway path, environmental laws forbidding pollution of the land, limitations on the height of buildings on the land, etc.
4. Present and defend your view concerning the relative strengths and weaknesses of the views of libertarian rights ethicists and those rights ethicists who believe in both liberty and welfare rights. In doing so, comment on why libertarianism is having considerable influence today, and yet why the Libertarian Party repeatedly cannot win widespread support for its goals to dismantle all welfare programs, such as guaranteed public education from kindergarten to twelfth grade and health care for the elderly and low-income families.

5. Write down a list of duties that you believe all reasonable persons should recognize as absolute, that is, as having no justified exceptions. Is the list very long? Explain why your list is short or long, and defend your view against possible criticism.

6. Americans are sometimes criticized for being too individualistic, and in particular for approaching moral issues with too great an emphasis on rights. Although we said that rights and duties are usually correlated with each other, what difference (if any) do you think would occur if Jefferson had written, "We hold these truths to be self-evident; that all people are created equal; that they owe duties of respect to all other persons, and are owed these duties in return"?

7. What does the Golden Rule imply concerning how engineers and corporations should behave toward customers in designing and marketing products? As a focus, discuss whether crash-test information should be made available to customers concerning the possibly harmful side effects of a particular automobile. Does it matter whether the negative or positive version of the Golden Rule is used? And does either version provide an answer that everyone might find morally reasonable?

3.3 VIRTUE ETHICS

Virtue ethics emphasizes character more than rights and rules. Character is the pattern of virtues (morally desirable features) and vices (morally undesirable features) in an individual. Virtues are desirable habits or tendencies in action, commitment, motive, attitude, emotion, ways of reasoning, and ways of relating to others. Vices are morally undesirable habits or tendencies. Words for specific virtues are familiar, both in engineering and in everyday life—for example, competence, honesty, courage, fairness, loyalty, and humility. Words for specific vices are also familiar: incompetence, dishonesty, cowardice, unfairness, disloyalty, and arrogance.

3.3.1 Virtues in Engineering

As noted in chapter 1, the Greek word *arete* can be translated as "virtue," "ethics," and "excellence," an etymological fact that reinforces our theme of ethics and excellence going together in engineering. The most comprehensive virtue of engineers is responsible professionalism. This umbrella virtue implies four (overlapping) categories of virtues: public well-being, professional competence, cooperative practices, and personal integrity.

Public-spirited virtues are focused on the good of clients and the wider public. The minimum virtue is nonmaleficence, that is, the tendency not to harm others intentionally. As Hippocrates reportedly said in connection with medicine, "Above all, do no harm." Engineering codes of professional conduct also call for beneficence, which is preventing or removing harm to others and, more positively, promoting the public safety, health, and welfare. Also important is a sense of community, manifested in faith and hope in the prospects for meaningful life within professional and public communities. Generosity, which means going beyond the minimum requirements in helping others, is shown by engineers who voluntarily give their time, talent, and money to their professional societies and local communities. Finally, justice within corporations, government, and economic practices is an essential virtue in the profession of engineering.

Proficiency virtues are the virtues of mastery of one's profession, in particular mastery of the technical skills that characterize good engineering practice. Following Aristotle, some thinkers regard these values as intellectual virtues rather than distinctly moral ones. As they contribute to sound engineering, however, they are morally desirable features. The most general proficiency virtue is competence: being well prepared for the jobs one undertakes. Also important is diligence: alertness to dangers and careful attention to detail in performing tasks by, for example, avoiding the deficiency of laziness and the excess of the workaholic. Creativity is especially desirable within a rapidly changing technological society.

Teamwork virtues are those that are especially important in enabling professionals to work successfully with other people. They include collegiality, cooperativeness, loyalty, and respect for legitimate authority. Also important are leadership qualities that play key roles within authority-structured corporations, such as the responsible exercise of authority and the ability to motivate others to meet valuable goals. More recently, philosopher William J. Frey conceptualizes four virtues that are critical for effective and ethical teamwork: (1) justice (in the distribution of work); (2) responsibility (in specifying tasks, assigning blame, and awarding credit); (3) reasonableness (ensuring participation, resolving conflict, and reaching consensus); and (4) honesty (avoiding deception, corruption, and impropriety).[20]

Finally, *self-governance virtues* are those necessary in exercising moral responsibility.[21] Some of them center on moral understanding and perception: for example, self-understanding and good moral judgment—what Aristotle called practical wisdom. Other self-governance virtues center on commitment and on putting understanding into action: for example, courage, self-discipline, perseverance, fidelity to commitments, self-respect, and integrity. Honesty falls into both groups of self-direction virtues, for it implies truthfulness in speech and belief and trustworthiness in commitments.

3.3.2 Florman: Competence and Conscientiousness

Like rights ethics, duty ethics, and utilitarianism, virtue ethics takes alternative forms, especially in the particular virtues emphasized and their roles in morally good lives. As an illustration, let us contrast Samuel Florman's emphasis on loyalty to employers with Aristotle's emphasis on loyalty to community, referring as well to Alasdair MacIntyre, who applied Aristotle's perspective to contemporary professions.

Florman is most famous for his celebration of the "existential pleasures" of engineering—the deeply rooted and elemental satisfactions in engineering that contribute to happiness.[22] These pleasures have many sources. There is the desire to improve the world, which engages individuals' sense of personal involvement and power. There is the challenge of practical and creative effort, including planning, designing, testing, producing, selling, constructing, and maintaining, all of which bring pride in achieving excellence in the technical aspects of one's work. There is the desire to understand the world—an understanding that brings wonder, peace, and sense of being at home in the universe. There is the sheer magnitude of natural phenomena—oceans, rivers, mountains, and prairies—that both inspires and challenges the design of immense ships, bridges, tunnels, communication links, and other vast undertakings. There is the presence of machines that can generate a comforting and absorbing sense of a manageable, controlled, and ordered world. Finally, engineers live with a sense of helping, of contributing to the well-being of other human beings.

In elaborating on these pleasures, Florman implicitly sets forth a virtue ethics. In his view, "the essence of engineering ethics" is best captured by the word *conscientiousness*.[23] Engineers who do their jobs well are morally good engineers, and doing their jobs well is to be understood in terms of the more specific virtues of competence, reliability, inventiveness, loyalty to employers, and respect for laws and democratic processes. Competence and loyalty are the two virtues Florman most emphasizes.

On the one hand, conscientious engineers are competent. Florman estimates that 98 percent of engineering failures are caused by incompetence. The other 2 percent involve greed, fraud, dishonesty, and other conventional understandings of wrongdoing, often in addition to sloppiness. "Competent" does not mean minimally adequate, but instead performing with requisite skill and experience. It implies exercising due care, persistence and diligence, and attention to detail and avoiding sloppiness. In addition to competence, conscientious engineering often requires creative problem solving and innovative thinking.

On the other hand, conscientious engineers are loyal to their employers, within the boundaries of laws and democratic institutions. Within a democratic setting in which laws express a public consensus, economic competition among corporations makes possible technological achievements that benefit the public. Competition depends on engineers who are loyal to their organizations, which is analogous to how members of a baseball team work together in competition. Like attorneys defending clients, engineers need not believe that their company is always best serving the interests of humanity at large. In fact, engineers should keep their personal commitments largely to themselves, although it is gratifying when they can match their personal convictions to the goals of their companies. Professional restraints should be laws and government regulations rather than personal conscience. In this view, even professional codes of ethics are largely ceremonial expressions, and "a code with real meaning and teeth is beyond the realm of possibility."[24]

We can agree that engineers should be conscientious in meeting their responsibilities, but the question is which responsibilities take priority. Florman defends the priority of duties to employers, in opposition to professional codes that require engineers to hold "paramount" the safety, health, and welfare of the public. His competence-and-loyalty credo could easily be used to encourage engineers to be passive in accepting the dictates of employers and relying on laws as sufficient to protect the public. Rather than "filtering their everyday work through a sieve of ethical sensitivity," he tells us, professionals have the task of meeting the expectations of their clients and employers.[25] Yet, in some important sense, such "filtering" is exactly what should be expected of engineers in exercising their professional judgment.

3.3.3   Aristotle: Community and the Golden Mean

Aristotle (384–322 B.C.) defined the moral virtues as habits of reaching a proper balance between extremes in conduct, emotion, desire, and attitude.[26] To use the phrase inspired by his theory, virtues are tendencies to find the *Golden Mean* between the extremes of too much (excess) and too little (deficiency) with regard to particular aspects of our lives. Thus, truthfulness is the appropriate middle ground (mean) between revealing all information in violation of tact and confidentiality (excess) and being secretive or lacking in candor (deficiency) in dealing with truth. Again, courage is the mean between foolhardiness (the excess of rashness) and cowardice (the deficiency of self-control) in confronting dangers. The most important virtue is practical wisdom, that is, morally good judgment, which enables one to discern the mean for all the other virtues.

What exactly is the morally good judgment required in discerning the mean in particular circumstances? Aristotle tells us it arises from the development of good habits as achieved through proper training within families and communities. This answer, however, merely pushes the question a step backward: How do we identify proper training, and how do we ensure that it results in good judgment? Aristotle's appeal to good judgment conceals the specific moral requirements and ideals, much like an appeal to "reasonable person" in the law conceals a great complexity of legal rules. The ultimate reference, however, is to goods made possible within particular communities.

More recently, Alasdair MacIntyre applied Aristotle's themes, including his emphasis on community and public goods, to the professions.[27] MacIntyre conceives of professions as valuable social activities, which he calls *social practices*. A social practice is

> any coherent and complex form of socially established cooperative human activity through which goods internal to that form of activity are realized in the course of trying to achieve those standards of excellence which are appropriate to, and partially definitive of, that form of activity, with the result that human powers to achieve excellence, and human conceptions of the ends and goods involved, are systematically extended.[28]

There are three key ideas in this definition: internal goods, standards of excellence, and human progress ("extension"). *Internal goods* are good things (products, activities, experiences, etc.) that are so essential to a social practice that they partly define it. Some internal goods are *public goods*—benefits provided to the community. Thus, health is the internal good of medicine, and legal justice the internal good of law. The internal goods of engineering, abstractly stated, are safe and useful technological products—products that can be further specified with regard to each area of engineering.

Other internal goods are *personal goods* connected with meaningful work. As an illustration, MacIntyre says that portrait painters discover "the good of a certain kind of life . . . *as a painter*" through "participation in the attempts to sustain progress and to respond creatively to problems," and more generally in the pursuit of excellence as an artist.[29] Similarly, personal meaning in working *as an engineer* connects with personal commitments to create useful and safe public goods and services.

Social practices produce *external goods,* which are goods that can be earned through engaging in a variety of practices. External goods include money, power, self-esteem, and prestige. External goods are, of course, vitally important to individuals and to organizations, and, although MacIntyre does not say so, sometimes they also partly define practices. Thus, we could not understand professions as forms of work without mentioning the money they make possible. Nevertheless, excessive concern for external goods, whether by individuals or organizations, threatens internal goods (both public and personal goods). In extreme instances, they thoroughly corrupt institutions and undermine social practices, as when managers use corporate resources for private gain or when engineers become so demoralized that they fail to maintain standards of professionalism.

Standards of excellence enable internal goods to be achieved (consistent with other important values within democracies). In professions like engineering, these standards include technical guidelines that specify state-of-the-art quality. Most important, they also include the requirements stated in professional codes of ethics, which are incumbent on all members of a profession. The codes promote internal goods positively by encouraging engineers to commit themselves to codified standards of conduct. Codes are also used to impose penalties for dishonesty, destructive types of conflicts of interest, and other failures of professionalism.

The virtues enable engineers to meet standards of excellence and thereby achieve internal goods, especially public or community goods, without allowing external goods such as money and power to distract their public commitments. The virtues thereby add to the personal meaning that engineers find in their work by linking individual lives to wider communities. All four categories of the virtues play key roles in engineers' commitments to the safety, health, and welfare of the public. That is obviously true of the public-spirited, proficiency, and self-governance virtues, but it is equally true of the teamwork virtues required within the organizations that make possible contemporary technological development.

Finally, *progress* is made possible through social practices. Nowhere is this truer than in the professions, which systematically expand our understanding and achievement of public and private goods. Think how dramatically engineers have improved human life by developing the internal combustion engine, computers, the Internet, and a host of consumer products. In this way, engineering and other professions are embedded in wider circles of meaning, in particular within communities and traditions.

We conclude by noting two challenges to virtue ethics from most recent studies in moral psychology. First, moral psychologists challenge the idea of "global virtues" in virtue ethics. Classical virtue ethicists assume that virtues are all *global* traits (e.g., virtues function in all similar circumstances). A person who holds the virtue of honesty can be honest across all situations. Therefore, such a person is reliable and predictable as their virtue of honesty is stable. Moral psychologists argue that in reality most of us only hold "*local* virtues." In other words, virtues only function in one or a few specific (not necessarily) all situations. It is not too difficult to imagine that a mechanic can be professional and caring about their clients while being an irresponsible parent at home (e.g., never caring about their children).

Second, moral psychologists further challenge whether virtues in fact exist at all. Or, instead, those virtuous predispositions we have observed in moral exemplars are in fact activated by some characteristics of situations. We may praise a brave solider on the battleground. However, as suggested by moral psychologists, it is also possible that the soldier's brave behavioral traits are activated by the intense climate on the battleground.[30]

3.3.4 Confucian Role Ethics

A most important non-Western approach to virtue ethics is Confucian ethics. In recent decades, philosophers have employed various approaches to engaging Confucian ethics, the most influential school of thought in the Chinese history, ranging from overtly historical or textual approaches to comparative approaches that put ideas from the classical period into conversation with contemporary Western ethical, social, and scientific theories.[31] Until very recently, scholars have attempted to theorize Confucian ethics as a kind of role-based moral theory.[32] The role-based approach to Confucian ethics is a most recent innovative effort to reinterpret and rediscover the value of Confucian ethics.

In Confucian role ethics, our moral actions in different situations are shaped by the specific roles we take in these situations. We as humans all assume various roles that are determined by the relationships we have with others. These different social relationships and roles affect the ways we choose to interact with others. The tone we use to speak to our parents is different from the one we use to communicate with strangers. The nature of a particular role relationship often evokes feelings and expectations characteristic of that relationship.[33] Roles do not simply describe the social relationships we have with others but also provide normative expectations about the ideal forms of these relationships. In the first place, the roles in Confucian ethics are family-based roles, such as son, daughter, mother, older, sibling, and grandfather. Other social roles discussed in classic Confucianism such as ruler, subject, husband, wife, minister, and friend are also of interest to Confucian scholars.[34] The discussion of role-based morality can be further extended from family roles to social or even professional roles such as engineer, doctor, teacher, and nurse.

Differentiation and fulfillment of different social roles is critical for a harmonious and flourishing society. Through living and reflecting on these social roles, we get to cultivate virtues that are necessitated by the ideal forms of these social roles in particular contexts. For example, to live and reflect on the role as a medical doctor, one gets to cultivate virtues (e.g., benevolence) that are required by an ideal medical doctor. Nevertheless, such process of cultivating virtues cannot be solely completed by the doctor themselves. It needs to be done by both the doctor and the patients they take care of. Therefore, Confucian role ethics advocates a kind of relational moral epistemology: becoming benevolent is something we either do together, or not at all.[35] Confucian role ethics acknowledges the value of social roles in making an agent a true person.[36] What characterizes the personhood is not so much about one's innate and inalienable individual human rights as most Western political and ethical theorists would emphasize. Instead, Confucian role ethicists insist that it is one's intentional efforts to actively live their social roles that defines their personhood.

Therefore, Confucian role ethics defines humans as "the sum of the roles we live in consonance with our fellows."[37] Confucian role ethics appeals to the actual life experience we are living with others both cognitively and affectively. A critical way of becoming virtuous persons in the Confucian tradition is to observe how others practice *li* (rituals, 礼) that are required by the social roles they live. Practicing rituals appropriately can be conducive to the reinforcement of human relationships and associated communal roles. Ritual practices require us to remain both physically and emotionally engaged.[38] Emotions and feelings are thus critical for us to demonstrate our commitment to the practice of rituals and the fulfillment of our role-based moral obligations. A truly caring nurse can never be one who only knows how to follow rules. They develop the virtue of benevolence by feeling what their patients are suffering. Arguably, their emotional engagement with patients' experience allows and encourages them to develop qualities and dispositions that define a truly caring nurse.

From the Confucian perspective, technology is never value neutral. Good technologies should always help promote the values respected and maintained in the communities. Therefore, a Confucian approach to the ethics of technology evaluates to what extent and in what ways technology contributes to a process of harmonization. Reliable technological development often leads to "a continuous negotiation and adjustment of relationships between human beings, society and technology."[39]

More specifically, a major task for the Confucian ethics of technology is to investigate whether practices engendered by technology "are conducive or detrimental to our performance of the social roles."[40] In this sense, the development of technologies such as artificial intelligence can and should be encouraged by our political communities if these technologies help us realize our constitutive commitments or moral obligations prescribed by our social roles (e.g., child, parent). Similarly, technologies that undermine the realization of our constitutive commitments should be restricted.[41] Therefore, if an AI-enabled technology can free us from socially necessary work so that we can be easier to spend time caring for our parents with love and compassion, then such technology should be supported. In contrast, if a cute-looking robot relives all of our caring obligations and our parents are convinced that the robot truly cares about their well-being, then the parents care more about the robots than about their own children. Such robotic technology should be restricted, from the Confucian perspective.

DISCUSSION QUESTIONS

1. Apply Aristotle's idea of the Golden Mean in understanding these virtues of engineers: (a) loyalty to employers, (b) courage in serving the public. In each case, provide an illustration of excess (too much) and defect (too little). Do you find his idea of the Golden Mean illuminating, or does it provide insufficient guidance? Also, Aristotle acknowledged that some virtues have no excess, that is, the more the better. Is that true of the virtue of engineering competence?

2. Review the NSPE Code of Ethics. To what extent do its "Fundamental Canons" rely on the language of the virtues? Try rewriting the canons entirely in terms of the virtues, and comment on what is lost or gained in doing so.

3. In your own words, explain the key ideas in Alasdair MacIntyre's conception of social practices (as they apply to engineering) and Samuel Florman's credo. Discuss similarities and differences in their views, and what you find insightful and problematic in their views.

4. Wrongdoing by professionals is often due in part to pressures within their organizations, but character remains important in understanding why only some professionals succumb to those pressures and engage in wrongdoing. Return to LeMessurier in the Citicorp case in chapter 1 and discuss what kinds of character faults might tempt other engineers in his situation to simply ignore the problem. The faults might be general ones in an individual or those limited to the situation. Consider each of the following categories: (a) moral indifference and negligence, (b) intentional (knowing) wrongdoing, (c) professional incompetence, (d) bias or lack of objectivity, (e) fear, (f) lack of effort, (g) lack of imagination or perspective.

5. We defined virtues as desirable and undesirable habits of conduct, motive, attitude, etc. By extension, we also speak of the character of organizations, that is, the patterns of virtues and vices that are manifested by management, employees, and corporate policies. For example, what is meant when we call a company honest or fair? And which vices were manifested by Enron, as discussed in chapter 1? Or Ford, with regard to safety?

6. Self-respect is an essential virtue in engineering, as elsewhere. What is self-respect? Is it the same as self-esteem, which is an idea popularized in the self-esteem movement in education and in teaching values in recent decades?[42]

7. Imagine you are designing the following social robots inspired by Confucian role ethics, what features do you want to program into the robots? Or, what features you may not want to program? Explain why.

    a. A learning companion robot for university students enrolled in a Chemistry class
    b. An elder care robot that assists nurses at an assisted living center
    c. A therapy robot teaching social skills to children with autism
    d. A robot that takes care of your daily schedule and helps you manage time
    e. A robot nanny that takes care of your children while you are busy at work
    f. A nutritionist robot who gives you advice on your dietary habit

3.4   SELF-REALIZATION AND SELF-INTEREST

Each of the preceding ethical theories leaves considerable room for self-interest, that is, for pursuing what is good for oneself. Utilitarians believe that self-interest should enter into our calculations of the overall good; rights ethics says we have rights to pursue our legitimate interests; duty ethics says we have duties to ourselves; and virtue ethics links our personal good with participating in communities and social practices. Self-realization ethics, however, gives greater prominence to self-interest and to personal commitments that individuals develop.

As with the other ethical theories, we will consider two versions, this time depending on how the self (the person) is conceived. In one version, called ethical egoism, the self is conceived in a highly individualistic manner. In a second version, the self to be realized is understood in terms of caring relationships and communities.

3.4.1   Ethical Egoism

Ethical egoism says that each of us ought always and only to promote our own self-interest. The theory is *ethical* because it is a theory about morality, and it is *egoistic* because it says the sole duty of each of us is to maximize our well-being. Self-interest is understood as our long-term and enlightened well-being (good, happiness), rather than a narrow, short-sighted pursuit of immediate pleasures that leaves us frustrated or damaged in the long run. Thus, Thomas Hobbes (1588–1679) and Ayn Rand (1905–1982) recommend a "rational" concern for one's long-term interests. Hobbes says that rational persons will agree to abide by a "social contract" in which one obeys the laws when others are willing to do so, thereby lifting them from a "state of nature" in which constant war makes life "solitary, poor, nasty, brutish, and short."[43] Rand celebrates a host of virtues exercised on behalf of oneself: self-respect, honesty with oneself, courage and excellence in pursuing personal projects, and even respect for others insofar as it tends to promote one's endeavors.[44]

Nevertheless, these and other ethical egoists do not assume that well-being must involve community and caring for others. Indeed, ethical egoists deny the value of altruism, of caring about others for their sake. Their ethical standard is that each of us should care about our self-interest—period. As such, ethical egoism sounds like an endorsement of selfishness. It implies that engineers should think first and last about what is beneficial to themselves, an implication at odds with the injunction to keep paramount the public health, safety, and welfare. As such, ethical egoism is an alarming view.

Are there any arguments to support ethical egoism? Rand offers three arguments. First, she emphasizes the importance of self-respect, and then portrays altruism toward others as incompatible with valuing oneself. She contends that acts of altruism are degrading, both to others and to oneself: "altruism permits no concept of a self-respecting, self-supporting man—a man who supports his life by his own effort and neither sacrifices himself nor others."[45] This argument contains one important premise: Independence is a value of great importance, especially in democratic and capitalistic economies. Yet independence is not the only important value. In infancy, advanced age, and various junctures in between, each of us is vulnerable. We are also interdependent, as much as independent. Self-respect includes recognition of our vulnerabilities and interdependencies, and certainly it is compatible with caring about other persons as well as about ourselves.

Rand's second argument is that the world would be a better place if all or most people embraced ethical egoism. Especially in her novels, Rand portrays heroic individuals who by pursuing their self-interest indirectly contribute to the good of others. She dramatizes Adam Smith's "invisible hand" argument, set forth in 1776 in *The Wealth of Nations*. According to Smith, in the marketplace individuals do and should seek their own economic interests, but in doing so it is as if each businessperson were "led by an invisible hand to promote an end which was no part of his intention."[46] To be sure, Smith had in mind the invisible hand of God, whereas Rand is an atheist, but both appeal to the general good for society of self-seeking in the professions and business. This argument, too, contains an enormously important truth (although it is doubtful that unrestrained capitalism always maximizes the general good). Nevertheless, contrary to Rand, this argument does not support ethical egoism. For notice that it assumes we ought to care about the well-being of others, for their sake—something denied by ethical egoism itself! And once the general good becomes the moral touchstone, we are actually dealing with a version of utilitarianism.

Rand's third argument is more complex, and it leads to a discussion of human nature and motivation. It asserts that ethical egoism is the only psychologically realistic ethical theory. By nature, human beings are exclusively self-seeking; our sole motives are to benefit ourselves. More fully, *psychological egoism* is true: all people are always and only motivated by what they believe is good for them in some respect. Psychological egoism is a theory about psychology, about what actually motivates human beings, whereas ethical egoism is a statement about how they ought to act. But if psychological egoism is true, ethical egoism becomes the only plausible ethical theory. If by nature we can only care about ourselves, we should at least adopt an enlightened view about how to promote our well-being.

Is psychological egoism true? Is the only thing an engineer or anyone else cares about, ultimately, their own well-being? Psychological egoism flies in the face of common sense, which discerns motives of human decency, compassion, and justice. It is difficult to refute psychological egoism directly, because it radically reinterprets both common sense and experimental data. But we can show that most arguments for psychological egoism are based on seductive and simple confusions. Here are four such arguments for psychological egoism.[47]

Argument 1. We always act on our own desires; therefore, we always and only seek something for ourselves, namely the satisfaction of our desires.

—In reply, the premise is true: we always act on our own desires. By definition, *my* actions are motivated by *my* desires together with *my* beliefs about how to satisfy those desires. But the conclusion does not follow. There are many different kinds of desires, depending on what the desire is for—the object of the desire. When we desire goods for ourselves, we are self-seeking; but when we desire goods for other people (for their sake), we are altruistic. The mere fact that in both instances we act on our own desires does nothing to support psychological egoism.

Argument 2. People always seek pleasures; therefore they always and only seek something for themselves, namely their pleasures.

—In reply, there are different sources of pleasures. Taking pleasure in seeking and getting a good solely for oneself is different from taking pleasure in helping others.

Argument 3. We can always imagine there is an ulterior, exclusively self-seeking motive present whenever a person helps someone else; therefore people always and only seek goods for themselves.

—In reply, there is a difference between imagination and reality. We can also imagine that people who help others have an ulterior desire to eat ants, but it does not follow that altruists are anteaters!

Argument 4. When we look closely, we invariably discover an element of self-interest in any given action; therefore people are solely motivated by self-interest.

—In reply, there is an enormous difference between the presence of "an element" of self-interest (asserted in the premise) and inferring the element is the only motive (asserted in the conclusion). Many actions have multiple motives, with an element of self-interest mixed in with concern for others.

We conclude that there are no sound reasons for believing psychological egoism, nor for believing ethical egoism. In preparation for discussing the second version of self-realization ethics, however, let us comment more fully on the question of what motivates engineers.

3.4.2 Motives of Engineers

Having emphasized that self-seeking is not the only human motive, we now grant that it is a very strong motive. Indeed, it is probably the strongest motive in most of us most of the time. Following Gregory Kavka, let us dub this commonsense view *predominant egoism:* the strongest desire for most people most of the time is self-seeking.[48] Predominant egoism is plausible and open to scientific confirmation. It is also plausible to believe that most acts of helping and service to others involve *mixed motives,* that is, a combination of self-concern and concern for others.

Unlike psychological egoism, predominant egoism acknowledges human capacities for love, friendship, and community involvement. It also acknowledges engineers' capacities for genuinely caring about the public safety, health, and welfare. Engineers are strongly motivated by self-interest, but they are also capable of responding to moral reasons in their own right, as well as additional motives concerned with the particular nature of their work. Their motives are as many and varied as the existential pleasures cited by Samuel Florman.

As just one illustration, consider the motives of Jack Kilby in inventing the microchip.[49] The invention has had momentous importance in making possible the development of today's powerful computers, so much so that in 2000 Kilby was awarded a Nobel Prize—a rare event for an engineer, since Nobel Prizes are usually given for fundamental contributions to science, not engineering.[50] In retrospect, the idea behind the microchip seems simple, as do many creative breakthroughs. During the 1950s the miniaturization of transistors was being pursued at a relentless pace, but it was clear there would soon be a limit to the vast number of minute components that could be wired together. Kilby was well aware of the problem and sensed the need for a fundamentally new approach. In July 1958, only a few weeks after starting a new job at Texas Instruments, he discovered the solution: make all parts of the circuit out of one material integrated on a piece of silicon, thereby removing the need to wire together miniature components.

In making his discovery, Kilby was not pursuing a grand humanitarian intention to provide humanity with the remarkable goods the microchip would make possible, although it is true he was known for his everyday kindness to colleagues. When he was about to give his Nobel lecture, he was introduced as having made the invention that "launched the global digital revolution, making possible calculators, computers, digital cameras, pacemakers, the Internet, etc., etc."[51] In response, he told a story borrowed from another Nobel laureate: "When I hear that kind of thing, it reminds me of what the beaver told the rabbit as they stood at the base of Hoover Dam: 'No, I didn't build it myself, but it's based on an idea of mine.'"

Was Kilby merely seeking money, power, fame, and other rewards just for himself? No, although these things mattered to him. As one biographer suggests, "we see nothing extraordinary in Jack Kilby's private ambition or in his aim to find personal fulfillment through professional achievement. In that regard he was the same as the rest of us: We all pick professions with a mind to fulfilling ourselves."[52] Primarily, Kilby was pursuing interests he had developed years earlier in how to solve technical problems in engineering. In this regard he was exceptional only in his passion for engineering work. Like many creative individuals, he was persistent to the point of being driven, and he found great joy in making discoveries. But even saying this by itself would be misleading. The accurate observation is that he had multiple motives, including motives to advance technology, to be compensated for his work, and to do some good for others.

Building on this observation, we might sort the motives of professionals into three categories: proficiency, compensation, and moral.

Proficiency motives, and their associated values, center on excellence in meeting the technical standards of a profession, together with related aesthetic values of beauty. The undergraduate curriculum for engineering is generally acknowledged to be more rigorous and difficult than the majority of academic disciplines. We might guess that students are attracted to engineering in part because of the challenge it offers to intelligent people. Do empirical studies back up this somewhat flattering portrayal? To a significant extent, yes. Typically, students are motivated to enter engineering primarily by a desire for interesting and challenging work. They have an "activist orientation" in the sense of wanting to create concrete objects and systems—to build them and to make them work. They are more skilled in math than average college students, although they tend to have a low tolerance for ambiguities and uncertainties that cannot be measured and translated into figures.[53]

Compensation motives are for social rewards such as income, power, recognition, and job or career stability. We tend to think of these motives and values as self-interested, and in large degree they are. Yet most people seek money for additional reasons, such as to benefit family members or even to be able to help others in need. In addition, financial independence prevents one from becoming a burden on others. In general, due regard for one's self-interest is a moral virtue—the virtue of *prudence*—assuming it does not crowd out other virtues.

Moral motives include desires to meet one's responsibilities and to respect the rights of others. Such motives of moral respect and caring involve affirming that other people have inherent moral worth. In addition, moral concern involves maintaining self-respect and integrity—valuing oneself as having equal moral worth.

For the most part, these motives are interwoven and mutually supportive. All of them, not only moral motives, contribute to providing valuable services to the community, as well as professional relationships among engineers, other involved workers, and clients. Engineering is demanding, and it requires engineers to summon and to integrate a wide range of motivations. Indeed, life itself is demanding, and it can be argued that our survival requires constant interweaving and cross-fertilization of motives. As Mary Midgley observed, human nature "must consist of a number of motives which are genuinely distinct and autonomous, but which are adapted to fit together, in the normal maturing of the individual, into a life that can satisfy."[54]

3.4.3 Self-Realization, Personal Commitments, and Communities

We turn now to the more community-oriented version of self-realization ethics. This version says that each individual ought to pursue self-realization, but it emphasizes the importance of caring relationships and communities in understanding self-realization and in defining the "self" to be fulfilled. It also highlights personal commitments, such as those of Jack Kilby, which express and develop individual talents while enriching communities.

On the one hand, this version of self-realization ethics emphasizes that we are social beings whose identities and meaning are linked to the communities in which we participate. This theme is expressed by F. H. Bradley (1826–1924): "The 'individual' apart from the community is an abstraction. It is not anything real, and hence not anything that we can realize. . . . I am myself by sharing with others, by including in my essence relations to them, the relations of the social state."[55]

On the other hand, self-realization ethics points to the particular commitments individuals make in their work, as well as in their personal lives. Indeed, a central theme is how personal commitments motivate, guide, and give meaning to the work of engineers and other professionals.[56] They also form the core of an individual's character.[57] As such, they reflect what engineers care about deeply in ways that evoke our interest and energy, shape our identities, and generate pride or shame in our work. These commitments contribute to both public goods and personal fulfillment.

As noted in chapter 1, personal commitments are commitments that might not be incumbent on every member of a profession, including humanitarian, environmental, religious, political, aesthetic, supererogatory, and family commitments. They also include, however, voluntary commitments to obligatory professional standards, especially when these are linked to an individual's broader value perspective.

Personal commitments are often neglected in thinking about professional ethics because we associate professionalism with setting aside personal values in order to be objective and to meet shared standards of the profession. Of course, professionalism does require that personal biases not be allowed to undermine objectivity and shared standards. In general, there are limits to how these commitments are exercised in professional life—limits established primarily by the mandatory requirements expressed in codes of ethics, as well as by common decency and justice.[58] Yet the passion for objectivity and the reasoned devotion to professional standards are themselves personal commitments essential in engineering and science.

Personal commitments are relevant in many ways to professional life.[59] Most important, they create meaning; thereby they motivate professionalism throughout long careers. Professions offer special opportunities for meaningful work, which explains much of their attraction to talented individuals. The relevant idea of meaning has subjective aspects—a "sense of meaning" that enlivens one's daily work and life. It also has objective aspects—the justified values that make work worthwhile and help make life worth living. In the following passage Joanne B. Ciulla has in mind both subjective and objective meaning.

> Meaningful work, like a meaningful life, is morally worthy work undertaken in a morally worthy organization. Work has meaning *because* there is some good in it. The most meaningful jobs are those in which people directly help others or create products that make life better for people. Work makes life better if it helps others; alleviates suffering; eliminates difficult, dangerous, or tedious toil, makes someone healthier and happier; or aesthetically or intellectually enriches people and improves the environment in which we live.[60]

Ciulla emphasizes meaning derived from public-spirited commitments, but equally important is the meaning derived from the technical challenges in work and the relationships among coworkers.

Again, personal commitments shape the kinds of work individuals undertake, including career choices, decisions about particular jobs, and discretionary choice in work assignments. Weapons development is a poignant example. Gene Moriarty reports that his first job prospect after college was a large aerospace company.

> The engineers in my prospective group were excitedly telling me about a system they were developing. It sensed the terrain with an ingenious radar mechanism, employed an elaborate feedback control structure, and made determinations on the basis of statistical decision rules. The job offered fascinating prospects for sophisticated engineering designs. But then I took a wider look at the project and realized that the system I'd be working on was to form part of the signal processing unit of what came to be the Cruise Missile.[61]

Moriarty decided not to pursue the job because, while it offered "a technically sweet project," since childhood he had believed that "war was good for nothing, generally speaking, except making the rich people richer."[62]

In contrast, engineers with commitments to a strong national defense, as essential in safeguarding democratic values, might have responded quite differently, especially if they saw in the cruise missile an accurate weapon that could minimize civilian casualties. Such personal convictions and commitments should not be dismissed as mere subjective matters lacking relevance to engineering ethics. On the contrary, they enter centrally into individuals' understanding of their responsibilities to the public affected by their work.

As a different type of example, consider supererogatory conduct—admirable conduct beyond the minimum duties incumbent on all members of a profession. A dramatic example, chronicled by Loren R. Graham in *The Ghost of the Executed Engineer,* is the courageous and creative life of Peter Palchinsky, who literally sacrificed his life for his ideals.[63] Although he was officially a Marxist, Palchinsky crusaded for ideals such as the rights of workers and the safety of the public affected by technology. Educated as a mining engineer, his first job was studying workers in the coal mining operation in the Ukraine's Don Basin. He immediately saw that the efficiency and productivity of the mines was linked to the workers' living conditions, and he developed the first quantitative information about their poor housing and transportation. The experience was formative. During the next three decades of his career he persistently connected engineering with the people it affects, understanding technical matters as interwoven with social, economic, and political issues. Gradually moving into top leadership positions, he lobbied for engineers to become more broadly educated and to accept wider responsibilities for the human dimensions of their work. Clearly, his personal commitments were not reducible to the shared requirements stated in a professional code of ethics—if only because one of his goals was to win recognition for professional societies that could write such codes! Although his only crime was vigor in pursuing commitments to humane industry and humanitarian engineering, Joseph Stalin had him executed for treason. It is no exaggeration to say that he sacrificed his life in seeking to advance professional standards.

Palchinsky is an extreme case. A more immediate concern is whether the engineering profession should do more to encourage engineers to apply their skills in offering voluntary service to others.[64] There are many essential needs not met in our society. Other professions, especially law, have strong traditions of encouraging pro bono service, that is, services provided free or at reduced fees. Should engineering do the same?

3.4.4 Religious Commitments

These examples barely hint at the myriad ways in which personal commitments, ideals, and meaning enter into professional ethics, including how individuals construe codified responsibilities.[65] Later we offer additional illustrations; for example, in chapter 6 we comment on personal commitments in connection with whistleblowing, and in chapter 8 we comment on environmental commitments. Here we will discuss how personal religious beliefs have relevance to the professional lives of many engineers. Doing so will also provide the opportunity to reflect more widely on the relationship of religion to morality.

For many individuals, religious beliefs and spiritual attitudes are especially important personal commitments relevant to all aspects of their lives, including their professions. Here are two examples.

Egbert Schuurman is a Dutch Calvinist engineer who has written extensively on technology.[66] Highlighting the dangers of technology, he calls for redirecting technology to serve morally worthy aims, both human liberation and respect for the environment. He and his coauthors of *Responsible Technology* articulate normative principles for design. They include: cultural appropriateness (preserving valuable institutions and practices within a particular society); openness (divulging to the public the value judgments expressed in products and also their known effects); stewardship (frugality in the use of natural resources and energy); harmony (effectiveness of products together with promoting social unity); justice (respect for persons); caring (for colleagues and workers); and trustworthiness (deserving consumers' trust).[67]

Mark Pesce is the principal engineer for Shiva Corporation, which invented dial-up networking. In 1994, Pesce and a colleague developed the Virtual Reality Modeling Language (VRML), which allowed three-dimensional models to be placed on the World Wide Web.[68] Emphasizing the importance of spiritual attitudes in his work, he makes it clear that his beliefs are neither orthodox nor associated solely with any one world religion. He characterizes his beliefs as "a mélange of a lot of different religious traditions, including Christian, pre-Christian, Buddhist, Taoist and so on," integrated into a type of "paganism" which is "a practice of harmony, a religion of harmony with yourself and the environment."[69] He is aware that his contributions to technology can be used as tools of communication or weapons of domination. Spiritual attitudes seek ways to allow aspects of the sacred into technology, to find ways for technology to make human life more interconnected through global communication, as well as attuned to nature, and to allow individuals to express themselves in more broadly creative ways through the Web.

These two examples of religious faith, traditional and nontraditional, underscore the highly personal nature of religious belief. They also remind us of the enormous diversity of religious belief. William James suggested that when we examine the full range of religious beliefs "we may very likely find no one essence, but many characters [that is, features] which may alternatively be equally important to religion."[70] Some religions make central belief in one deity (Judaism, Christianity, Islam), others are polytheistic (Hinduism), and still others are nontheistic (Zen Buddhism). Most religions endorse particular worldviews (about human destinies and the origin of the universe), moral perspectives, scriptures, ways of structuring religious communities, and rituals such as prayer and fasting—but the variations are enormous, not only among religions but even within a particular world religion.

Despite their diversity, religious beliefs can support morally responsible conduct in several ways. One way is by providing supporting motivation for being moral. We are not referring primarily to self-interested motives such as the fear of punishment, but rather inspiration rooted in religious faith. Another way religions support moral conduct is by stimulating moral reflection and offering practical guidance, often through stories, parables, and the celebration of moral exemplars such as the lives of prophets and saints. In addition, religions sometimes set a higher moral standard than is conventional. In doing so, many religions emphasize particular ideals of character, which as we noted have permissible variations within a framework of ethical pluralism. For example, the ethics of Christianity centers on the virtues of hope, faith, and especially love; Judaism emphasizes the virtue of *tsedakah* (righteousness); Buddhism emphasizes compassion; Islam emphasizes *ihsan* (translated as either piety or the pursuit of excellence); and Navajo ethics centers on *hozho* (translated variously as harmony, peace of mind, beauty, health, or well-being).

To be sure, sometimes sects employ moral standards below what most of us view as acceptable, for instance by not recognizing the equal rights of women, or by treating children in ways that health professionals see as harmful.[71] Tragically, some religious subgroups engage in terrorism, reminding us that some personal commitments are unjustified by both professional codes and common decency. Religious views are themselves open to moral scrutiny.

3.4.5   Which Ethical Theory Is Best?

Just as ethical theories are used to evaluate actions, rules, and character, ethical theories can themselves be evaluated. In this chapter, our concern has been to introduce some of the most influential ethical theories rather than to try to determine which is preferable. Nevertheless, we sometimes argue against particular versions of each type of theory. For example, we argued against act-utilitarianism, as compared with rule-utilitarianism, and we argued against ethical egoism. We hinted at our preference, as authors, for nonlibertarian versions of rights ethics. And we suggested that few duties are absolute, contrary to Kant.

Which criteria can be used in assessing ethical theories, and which criteria did we use? The criteria follow from the very definition of what ethical theories are. Ethical theories are attempts to provide clarity and consistency, systematic and comprehensive understanding, and helpful practical guidance in moral matters. Sound ethical theories succeed in meeting these aims.

First, sound ethical theories are clear and coherent. They rely on concepts (ideas) that are sufficiently clear to be applicable, and their various claims and principles are internally consistent.

Second, sound ethical theories organize basic moral values in a systematic and comprehensive way. They highlight important values and distinguish them from what is secondary. And they apply to all circumstances that interest us, not merely to a limited range of examples.

Third, and most important, sound ethical theories provide helpful guidance that is compatible with our most carefully considered moral convictions (judgments, intuitions) about concrete situations. Who does "our" refer to? It refers to each of us, in moral dialogue with others. To take an extreme case, if an ethical theory said it was all right for engineers to create extremely dangerous products without the public's informed consent, then that would show the theory is inadequate.

Of course, even our most carefully considered convictions can be mistaken, sometimes flagrantly so as with racists and other bigots. An important role of a sound ethical theory is to improve our moral insight into particular problems. Hence, there is an ongoing checking of an ethical theory (or general principles and rules) against the judgments about specific situations (cases, dilemmas, issues) that we are most confident are correct, and, in reverse, a checking of our judgments about specific situations by reference to the ethical theory. Theories and specific judgments are continually adjusted to each other in a back-and-forth process until we reach what John Rawls calls a *reflective equilibrium:* "It is an equilibrium because at last our principles and judgments coincide; and it is reflective since we know to what principles our judgments conform and the premises of their derivation."[72]

Which of the ethical theories most fully satisfies these criteria? In our view, some versions of rule-utilitarianism, rights ethics, duty ethics, virtue ethics, and self-realization ethics all satisfy the criteria in high degrees. We find ourselves more impressed by the similarities and connections, rather than the differences, among the general types of theories.

Thus, we suggested that duty ethics and rights ethics largely differ in emphasis. We also suggested that virtue ethics needs to be complemented by the other theories. There are many other connections among the theories that might be pursued. For example, the community-oriented version of self-realization ethics can be linked to Kant's idea of duties to oneself, Mill's emphasis on personal liberty, and to the Aristotelian pursuit of excellence. In any case, the differences within each of the moral traditions are at least as striking as the differences between the types of theories themselves.[73] For example, the internal differences between libertarianism and most rights ethics, or between ethical egoism and community-oriented self-realization theories, reflect the broader differences in moral perspectives about the relationships between individuals and communities.

DISCUSSION QUESTIONS

1. The following widely discussed case study was written by Bernard Williams (1929–2003). The case is about a chemist, but the issues it raises are equally relevant to engineering. What should George do in order to best preserve his integrity? Is it permissible for him to take the job and "compartmentalize" so as to separate his work and his personal commitments? In your answer, discuss whether in taking the job George would be compromising in either of the two senses of "compromise": (1) undermine integrity by violating one's fundamental moral principles; (2) settle moral dilemmas and differences by mutual concessions or to reconcile conflicts through adjustments in attitude and conduct.[74]

   "George, who has just taken his Ph.D. in chemistry, finds it extremely difficult to get a job. He is not very robust in health, which cuts down the number of jobs he might be able to do satisfactorily. His wife has to go out to work to keep [i.e., to support] them, which itself causes a great deal of strain, since they have small children and there are severe problems about looking after them. The results of all this, especially on the children, are damaging. An older chemist, who knows about this situation, says that he can get George a decently paid job in a certain laboratory, which pursues research into chemical and biological warfare."[75]

2. With regard to each of the following cases, first discuss what morality requires and then what self-interest requires. Is the answer the same or different?

   a. Bill, a process engineer, learns from a former classmate who is now a regional compliance officer with the Occupational Safety and Health Administration (OSHA) that there will be an unannounced inspection of Bill's plant. Bill believes that unsafe practices are often tolerated in the plant, especially in the handling of toxic chemicals. Although there have been small spills, no serious accidents have occurred in the plant during the past few years. What should Bill do?[76]

   b. On a midnight shift, a botched solution of sodium cyanide, a reactant in an organic synthesis, is temporarily stored in drums for reprocessing. Two weeks later, the day shift foreperson cannot find the drums. Roy, the plant manager, finds out that the batch has been illegally dumped into the sanitary sewer. He severely disciplines the night shift foreperson. Upon making discreet inquiries, he finds out that no apparent harm has resulted from the dumping.[77] Should Roy inform government authorities, as is required by law in this kind of situation?

3. A *work ethic* is a set of attitudes, which implies a motivational orientation, concerning the value of work.[78] Which, if any, of the following work ethics do you find attractive, and why? Which of them, as applied to engineering, are compatible or incompatible with the kinds of commitments desirable for professionals?

   a. The Protestant work ethic, as named and analyzed by sociologist Max Weber in *The Protestant Ethic and the Spirit of Capitalism,* was the idea that financial success is a sign that predestination has ordained one as favored by God. This was thought to imply that making maximal profits is a duty mandated by God. Profit becomes an end in itself rather than a means to other ends. It is to be sought rationally, diligently, and perhaps without compromise with other values such as spending time with one's family.

b. Work is a necessary evil. It is the sort of thing one must do in order to avoid worse evils, such as dependency and poverty. But it is mind-numbing, degrading, and a major source of anxiety and unhappiness.

c. Work is the major instrumental good in life. It is the central means for providing the income needed to avoid economic dependence on others, for obtaining desired goods and services, and for achieving status and recognition from others.

d. Work is intrinsically valuable to the extent that it is enjoyable or meaningful in allowing personal expression and self-fulfillment. Meaningful work is worth doing for its own sake and for the sense of personal identity and self-esteem it brings.

4. Discuss the following claim: "It is irrelevant what the motives of professionals are; what matters is that they do what is right." In your answer, distinguish questions about the motives for a specific right action and questions about habits or patterns of motivation throughout a career.

5. One argument against ethical egoism is that it is self-defeating. In stating "the paradox of happiness," John Stuart Mill wrote: "Those only are happy . . . who have their minds fixed on some object other than their own happiness; on the happiness of others, on the improvement of mankind, even on some art or pursuit, followed not as a means, but as itself an ideal end. Aiming thus at something else, they find happiness by the way."[79] The idea is that self-absorption tends to narrow our interests and shut us off from rewarding relationships. Most of life's deepest satisfactions, whether in one's work or in personal relationships, come from developing commitments to other persons and activities. Assess Mill's argument, and discuss whether it provides a refutation of ethical egoism.

6. Psychologist Carol Gilligan, in her book *In a Different Voice,* argues that women tend to define themselves more in terms of caring relationships with others, whereas men tend to think of themselves more individualistically.[80] Based on your experience, is that true? If so, what implications might it have in thinking about engineering ethics?

7. Long before H. G. Wells wrote *The Invisible Man,* Plato (428–348 B.C.) in *The Republic* described a shepherd named Gyges who, according to a Greek legend, discovers a ring that enables him to become invisible when he turns its bezel. Gyges uses his magical powers to seduce the queen, kill the king, and take over an empire. If we have similar powers, why should we feel bound by moral constraints? In particular, if professionals are sufficiently powerful to pursue their desires without being caught for malfeasance, why should they care about the good of the wider public?

In your answer, reflect on the question "Why be moral?" Is the question asking for self-interested reasons for being moral, and if so does it already presuppose that only self-interest, not morality, provides valid reasons for conduct?

KEY CONCEPTS

—**Utilitarianism:** Right action consists in producing the most good for the most people, giving equal consideration to everyone affected. *Act-utilitarianism* says maximize the overall good of each action, in each situation. *Rule-utilitarianism* says live by a set of rules that maximize the overall good.

—**Theories of good** specify *intrinsic goods,* that is, things inherently worth seeking, perhaps such things as pleasure, happiness, a list of desirable activities and relationships, satisfaction of preferences, or satisfaction of rational desires.

—**Rights ethics:** Right action consists in respecting human rights. *Most rights ethicists* believe there are both liberty rights (right not to be interfered with) and welfare rights (right to benefits needed for a decent human life when one cannot earn those benefits on one's own and when the community has them available). In contrast, *libertarians* believe there are only liberty rights. In addition to human rights, which we have because we are human beings, there are special moral rights that arise because of contracts and other special relationships.

—**Duty ethics:** Right actions are those required by principles of duty to respect the autonomy (self-determination) of individuals.

—**Prima facie duties** are duties that have some permissible exceptions when they conflict with more pressing duties, as distinct from *absolute duties* that never have justified exceptions. (In similar senses, "prima facie" is sometimes applied to rights, rules, principles, etc.)

—**Virtue ethics:** We should develop and manifest good character as defined by the *virtues*—desirable habits or tendencies in action, commitment, motive, attitude, emotion, ways of reasoning, and ways of relating to others.

—**Self-realization ethics:** Right action consists in seeking self-fulfillment. In one version, the self to be realized is defined by caring relationships with other individuals and communities. In another version, called *ethical egoism,* right action consists in always promoting what is good for oneself, with no presumption that the self is defined in terms of caring and community relationships.

—**Theories about motivation:** General perspectives on what motivates engineers and others. *Psychological egoism* says that all people are only motivated by self-seeking, that is, by what they believe is good for them (at least in some respect). More plausibly, *predominant egoism* says that the strongest desire for most people most of the time is self-seeking. This view allows that engineers are motivated by combinations of *proficiency motives* (skill, excellence), *compensation motives* (money, power, recognition), and *moral motives* (respect and caring for others).

REFERENCES

1. Matthew D. Adler and Eric A. Posner (eds.), *Cost-Benefit Analysis: Legal, Economic, and Philosophical Perspectives* (Chicago: University of Chicago Press, 2001).
2. Frank Camps, "Warning an Auto Company about an Unsafe Design," in Alan F. Westin (ed.), *Whistle-Blowing!: Loyalty and Dissent in the Corporation* (New York: McGraw-Hill, 1981), pp. 119–29; W. Michael Hoffman, "The Ford Pinto," in W. Michael Hoffman, Robert E. Frederick, and Mark S. Schwartz (eds.), *Business Ethics,* 4th ed. (Boston: McGraw-Hill, 2001), pp. 497–503.
3. Mark Dowie, "Pinto Madness," *Mother Jones* (September-October 1977). For an especially helpful collection of essays, see Douglas Birsch and John H. Fielder (eds.), *The Ford Pinto Case* (Albany, NY: State University of New York Press, 1994).
4. John Stuart Mill, *Utilitarianism, with Critical Essays,* ed. Samuel Gorovitz (Indianapolis, IN: Bobbs-Merrill, 1971).
5. Richard B. Brandt, *A Theory of the Good and the Right* (Oxford: Clarendon Press, 1979).
6. See Sterling Harwood, "Eleven Objections to Utilitarianism," in Louis Pojman (ed.), *Moral Philosophy: A Reader,* 2nd ed. (Indianapolis: Hackett Publishing, 1998), pp. 179–92.
7. Benjamin Todd, "*Is It Ever Okay to Take a Harmful Job in Order to Do More Good? An In-depth Analysis,*" 8000 Hours (August 2017).
8. Such a theory of good, called "ideal utilitarianism," was set forth by G. E. Moore (1873–1958) in *Principia Ethica* (Cambridge: Cambridge University Press, 1959 [1903]).

9. Roger Rosenblatt, "How Do Tobacco Executives Live with Themselves?" *New York Times Magazine,* March 20, 1994, pp. 34–41, 55.

10. Alan H. Goldman, *The Moral Foundations of Professional Ethics* (Totowa, NJ: Rowman and Littlefield, 1980); and Alan Gewirth, "Professional Ethics: The Separatist Thesis," *Ethics* 96 (1986).

11. A. I. Melden, *Rights and Persons* (Berkeley, CA: University of California Press, 1977). Other rights ethicists include Ronald Dworkin, author of *Taking Rights Seriously* (Cambridge, MA: Harvard University Press, 1978) and many additional books; Alan Gewirth, author of *Human Rights* (Chicago: University of Chicago Press, 1982); and other authors anthologized in Patrick Hayden (ed.), *The Philosophy of Human Rights* (St. Paul, MN: Paragon House, 2001).

12. John Locke, *Two Treatises of Government* (Cambridge: Cambridge University Press, 1960).

13. Heather Widdows, *Global Ethics: An Introduction* (Abingdon, UK: Routledge, 2014).

14. Robin R. Wang, "Globalizing the Heart of the Dragon: The Impact of Technology on Confucian Ethical Values," *Journal of Chinese Philosophy* 29, no (4): 553–569.

15. Bernard Gert, *Morality* (New York: Oxford University Press, 1988), p. 157.

16. Immanuel Kant, *Foundations of the Metaphysics of Morals,* trans. L. W. Beck (New York: Liberal Arts Press, 1959). Anthologized widely, for example in A. I. Melden, *Ethical Theories,* 2nd ed. (Englewood Cliffs, NJ: Prentice Hall, 1967), p. 345. Also translated as *Grounding for the Metaphysics of Morals,* trans. James W. Ellington (Indianapolis: Hackett, 1981). On interpreting Kant, see Roger J. Sullivan, *An Introduction to Kant's Ethics* (Cambridge: Cambridge University Press, 1994).

17. Immanuel Kant, *Foundations of the Metaphysics of Morals,* p. 339.

18. Jeffrey Wattles, *The Golden Rule* (New York: Oxford University Press, 1996).

19. Glen Miller, "Toward Lifelong Excellence: Navigating the Engineering-Business Space," in Steen Hyldgaard Christensen, Bernard Delahouse, Christelle Didier, Martin Meganck, and Mike Murphy (eds.), *The Engineering-Business Nexus: Symbiosis, Tension and Co-Evolution* (Cham, Switzerland: Springer, 2019): 81–101.

20. William J. Frey, "Ethics of Team Work," *Ethics in Science and Engineering National Clearinghouse.* 334. Available at: https://scholarworks.umass.edu/esence/334

21. John Kekes, *The Examined Life* (Lewisburg, PA: Buckness University Press, 1988).

22. Samuel C. Florman, *The Existential Pleasures of Engineering,* 2nd ed. (New York: St. Martin's Griffin, 1994).

23. Samuel C. Florman, *The Civilized Engineer* (New York: St. Martin's Press, 1987), p. 101. Here we offer an interpretation of Florman, who would possibly identify himself as a virtue ethicist were he to orient himself to the history of philosophy.

24. Ibid., p. 83.

25. Samuel C. Florman, "Moral Blueprints," *Harpers* 257 (October 1978): 32. Reprinted in Samuel C. Florman, *Blaming Technology* (New York: St. Martin's Press, 1981), pp. 162–80.

26. Aristotle, *Ethics,* trans. J. A. K. Thomson and Hugh Tredennick (New York: Penguin, 1976).

27. Some of the material that follows is from Mike W. Martin, "Personal Meaning and Ethics in Engineering," *Science and Engineering Ethics* 8 (2002): 545–60. It is used with the permission of the publisher. Albert Flores also applies MacIntyre's thinking to the professions in "What Kind of Person Should a Professional Be?" in Albert Flores (ed.), *Professional Ideals* (Belmont, CA: Wadsworth, 1988), p. 6.

28. Alasdair MacIntyre, *After Virtue,* 2nd ed. (South Bend, IN: University of Notre Dame Press, 1984), p. 187.

29. Ibid., pp. 189–90.

30. Edward Slingerland, "The Situationist Critique and Early Confucian Virtue Ethics," *Ethics* 121, no. 2: 390–419.

31. Sarah Mattice, "Confucian Role Ethics: Issues of Naming, Translation, and Interpretation," in Alexus Mcleod (ed.), *The Bloomsbury Research Handbook of Early Chinese Ethics and Political Philosophy* (London, UK: Bloomsbury Academic, 2019), pp. 25–44.

32. Roger T. Ames, *Confucian Role Ethics: A Vocabulary* (Honolulu, HI: The University of Hawai'i Press, 2011).

33. Roger T. Ames, *Confucian Role Ethics.*

34. Stephen Angel, "Building Bridges to Distinct Shores: Pragmatic Problems with Confucian Role Ethics," in Jim Behuniak (ed.), *Appreciating the Chinese Difference: Engaging Roger T. Ames on Methods, Issues, and Roles* (Albany, NY: SUNY Press, 2018), pp. 159–82.

35. Roger T. Ames, *Confucian Role Ethics.*

36. A. T. Nuyen, "Confucian Ethics as Role-based Ethics," *International Philosophical Quarterly* 47, no. 3 (2007): 315–28.

37. Henry Rosemont and Roger Ames, *Confucian Role Ethics: A Moral Vision for the 21st Century?* (Taipei: National Taiwan University, 2016), p. 112.

38. Kurtis Hagen, "The Propriety of Confucius: A Sense-of-Ritual," *Asian Philosophy* 20, no. 1 (2010): 1–25.

39. Pak-hang Wong, "Dao, Harmony and Personhood: Towards a Confucian Ethics of Technology," *Philosophy and Technology* 25, no. 1 (2012): 67–86.

40. Ibid., p. 83.

41. Daniel A. Bell and Wang Pei, *Just Hierarchy: Why Social Hierarchies Matter in China and the Rest of the World* (Princeton, NJ: Princeton University Press, 2020).

42. See Andrew M. Mecca, Neil J. Smelser, and John Vasconcellos (eds.), *The Social Importance of Self-Esteem* (Berkeley, CA: University of California Press, 1989); and Robin S. Dillon (ed.), *Dignity, Character, and Self-Respect* (New York: Routledge, 1995).

43. Thomas Hobbes, *Leviathan*, chapter 13. For an especially insightful discussion, see Gregory S. Kavka, *Hobbesian Moral and Political Theory* (Princeton, NJ: Princeton University Press, 1986).

44. Ayn Rand, *The Virtue of Selfishness* (New York: New American Library, 1964). Many of Rand's themes derive from Friedrich Nietzsche, who developed them with a more elitist emphasis on the exceptionally talented individual. See especially *Thus Spoke Zarathustra,* trans. Walter Kaufmann (New York: Penguin, 1966).

45. Ayn Rand, *The Virtue of Selfishness*, p. ix, italics removed.

46. Adam Smith, *An Inquiry Into the Nature and Causes of the Wealth of Nations* (New York: Oxford University Press, 1976), vol. 1, p. 456.

47. See James Rachels, *The Elements of Moral Philosophy,* 4th ed. (Boston: McGraw-Hill, 2003), pp. 63–75.

48. Gregory S. Kavka, *Hobbesian Moral and Political Theory* (Princeton, NJ: Princeton University Press, 1986).

49. T. R. Reid, *The Chip* (New York: Random House, 2001); and Jeffrey Zygmont, *Microchip: An Idea, Its Genesis, and the Revolution It Created* (Cambridge, MA: Perseus Publishing, 2003).

50. Another engineer, Robert Noyce, would no doubt have shared the prize had he not died 10 years earlier (the award is not given posthumously) because he also invented the integrated circuit about the same time as Kilby.

51. T. R. Reid, *The Chip,* p. 265.

52. Jeffrey Zygmont, *Microchip,* p. 3.

53. Robert Perrucci and Joel E. Gerstl, *Profession Without Community: Engineers in American Society* (New York: Random House, 1969), pp. 27–52.

54. Mary Midgley, *Beast and Man: The Roots of Human Nature,* rev. ed. (New York: New American Library, 1994), p. 331.

55. F. H. Bradley, *Ethical Studies* (New York: Oxford University Press, 1962), p. 173. The quotation is from "My Station and Its Duties," Bradley's most famous essay in ethics. Bradley is an Hegelian, that is, a philosopher strongly influenced by Hegel (1770–1831).

56. Some of the material that follows is taken from Mike W. Martin, "Personal Meaning and Ethics in Engineering," *Science and Engineering Ethics* 8 (2002): 545–60. It is used with permission of the publisher.

57. Bernard Williams, "Persons, Character and Morality," in Bernard Williams, *Moral Luck* (New York: Cambridge University Press, 1981), p. 5.

58. For an especially illuminating study of immoral personal commitments by engineers and scientists, forbidden by professional codes and common decency alike, see Robert Jay Lifton,

Destroying the World to Save It: Aum Shinrikyo, Apocalyptic Violence, and the New Global Terrorism (New York: Henry Holt, 2000).

59. See Arnold Pacey, *Meaning in Technology* (Cambridge, MA: MIT Press, 1999); and Mike W. Martin, "Personal Meaning and Ethics in Engineering," *Science and Engineering Ethics* 8 (2002): 545–60.

60. Joanne B. Ciulla, *The Working Life: The Promise and Betrayal of Modern Work* (New York: Times Books, 2000), pp. 225–26.

61. Gene Moriarty, "Ethics, Ethos and the Professions: Some Lessons from Engineering," *Professional Ethics* 4 (1994): 77.

62. Ibid., p. 77.

63. Loren R. Graham, *The Ghost of the Executed Engineer: Technology and the Fall of the Soviet Union* (Cambridge, MA: Harvard University Press, 1993).

64. Robert J. Baum, "Engineering Services," *Business and Professional Ethics Journal* 4 (1985): 117–35.

65. For additional examples, see: Caroline Whitbeck, *Ethics in Engineering Practice and Research* (New York: Cambridge University Press, 1998), pp. 306–12; Stephen V. Monsma (ed.), *Responsible Technology: A Christian Perspective* (Grand Rapids, MI: William B. Eerdmans Publishing Company, 1986); and Mary Tiles and Hans Oberdiek, *Living in a Technological Culture: Human Tools and Human Values* (New York: Routledge, 1995), pp. 172–75.

66. Egbert Schuurman, *Technology and the Future* (Toronto: Wedge Publishing, 1980); and "The Modern Babylon Culture," in *Technology and Responsibility,* ed. Paul Durbin (Dordrecht, Holland: D. Reidel, 1987).

67. Stephen V. Monsma et al., *Responsible Technology* (Grand Rapids, MI: William B. Eerdmans Publishing, 1986), pp. 171–77.

68. Robert H. Reid, *Architects of the Web* (New York: John Wiley & Sons, 1997), pp. 167–209.

69. "Virtually Sacred," an interview in W. Mark Richardson and Gordy Slack, *Faith in Science* (London: Routledge, 2001), p. 104.

70. William James, *The Varieties of Religious Experience* (New York: Modern Library, 1902), p. 27.

71. Margaret P. Battin, *Ethics in the Sanctuary: Examining the Practices of Organized Religion* (New Haven: Yale University Press, 1990).

72. John Rawls, *A Theory of Justice,* rev. ed. (Cambridge, MA: Harvard University Press, 1999), p. 18.

73. For example, Mill presents his famous defense of freedom in *On Liberty* as a utilitarian perspective, but libertarians generally cite it as a compelling defense of their viewpoint.

74. Martin Benjamin, *Splitting the Difference: Compromise and Integrity in Ethics and Politics* (Lawrence, KS: University of Kansas Press, 1990).

75. Bernard Williams, *Utilitarianism: For and Against* (New York: Cambridge University Press, 1973), pp. 97–98.

76. Jay Matley, Richard Greene, and Celeste McCauley, "Health, Safety and Environment," *Chemical Engineering* 28 (September 1987), p. 115.

77. Ibid., p. 117.

78. David J. Cherrington, *The Work Ethic* (New York: AMACOM, 1980), pp. 19–30, 253–74.

79. John Stuart Mill, *The Autobiography of John Stuart Mill* (Garden City, NY: Doubleday), p. 110.

80. Carol Gilligan, *In a Different Voice: Psychological Theory and Women's Development* (Cambridge, MA: Harvard University Press, [1982] 1993). On possible applications to engineering ethics, see Alison Adam, "Heroes or Sibyls? Gender and Engineering Ethics," *IEEE Technology and Society Magazine* 20, no. 3 (Fall 2001): 39–46.

CHAPTER
4

ENGINEERING
AS SOCIAL
EXPERIMENTATION

As it departed on its maiden voyage in April 1912, the *Titanic* was proclaimed the greatest engineering achievement ever. Not merely was it the largest ship the world had seen, having a length of almost three football fields; it was also the most glamorous of ocean liners, complete with a tropical vinegarden restaurant and the first seagoing masseuse. It was touted as the first fully safe ship. Since the worst collision envisaged was at the juncture of two of its sixteen watertight compartments, and since it could float with any four compartments flooded, the *Titanic* was believed to be virtually unsinkable.

Buoyed by such confidence, the captain allowed the ship to sail full speed at night in an area frequented by icebergs, one of which tore a large gap in the ship's side, flooding five compartments. Time remained to evacuate the ship, but there were not enough lifeboats to accommodate all the passengers and crew. British regulations then in effect did not foresee vessels of this size. Accordingly only 825 places were required in lifeboats, sufficient for a mere one-quarter of the *Titanic*'s capacity of 3,547 passengers and crew. No extra precautions had seemed necessary for an unsinkable ship resulting in the deaths of 1,522 people out of the 2,227 on board.[1]

The *Titanic* remains a haunting tragedy of technological complacency. So many products of technology present some potential dangers that engineering should be regarded as an inherently risky activity. In order to underscore this fact and help to explore its ethical implications, we suggest that engineering should be viewed as an *experimental* process. It is not, of course, an experiment conducted solely in a laboratory under controlled conditions. Rather, it is an experiment on a social scale involving human subjects.

There are conjectures that the *Titanic* left England with a coal fire on board, that this made the captain rush the ship to New York, and that water entering the coal bunkers through the gash caused an explosion and greater damage to the compartments. Others maintain that embrittlement of the ship's steel hull in the icy waters caused a much larger crack than a collision would otherwise have produced. Shipbuilders have argued that carrying the watertight bulkheads up higher on such a big ship instead of allowing less obstructed space on the passenger decks for arranging cabins would have kept the ship afloat. However, what matters most is that the lack of lifeboats and the difficulty of launching those available from the listing ship prevented a safe exit for two-thirds of the persons on board, where a *safe exit* is a mechanism or procedure for escape from harm in the event a product fails.

4.1   ENGINEERING AS EXPERIMENTATION

Experimentation is commonly recognized as playing an essential role in the design process. Preliminary tests or simulations are conducted from the time it is decided to convert a new engineering concept into its first rough design. Materials and processes are tried out, usually employing formal experimental techniques. Such tests serve as the basis for more detailed designs, which in turn are tested. At the production stage further tests are run, until a finished product evolves. The normal design process is thus iterative, carried out on trial designs with modifications being made on the basis of feedback information acquired from tests. Beyond those specific tests and experiments, however, each engineering project taken as a whole may be viewed as an experiment.

4.1.1   Similarities to Standard Experiments

Several features of virtually every kind of engineering practice combine to make it appropriate to view engineering projects as experiments. First, any project is carried out in partial ignorance. There are uncertainties in the abstract model used for the design calculations; there are uncertainties in the precise characteristics of the materials purchased; there are uncertainties in the precision of materials processing and fabrication; there are uncertainties about the nature of the stresses the finished product will encounter. Engineers often have to commence working before all the relevant facts are received. Indeed, one talent crucial to an engineer's success lies precisely in the ability to accomplish tasks safely with only a partial knowledge of scientific laws about nature and society.

Second, the final outcomes of engineering projects, like those of experiments, are generally uncertain. Often in engineering it is not even known what the possible outcomes may be, and great risks may attend even seemingly benign projects. A reservoir may do damage to a region's social fabric or to its ecosystem. It may not even serve its intended purpose if the dam leaks or breaks. An aqueduct may bring about a population explosion in a region where it is the only source of water, creating dependency and vulnerability without adequate safeguards. A special-purpose fingerprint reader may find its main application in the identification and surveillance of dissidents by totalitarian regimes. A nuclear reactor, the scaled-up version of a successful smaller model, may exhibit unexpected problems that endanger the surrounding population, leading to its untimely shutdown at great cost to owner and consumers alike. In the past, a hair dryer may have exposed users to lung damage from the asbestos insulation in its barrel.

Third, effective engineering relies upon knowledge gained about products both before and after they leave the factory—knowledge needed for improving current products and creating better ones. That is, ongoing success in engineering depends upon gaining new knowledge, as does ongoing success in experimentation. Monitoring is thus as essential to engineering as it is to experimentation in general. To monitor is to make periodic observations and tests in order to check for both successful performance and unintended side effects. But since the ultimate test of a product's efficiency, safety, cost-effectiveness, environmental impact, and aesthetic value lies in how well that product functions within society, monitoring cannot be restricted to the in-house development or testing phases of an engineering venture. It also extends to the stage of client use. Just as in experimentation, both the intermediate and final results of an engineering project deserve analysis if the correct lessons are to be learned from it.

4.1.2 Learning from the Past

Usually engineers learn from their own earlier design and operating results, as well as from those of other engineers, but unfortunately that is not always the case. A lack of established channels of communication, misplaced pride in not asking for information, embarrassment at failure or fear of litigation, and neglect often impede the flow of such information and lead to many repetitions of past mistakes. Here are a few examples:

1. The *Titanic* lacked a sufficient number of lifeboats decades after most of the passengers and crew on the steamship *Arctic* had perished because of the same problem.[2]
2. "Complete lack of protection against impact by shipping caused Sweden's worst ever bridge collapse on Friday as a result of which eight people were killed." Thus reported the *New Civil Engineer* on January 24, 1980. On May 15 of the same year it also reported the following: "Last Friday's disaster at Tampa Bay, Florida, was the largest and most tragic of a growing number of incidents of errant ships colliding with bridges over navigable waterways." As the empty phosphate freighter "Summit Venture" slammed into a pier of the bridge, it had knocked 1,261 feet of center span, cantilever approach, and roadway into the bay, along with 35 people on board a greyhound bus. While collisions of ships with bridges do occur—other well-known cases include the Maracaibo Bridge (Venezuela, 1964) and the Tasman Bridge (Australia, 1975)—Tampa's Sunshine Skyline Bridge was not designed with horizontal impact forces in mind because the code did not require it. Engineers now

recommend the use of floating concrete bumpers that can deflect ships, but that recommendation can go unheeded as seen by the 1993 collapse of the Bayou Canot Bridge which cost 43 passengers of the *Sunset Limited* their lives.

3. In June 1966 a section of the Milford Haven Bridge in Wales collapsed during construction. A bridge of similar design was being erected by the same bridge builder (Freeman Fox and Partners) in Melbourne, Australia, when it, too, partially collapsed, killing 33 people and injuring 19. This happened in October of the same year, shortly after chief construction engineer Jack Hindshaw (also a casualty) had assured worried workers that the bridge was safe.[3]

4. Valves are notorious for being among the least reliable components of hydraulic systems. It was a pressure relief valve, and a lack of definitive information regarding its open or shut state, which contributed to the nuclear reactor accident at Three Mile Island on March 28, 1979. Similar malfunctions had occurred with identical valves on nuclear reactors at other locations. The required reports had been filed with Babcock and Wilcox, the reactor's manufacturer, but no attention had been given to them.[4]

These examples illustrate why it is not enough for engineers to rely on handbooks and computer programs without knowing the limits of the tables and algorithms underlying their favorite tools. They do well to visit shop floors and construction sites to learn from workers and testers how well the customers' wishes were met. The art of back-of-the-envelope calculations to obtain ballpark values with which to quickly check lengthy and complicated computational procedures must not be lost. Engineering, just like experimentation, demands practitioners who remain alert and well informed at every stage of a project's history and who exchange ideas freely with colleagues in related departments.

4.1.3   Contrasts with Standard Experiments

To be sure, engineering differs in some respects from standard experimentation. It is worth noting that the "standard experiments" here we are comparing with engineering mainly refer to medical experiments which often involve human subjects. Some scientific experiments such as those in chemistry, physics, and geological sciences may not directly involve human subjects. Some of those very differences help to highlight the engineer's special responsibilities. Exploring the differences can also aid our thinking about the moral responsibilities of all those engaged in engineering.

EXPERIMENTAL CONTROL. One great difference arises with experimental control. In a standard experiment this involves the selection, at random, of members for two different groups. The members of one group receive the special, experimental treatment. Members of the other group, called the control group, do not receive that special treatment, although they are subjected to the same environment as the first group in every other respect.

In engineering, this is not the usual practice—unless the project is confined to laboratory experimentation—because the experimental subjects are human beings or finished and sold products out of the experimenter's control. Indeed, clients and consumers exercise most of the control as they choose the product or item they wish to use. This makes it impossible to obtain a random selection of participants from various groups. Nor can parallel control groups be established based on random sampling. Thus it is not possible to study the effects that changes in variables have on two or more comparison groups, and one must simply work with the available historical and retrospective data about various groups that use the product.

INFORMED CONSENT. Viewing engineering as an experiment on a societal scale places the focus where it should be: on the human beings affected by technology, for the experiment is performed on persons, in rare occasions could also be on inanimate objects such as geological engineering experiments or more recently geoengineering projects. In this respect, albeit on a much larger scale, engineering closely parallels medical testing of new drugs or procedures on human subjects.

A subject's safety and freedom of choice as to whether to participate in medical experiments is of the utmost importance. Ever since the revelations of the horrors conducted in prisons and concentration camps in the name of science and medicine, an increasing number of moral and legal safeguards were put in place to ensure that subjects in experiments participate on the basis of informed consent.

Contemporary engineering practice is only beginning to recognize that informed consent, which is so vital to the concept of a properly conducted experiment involving human subjects, should be the keystone in the interaction between engineers and the public. When a manufacturer sells a new device to a knowledgeable firm that has its own engineering staff, there is usually an agreement regarding the shared risks and benefits of trying out the technological innovation.

Informed consent is understood as including two main elements: knowledge and voluntariness. First, subjects should be given not only the information they request, but all the information needed to make a reasonable decision. Second, subjects must enter into the experiment without being subjected to coercion, fraud, or deception.

The mere purchase of a product does not constitute informed consent, any more than does the act of showing up on the occasion of a medical examination. The public and clients must be given information about the practical risks and benefits of the process or product in terms they can understand. Supplying complete information is neither necessary nor in most cases possible. In both medicine and engineering there may be an enormous gap between the experimenter's and the subject's understanding of the complexities of an experiment. But while this gap most likely cannot be entirely closed, it should be possible to convey all pertinent information needed for making a reasonable decision on whether to participate.

We do not propose a process resembling the preparation and release of environmental impact reports. Those reports should be carried out anyway when large projects are involved. We favor the kind of sound advice a responsible physician gives a patient when prescribing a course of drug treatment that has possible side effects. The physician must search beyond the typical sales brochures from drug manufacturers for adequate information; hospital management must allow the physician the freedom to undertake different treatments for different patients, as each case may constitute a different "experiment" involving different circumstances; finally, the patient must be readied to receive the information.

Likewise, engineers cannot succeed in providing essential information about a project or product unless there is cooperation by superiors and also receptivity on the part of those who should have the information. Management is often understandably reluctant to provide more information than current laws require, fearing disclosure to potential competitors and exposure to potential lawsuits. Moreover, it is possible that, paralleling the experience in medicine, clients or the public may not be interested in all of the relevant information about an engineering project, at least not until a crisis looms. It is important nevertheless that all avenues for disseminating such information be kept open and ready.

We note that the matter of informed consent is surfacing indirectly in the continuing debate over acceptable forms of energy. Representatives of the nuclear industry can be heard expressing their impatience with critics who worry about reactor malfunction while engaging in statistically more hazardous activities such as driving automobiles and smoking cigarettes. But what is being overlooked by those industry representatives is the common enough human readiness to accept *voluntarily undertaken risks* (as in daring sports), even while objecting to *involuntary risks* resulting from activities in which the individual is neither a direct participant nor a decision maker. In other words, we all prefer to be the subjects of our own experiments rather than those of somebody else. When it comes to approving a nearby oil-drilling platform or a nuclear plant, affected parties expect their consent to be sought no less than it is when a doctor contemplates surgery.

Prior consultation of the kind suggested can be effective. When Northern States Power Company (Minnesota) was planning a new power plant, it got in touch with local citizens and environmental groups before it committed large sums of money to preliminary design studies. The company was able to present convincing evidence regarding the need for a new plant and then suggested several sites. Citizen groups responded with a site proposal of their own. The latter was found acceptable by the company. Thus, informed consent was sought from and voluntarily given by those the project affected, and the acrimonious and protracted battles so common in other cases where a company has already invested heavily in decisions based on engineering studies alone was avoided.[5] Note that the utility company interacted with groups that could serve as proxy for various segments of the rate-paying public. Obviously it would have been difficult to involve the rate-payers individually.

We endorse a broad notion of informed consent, or what some would call *valid consent* defined by the following conditions:[6]

1. The consent was given voluntarily.
2. The consent was based on the information that a rational person would want, together with any other information requested, presented to them in understandable form.
3. The consenter was competent to process the information and make rational decisions.

Nevertheless, it is not always easy to acquire informed consent from all social groups who might be affected by engineering decisions. For instance, it will be very difficult to get informed consent from future generations who can be indirectly impacted by engineering designs. Therefore, we include two additional requirements for situations in which subjects cannot be readily identified as rational and autonomous individuals or they may not exist at the present moment such as future generations:

4. Information that a rational person would need, stated in understandable form, has been widely disseminated.
5. The subject's consent was offered in proxy by a group that collectively represents many subjects of like interests, concerns, and exposure to risk.

KNOWLEDGE GAINED. Scientific experiments are conducted to gain new knowledge, while engineering experiments may not always produce new scientific or fundamental knowledge, according to a valuable interpretation of our paradigm by Taft Broome.[7] When we carry out an engineering activity as if it were an experiment, we are primarily preparing ourselves for unexpected outcomes. The best outcome in this sense is one that tells us nothing new about fundamental scientific knowledge but merely affirms that we are right about something, although other forms of knowledge (e.g., procedure knowledge, knowledge necessary for translating scientific ideas to technical products) may be generated. Unexpected outcomes send us on a search for new knowledge—possibly involving an experiment of the first (scientific) type. For the purposes of our model the distinction is not vital because we are concerned about the manner in which the experiment is conducted, such as that informed consent of human subjects is sought, safety measures are taken, and means exist for terminating the experiment at any time and providing all participants a safe exit.

DISCUSSION QUESTIONS

1. On June 5, 1976, Idaho's Teton Dam collapsed, killing 11 people and causing $400 million in damage. The Bureau of Reclamation, which built the ill-fated Teton Dam, allowed it to be filled rapidly, thus failing to provide sufficient time to monitor for the presence of leaks in a project constructed with less-than-ideal soil.[8]

Drawing upon the concept of engineering as social experimentation, discuss the following facts uncovered by the General Accounting Office and reported in the press.

   a. Because of the designers' confidence in the basic design of Teton Dam, it was believed that no significant water seepage would occur. Thus sufficient instrumentation to detect water erosion was not installed.

   b. Significant information suggesting the possibility of water seepage was acquired at the dam site six weeks before the collapse. The information was sent through routine channels from the project supervisors to the designers, and arrived at the designers the day after the collapse.

   c. During the important stage of filling the reservoir, there was no around-the-clock observation of the dam. As a result, the leak was detected only five hours before the collapse. Even then, the main outlet could not be opened to prevent the collapse because a contractor was behind schedule in completing the outlet structure.

   d. Ten years earlier the Bureau's Fontenelle Dam had experienced massive leaks that caused a partial collapse, an experience the bureau could have drawn on.

2. Debates over responsibility for safety in regard to technological products often turn on who should be considered mainly responsible, the consumer ("buyer beware") or the manufacturer ("seller beware"). How might an emphasis on the idea of informed consent influence thinking about this question?

3. Thought models often influence thinking by effectively organizing and guiding reflection and crystallizing attitudes. Yet they usually have limitations and can themselves be misleading to some degree. With this in mind, critically assess the strengths and weaknesses you see in the social experimentation model.

One possible criticism you might consider is whether the model focuses too much on the creation of new products, whereas a great deal of engineering involves the routine application of results from past work and projects. Another point to consider is how informed consent is to be measured in situations where different groups are involved, as in the construction of a garbage incinerator near a community of people having mixed views about the advisability of constructing the incinerator.

4.2   ENGINEERS AS RESPONSIBLE EXPERIMENTERS

What are the responsibilities of engineers to society? Viewing engineering as social experimentation does not by itself answer this question. While engineers are the main technical enablers or facilitators, they are far from being the sole experimenters. Their responsibility is shared with management, the public, and others. Yet their expertise places them in a unique position to monitor projects, to identify risks, and to provide clients and the public with the information needed to make reasonable decisions.

From the perspective of engineering as social experimentation, four features characterize what it means to be a responsible person while acting as an engineer: a conscientious commitment to live by moral values, a comprehensive perspective, autonomy, and accountability.[9] Or, stated in greater detail as applied to engineering projects conceived as social experiments:

1. A primary obligation to protect the safety of human subjects and respect their right of consent.

2. A constant awareness of the experimental nature of any project, imaginative forecasting of its possible side effects, and a reasonable effort to monitor them.

3. Autonomous, personal involvement in all steps of a project.

4. Accepting accountability for the results of a project.

These features imply that engineers should also display technical competence and other attributes of professionalism. Inclusion of these four requirements as part of engineering practice would then earmark a definite "style" of engineering. In elaborating upon this style, we will note some of the contemporary threats to it.

4.2.1 Conscientiousness

People act responsibly to the extent that they conscientiously commit themselves to live according to moral values. But moving beyond this truism leads immediately to controversy over the precise nature of those values. Moral values transcend a consuming preoccupation with a narrowly conceived self-interest. Accordingly, individuals who think solely of their own good to the exclusion of the good of others are not moral agents. By conscientious moral commitment we mean a sensitivity to the full range of moral values and responsibilities relevant to a given situation, and the willingness to develop the skill and expend the effort needed to reach the best balance possible among those considerations. Conscientiousness implies consciousness: open eyes, open ears, and an open mind.

The contemporary working conditions of engineers tend to narrow moral vision solely to the obligations that accompany employee status. Over 90 percent of engineers are salaried employees, most of whom work within large bureaucracies under great pressure to function smoothly within the organization. There are obvious benefits in terms of prudent self-interest and concern for one's family that make it easy to emphasize as primary the obligations to one's employer. Gradually the minimal negative duties, such as not falsifying data, not violating patent rights, and not breaching confidentiality, may come to be viewed as the full extent of moral aspiration.

Conceiving engineering as social experimentation restores the vision of engineers as guardians of the public interest, whose professional duty is to hold paramount the safety, health, and welfare of those affected by engineering decisions. And this helps to ensure that such safety and welfare will not be disregarded in the quest for new knowledge, the rush for profits, a narrow adherence to rules, or a concern over benefits for the many that ignores harm to the few.

The role of social guardian should not suggest that engineers force, paternalistically, their own views of the social good upon society. As medical experimentation on humans, the social experimentation involved in engineering should be restricted by the participant's consent—voluntary and informed consent.

4.2.2 Comprehensive Perspective

Conscientiousness requires relevant factual information. Hence showing moral concern involves a commitment to obtain and properly assess all available information that is pertinent to meeting one's moral obligations. This means, as a first step, fully grasping the context of one's work, which makes it count as an activity having a moral import.

For example, in designing a heat exchanger, if we ignore the fact that it will be used in the manufacture of a potent, illegal hallucinogen, I am showing a lack of moral concern. It is this requirement that one be aware of the wider implications of one's work that makes participation in, say, a design project for a super-weapon morally problematic—and that makes it sometimes convenient for engineers self-deceivingly to ignore the wider context of their activities, a context that may rest uneasily with conscience.

Another way of blurring the context of one's work results from the ever-increasing specialization and division of labor that makes it easy to think of someone else in the organization as responsible for what otherwise might be a bothersome personal problem. For example, a company may produce items with obsolescence built into them, or the items might promote unnecessary energy usage. It is easy to place the burden on the sales department: "Let them inform the customers—if the customers ask." It may be natural to thus rationalize one's neglect of safety or cost considerations, but it shows no moral concern. More convenient is a shifting of the burden to the government and voters: "We will attend to this when the government sets standards so our competitors must follow suit," or "Let the voters decide on the use of super-weapons; we just build them."

These ways of losing perspective on the nature of one's work also hinder acquiring a full perspective along a second dimension of factual information: the consequences of what one does. And so while regarding engineering as social experimentation points out the importance of context, it also urges the engineer to view their specialized activities in a project as part of a larger whole having a social impact—an impact that may involve a variety of unintended effects. Accordingly, it emphasizes the need for wide training in disciplines related to engineering and its results, as well as the need for a constant effort to imaginatively foresee dangers.

No amount of disciplined and imaginative foresight, however, can anticipate all dangers. Because engineering projects are inherently experimental in nature, they need to be monitored on an ongoing basis from the time they are put into effect. Individual practitioners cannot privately conduct full-blown environmental and social impact studies, but they can choose to make the extra effort needed to keep in touch with the course of a project after it has officially left their hands. This is a mark of personal identification with one's work, a notion that leads to the next aspect of moral responsibility.

4.2.3 Moral Autonomy

People are morally autonomous when their moral conduct and principles of action are their own, in a special sense derived from Kant: Moral beliefs and attitudes should be held on the basis of critical reflection rather than passive adoption of the particular conventions of one's society, church, or profession. This is often what is meant by "authenticity" in one's commitment to moral values. Those beliefs and attitudes, moreover, must be integrated into the core of an individual's personality in a manner that leads to committed action.

It is a comfortable illusion to think that in working for an employer, and thereby performing acts directly serving a company's interests, one is no longer morally and personally identified with one's actions. Selling one's labor and skills may make it seem that one has thereby disowned and forfeited power over one's actions.[10]

Viewing engineering as social experimentation can help overcome this tendency and restore a sense of autonomous participation in one's work. As an experimenter, an engineer is exercising the sophisticated training that forms the core of their identity as a professional. Moreover, viewing an engineering project as an experiment that can result in unknown consequences should help inspire a critical and questioning attitude about the adequacy of current economic and safety standards. This also can lead to a greater sense of personal involvement with one's work.

Philosopher Charles E. Harris suggests that good engineers need to develop both technical and non-technical "excellences" or virtues that can be helpful for capturing and navigating unexpected consequences or risks in highly complicated technical systems.[11] Two technical virtues suggested by Harris are: (1) sensitivity to risk: the sensitivity to capture that in a technical organization some risks may be reconsidered as safe as these risks have not led to any disasters for a long time; and (2) sensitivity to the tight coupling effects in a complex technical system: the sensitivity to capture that the interactions between different components of a highly complex technical system can lead to unexpected consequences.

The attitude of management plays a decisive role in how much moral autonomy engineers feel they have. It would be in the long-term interest of a high-technology firm to grant its engineers a great deal of latitude in exercising their professional judgment on moral issues relevant to their jobs (and, indeed, on technical issues as well). But the yardsticks by which a manager's performance is judged on a quarterly or yearly basis often discourage this. This is particularly true in our age of conglomerates, when near-term profitability is more important than consistent quality and long-term retention of satisfied customers.

In government-sponsored projects it is often a deadline that becomes the ruling factor, along with fears of interagency or foreign competition. Tight schedules contributed to the loss of the space shuttle *Challenger,* as we shall cover later.

Accordingly, engineers are compelled to look to their professional societies and other outside organizations for moral support. Yet it is no exaggeration to claim that the blue-collar worker with union backing has greater leverage at present in exercising moral autonomy than do many employed professionals. Professional societies, originally organized as learned societies dedicated to the exchange of technical information, lack comparable power to protect their members, although most engineers have no other group to rely on for such protection.

Only now is the need for moral and legal support of members in the exercise of their professional obligations being slowly recognized by those societies.[12]

4.2.4 Accountability

Finally, responsible people accept moral responsibility for their actions. Too often "accountable" is understood in the overly narrow sense of being culpable and blameworthy for misdeeds. But the term more properly refers to the general disposition of being willing to submit one's actions to moral scrutiny and be open and responsive to the assessments of others. It involves a willingness to present morally cogent reasons for one's conduct when called upon to do so in appropriate circumstances.

Submission to an employer's authority, or any authority for that matter, creates in many people a narrowed sense of accountability for the consequences of their actions. This was documented by some famous experiments conducted by Stanley Milgram during the 1960s.[13] Subjects would come to a laboratory believing they were to participate in a memory and learning test. In one variation, two other people were involved, the "experimenter" and the "learner." The experimenter was regarded by the subject as an authority figure, representing the scientific community. They would give the subject orders to administer electric shocks to the "learner" whenever the latter failed in the memory test. The subject was told the shocks were to be increased in magnitude with each memory failure. All this, however, was a deception. There were no real shocks, and the apparent "learner" and the "experimenter" were merely acting parts in a ruse designed to see how far the unknowing experimental subject was willing to go in following orders from an authority figure.

The results were astounding. When the subjects were placed in an adjoining room separated from the "learner" by a shaded glass window, more than half were willing to follow orders to the full extent: giving the maximum electric jolt of 450 volts. This was in spite of seeing the "learner," who was strapped in a chair, writhing in (apparent) agony. The same results occurred when the subjects were allowed to hear the (apparently) pained screams and protests of the "learner," screams and protests that became intense from 130 volts on. There was a striking difference, however, when subjects were placed in the same room within touching distance of the "learner." Then the number of subjects willing to continue to the maximum shock dropped by one-half.

Milgram explained these results by citing a strong psychological tendency in people to be willing to abandon personal accountability when placed under authority. He saw his subjects ascribing all initiative, and thereby all accountability, to what they viewed as legitimate authority. And he noted that the closer the physical proximity, the more difficult it becomes to divest oneself of personal accountability.

The divorce between causal influence and moral accountability is common in business and the professions, and engineering is no exception. Such a psychological schism is encouraged by several prominent features of contemporary engineering practice.

First, large-scale engineering projects involve fragmentation of work. Each person makes only a small contribution to something much larger. Moreover, the final product is often physically removed from one's immediate workplace, creating the kind of "distancing" that Milgram identified as encouraging a lessened sense of personal accountability.

Second, corresponding to the fragmentation of work is a vast diffusion of accountability within large institutions. The often massive bureaucracies within which so many engineers work are bound to diffuse and delimit areas of personal accountability within hierarchies of authority. Engineers working in large organizations often encounter the responsibility dilemma "the problem of many hands."[14] This idea refers to the difficulty to identify who is morally responsible for a consequence given that too many different persons contribute to the decision in various ways.

Third, there is often pressure to move on to a new project before the current one has been operating long enough to be observed carefully. This promotes a sense of being accountable only for meeting schedules.

Fourth, the contagion of malpractice suits currently afflicting the medical profession is carrying over into engineering. With this comes a crippling preoccupation with legalities, a preoccupation that makes one wary of becoming morally involved in matters beyond one's strictly defined institutional role.

We do not mean to underestimate the very real difficulties these conditions pose for engineers who seek to act as morally accountable people on their jobs. Much less do we wish to say engineers are blameworthy for all the harmful side effects of the projects they work on, even though they partially cause those effects simply by working on the projects. That would be to confuse accountability with *blameworthiness,* and also to confuse *causal* responsibility with *moral* responsibility. But we do claim that engineers who endorse the perspective of engineering as a social experiment will find it more difficult to divorce themselves psychologically from personal responsibility for their work. Such an attitude will deepen their awareness of how engineers daily cooperate in a risky enterprise in which they exercise their personal expertise toward goals they are especially qualified to attain, and for which they are also accountable.

4.2.5 A Balanced Outlook on Law

Hammurabi, as king of Babylon, was concerned with strict order in his realm, and he decided that the builders of his time should also be governed by his laws. In 1758 B.C.E. he decreed:

> If a builder has built a house for a man and has not made his work sound, and the house which he has built has fallen down and so caused the death of the householder, that builder shall be put to death. If it causes the death of the householder's son, they shall put that builder's son to death. If it causes the death of the householder's slave, he shall give slave for slave to the householder. If it destroys property he shall replace anything it has destroyed; and because he has not made sound the house which he has built and it has fallen down, he shall rebuild the house which

has fallen down from his own property. If a builder has built a house for a man and does not make his work perfect and the wall bulges, that builder shall put that wall into sound condition at his own cost.[15]

What should be the role of law in engineering, as viewed within our model of social experimentation? The legal regulations that apply to engineering and other professions are becoming more numerous and more specific as crises and special interest cases occur. This emphasis on law can cause problems in regard to ethical conduct quite aside from the more practical issues usually cited by those who favor deregulation.

For example, one of the greatest moral problems in engineering, and one fostered by the very existence of minutely detailed rules, is that of *minimal compliance*. This can find its expression when companies or individuals search for loopholes in the law that will allow them to barely keep to its letter even while violating its spirit. Or, hard-pressed engineers find it convenient to refer to standards with ready-made specifications as a substitute for original thought, perpetuating the "handbook mentality" and the repetition of mistakes. Minimal compliance led to the tragedy of the *Titanic:* Why should that ship have been equipped with enough lifeboats to accommodate all its passengers and crew when British regulations at the time required only a lower minimum, albeit with smaller ships in mind?

On the other hand, remedying the situation by continually updating laws or regulations with further specifications may also be counterproductive. Not only will the law inevitably lag behind changes in technology and produce a judicial vacuum, there is also the danger of overburdening the rules and the regulators.

Lawmakers cannot be expected always to keep up with technological development. Nor would we necessarily want to see laws changed upon each innovation. Instead we empower rule-making and inspection agencies to fill the void. The Food and Drug Administration (FDA), Federal Aviation Agency (FAA), and the Environmental Protection Agency (EPA) are examples of these in the United States. Though they are nominally independent in that they belong neither to the judicial nor the executive branches of government, their rules have, for all practical purposes, the effect of law, but they are headed by political appointees.

Industry tends to complain that excessive restrictions are imposed on it by regulatory agencies. But one needs to reflect on why regulations may have been necessary in the first place. Take, for example, the U.S. Consumer Product Safety Commission's rule for baby cribs, which specifies that "the distance between components (such as slats, spindles, crib rods, and corner posts) shall not be greater than $2\frac{3}{8}$ inches at any point." This rule came about because some manufacturers of baby furniture had neglected to consider the danger of babies strangling in cribs or had neglected to measure the size of babies' heads.[16]

Again, why must regulations be so specific when broad statements would appear to make more sense? When the EPA adopted rules for asbestos emissions in 1971, it was recognized that strict numerical standards would be impossible to promulgate. Asbestos dispersal and intake, for example, are difficult to measure in the field. So, being reasonable, the EPA many years ago specified a set of work practices to keep emissions to a minimum—for example that asbestos should be wetted down before handling and disposed of carefully. The building industry called for more specifics. Modifications in the Clean Air Act eventually permitted the EPA to issue enforceable rules on work practices, and now the Occupational Safety and Health Administration is also involved.

Society's attempts at regulation have indeed often failed, but it would be wrong to write off rule-making and rule-following as futile. Good laws, effectively enforced, clearly produce benefits. They authoritatively establish reasonable minimal standards of professional conduct and provide at least a self-interested motive for most people and corporations to comply. Moreover, they serve as a powerful support and defense for those who wish to act ethically in situations where ethical conduct might be less than welcome. By being able to point to a pertinent law, one can feel freer to act as a responsible engineer.

Engineering as social experimentation can provide engineers with a proper perspective on laws and regulations in that rules that govern engineering practice should not be devised or construed as rules of a game but as rules of responsible experimentation. Such a view places proper responsibility on the engineer who is intimately connected with their "experiment" and responsible for its safe conduct. Moreover, it suggests the following conclusions: Precise rules and enforceable sanctions are appropriate in cases of ethical misconduct that involve violations of well-established and regularly reexamined engineering procedures that have as their purpose the safety and well-being of the public. Little of an experimental nature is probably occurring in such standard activities, and the type of professional conduct required is most likely very clear. In areas where experimentation is involved more substantially, however, rules must not attempt to cover all possible outcomes of an experiment, nor must they force engineers to adopt rigidly specified courses of action. It is here that regulations should be broad, but written to hold engineers accountable for their decisions. Through their professional societies engineers should also play an active role in establishing (or changing) enforceable rules as well as in enforcing them, but with great care to forestall conflicts of interest. (See Discussion Question 4, on the Hydrolevel case.)

4.2.6 Industrial Standards

There is one area in which industry usually welcomes greater specificity, and that is in regard to standards. Product standards facilitate the interchange of components, they serve as ready-made substitutes for lengthy design specifications, and they decrease production costs.

Standards consist of explicit specifications that, when followed with care, ensure that stated criteria for interchangeability and quality will be attained. Examples range from automobile tire sizes and load ratings to computer protocols. Table 4-1 lists purposes of standards and gives some examples to illustrate those purposes.

TABLE 4-1
Types of standards

Criterion	Purpose	Selected examples
Uniformity of physical properties and functions	Accuracy in measurement, interchangeability, ease of handling	Standards of weights, screw dimensions, standard time, film size
Safety and reliability	Preparation of injury, death, and loss of income or property	National Electric Code, boiler code, methods of handling toxic wastes
Quality of product	Fair value for price	Plywood grades, lamp life
Quality of personnel and service	Competence in carrying out tasks	Accreditation of schools, professional licenses
Use of accepted procedures	Sound design, ease of communications	Drawing symbols, test procedures
Separability	Freedom from interference	Highway lane markings, radio frequency bands
Quality procedures approved by ISO, the International Standards Organization	Assurance of product acceptance in member countries	Quality of products, work, certificates, and degrees

A selection of modular telephone adaptors offered by Magellan's:

"Wouldn't it be nice if they could agree on a common telephone jack?"

Standards are established by companies for in-house use and by professional associations and trade associations for industry-wide use. They may also be prescribed as parts of laws and official regulations, for example, in mandatory standards, which often arise from lack of adherence to voluntary standards.

Standards not only help the manufacturers, they also benefit the client and the public. They preserve some competitiveness in industry by reducing overemphasis on name brands and giving the smaller manufacturer a chance to compete. They ensure a measure of quality and thus facilitate more realistic trade-off decisions. International standards are becoming a necessity in European and world trade. An interesting approach has been adopted by the International

Standards Organization (ISO) that replaces the detailed national specifications for a plethora of products with statements of procedures that a manufacturer guarantees to carry out to assure quality products.

Standards have been a hindrance at times. For many years they were mostly descriptive, specifying, for instance, how many joists of what size should support a given type of floor. Clearly such standards tended to stifle innovation. The move to performance standards, which in the case of a floor may specify only the required load-bearing capacity, has alleviated that problem somewhat. But other difficulties can arise when special interests (for example, manufacturers, trade unions, exporters, and importers) manage to impose unnecessary provisions on standards or remove important provisions from them in order to secure their own narrow self-interest. Requiring metal conduits for home wiring is one example of this problem. Modern conductor coverings have eliminated the need for metal conduit in many applications, but many localities still require it. Its use sells more conduit and labor time for installation. But standards did not foresee the dangers encountered when aluminum was substituted for copper as conductor in home wiring, as happened in the United States during the copper scarcity occasioned by the Vietnam War. Until better ways were devised for fastening aluminum conductors, many fires occurred due to the gradual loosening of screw terminals.

Nevertheless, there are standards nowadays for practically everything, it seems, and consequently we often assume that stricter regulation exists than may actually be the case. The public tends to trust the National Electrical Code in all matters of power distribution and wiring, but how many people realize that this code, issued by the National Fire Protection Association, is primarily oriented toward fire hazards? Only recently have its provisions against electric shock begun to be strengthened. Few consumers know that an Underwriter Laboratories seal prominently affixed to the cord of an electrical appliance may pertain only to the cord and not to the rest of the device. In a similar vein, a patent notation inscribed on the handle of a product may refer just to the handle, and then possibly only to the design of the handle's appearance.

Sometimes standards are thought to apply when in actuality there is no standard at all. Product appearances can be misleading in this respect. Years ago, when competing foreign firms were trying to corner the South American market for electrical fixtures and appliances, one manufacturing company had a shrewd idea. It equipped its lightbulbs with extra-long bases and threads. These would fit into the competitors' shorter lamp sockets and its own deep sockets. But the competitors' bulbs would not fit into the deeper sockets of its own fixtures (see figure 4-1). Yet so far as the unsuspecting consumer was concerned, all the lightbulbs and sockets continued to look alike.

During the introduction of novel products there is often a period during which the consumer is at a disadvantage, for example, in not knowing which word processing program or camera lens mount will win in the long run and make a recently purchased product prematurely obsolete or nonrepairable. Sometimes a

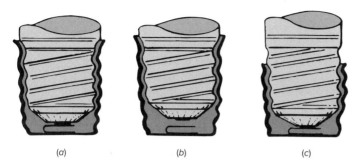

FIGURE 4-1
The light bulb story. (*a*) Long base, deep socket: firm contact. (*b*) Short
base, deep socket: no contact. (*c*) Long base, shallow socket: firm contact.

particular design stays in the front long enough that it becomes the standard, as
happened with Hayes modems and their command structure or to Sony and its tape
drives.

DISCUSSION QUESTIONS

1. A common excuse for carrying out a morally questionable project is, "If I don't do it
   somebody else will." This rationale may be tempting for engineers who typically work
   in situations where someone else might be ready to replace them on a project. Do you
   view it as a legitimate excuse for engaging in projects that might be unethical? In your
   answer, comment on the concept of responsible conduct developed in this section.
2. Another commonly used phrase, "I only work here," implies that one is not personally
   accountable for the company rules since one does not make them. It also suggests that
   one wishes to restrict one's area of responsibility within tight bounds as defined by
   those rules. In light of the discussion in this section, respond to the potential implications of this phrase and the attitude it represents when exhibited by engineers.
3. Threats to a sense of personal responsibility are neither unique to, nor more acute for,
   engineers than they are for others involved with engineering and its results. The reason
   is that, in general, public accountability also tends to lessen as professional roles
   become narrowly differentiated. With this in mind, critique each of the remarks made in
   the following dialogue. Is the remark true, or partially true? What needs to be added to
   make it accurate?

   ENGINEER: My responsibility is to receive directives and to create products within
   specifications set by others. The decision about what products to make and their general
   specifications are economic in nature and made by management.
   SCIENTIST: My responsibility is to gain knowledge. How the knowledge is applied
   is an economic decision made by management, or else a political decision made by
   elected representatives in government.
   MANAGER: My responsibility is solely to make profits for stockholders.
   STOCKHOLDER: I invest my money for the purpose of making a profit. It is up to our
   boards and managers to make decisions about the directions of technological development.

CONSUMER: My responsibility is to my family. Government should make sure corporations do not harm me with dangerous products, harmful side effects of technology, or dishonest claims.

GOVERNMENT REGULATOR: By current reckoning, government has strangled the economy through over-regulation of business. Accordingly, at present on my job, especially given decreasing budget allotments, I must back off from the idea that business should be policed, and urge corporations to assume greater public responsibility.

4. In 1975, Hydrolevel Corporation brought suit against the American Society of Mechanical Engineers (ASME), charging that two ASME volunteers, acting as agents of ASME, had conspired to interpret a section of ASME's Boiler and Pressure Vessel Code in such a manner that Hydrolevel's low-water fuel cutoff for boilers could not compete with the devices built by the employers of the two volunteers. On May 17, 1982, the Supreme Court upheld the lower courts that had found ASME guilty of violating antitrust provisions.

Writing on behalf of the six-to-three majority, Justice Harry A. Blackmun said:

> When ASME's agents act in its name, they are able to affect the lives of large numbers of people and the competitive fortunes of businesses throughout the country. By holding ASME liable under the antitrust laws for the antitrust violations of its agents committed with apparent authority, we recognize the important role of ASME and its agents in the economy, and we help to ensure that standard-setting organizations will act with care when they permit their agents to speak for them.

Acquaint yourself with the particulars of this case and discuss it as an illustration of the possible misuses of standards.[17]

5. Mismatched bumpers: Ought there to be a law? What happens when a passenger car rear-ends a truck or a sports utility vehicle (SUV)? The bumpers usually ride at different heights, so even modest collisions can result in major repair bills. (At high speed, with the front of the car nose down upon braking, people in convertibles have been decapitated upon contact devoid of protection by bumpers.) The people at Volvo recognized the problem long ago—we have observed that their trucks usually have low bumpers front and rear. Discuss how other companies building and selling trucks and high-riding vehicles can be induced to follow Volvo's example. Should older vehicles be retrofitted with lower bumpers or guards once a standard is established?

4.3 CHALLENGER

Several months before the destruction of *Challenger,* NASA historian Alex Roland wrote the following in a critical piece about the space shuttle program:

> The American taxpayer bet about $14 billion on the shuttle. NASA bet its reputation. The Air Force bet its reconnaissance capability. The astronauts bet their lives. We all took a chance.
>
> When John Young and Robert Crippen climbed aboard the orbiter *Columbia* on April 12, 1981, for the first shuttle launch, they took a bigger chance than any astronaut before them. Never had Americans been asked to go on a launch vehicle's maiden voyage. Never had astronauts ridden solid propellant rockets. Never had Americans depended on an engine untested in flight.[18]

Most of Alex Roland's criticism was directed at the economic and political side of what was supposed to become a self-supporting operation but never gave any indication of being able to reach that goal. Without a national consensus to back it, the shuttle program became a victim of year-by-year funding politics.

The *Columbia* and its sister ships, *Challenger, Discovery,* and *Endeavor,* were delta-wing craft with a huge payload bay. Early, sleek designs had to be abandoned to satisfy U.S. Air Force requirements when the Air Force was ordered to use the NASA shuttle instead of its own expendable rockets for launching satellites and other missions. As shown in figure 4-2, each orbiter has three main engines fueled by several million pounds of liquid hydrogen; the fuel is carried in an immense, external, divided fuel tank, which is jettisoned when empty. During liftoff the main engines fire for about 8.5 minutes, although during the first two minutes of the launch much of the thrust is provided by two booster rockets. These are of the solid-fuel type, each burning a one-million-pound load of a mixture of aluminum, potassium chloride, and iron oxide.

The casing of each booster rocket is about 150 feet long and 12 feet in diameter. It consists of cylindrical segments that are assembled at the launch site. The four field joints use seals composed of pairs of O-rings made of vulcanized rubber. The O-rings work in conjunction with a putty barrier of zinc chromide.

The shuttle flights were successful, although not as frequent as had been hoped. NASA tried hard to portray the shuttle program as an operational system that could pay for itself. Some Reagan administration officials had even suggested that the operations be turned over to an airline. Aerospace engineers intimately involved in designing, manufacturing, assembling, testing, and operating the shuttle still regarded it as an experimental undertaking in 1986. These engineers were employees of manufacturers, such as Rockwell International (orbiter and main rocket) and Morton-Thiokol (booster rockets), or they worked for NASA at one of its several centers: Marshall Space Flight Center, Huntsville, Alabama (responsible for the propulsion system); Kennedy Space Center, Cape Kennedy, Florida (launch operations); Johnson Space Center, Houston, Texas (flight control); and the office of the chief engineer, Washington, D.C. (overall responsibility for safety, among other duties).

After embarrassing delays, *Challenger*'s first flight for 1986 was set for Tuesday morning, January 28. But Allan J. McDonald, who represented Morton-Thiokol at Cape Kennedy, was worried about the freezing temperatures predicted for the night. As his company's director of the solid-rocket booster project, he knew of difficulties that had been experienced with the field joints on a previous cold-weather launch when the temperature had been mild compared to what was forecast. He therefore arranged a teleconference so that NASA engineers could confer with Morton-Thiokol engineers at their plant in Utah.

Arnold Thompson and Roger Boisjoly, two seal experts at Morton-Thiokol, explained to their own colleagues and managers as well as the NASA representatives how upon launch the booster rocket walls bulge and the combustion gases can blow past one or even both of the O-rings that make up the field joints (see figure 4-2).[19] The rings char and erode, as had been observed on many previous

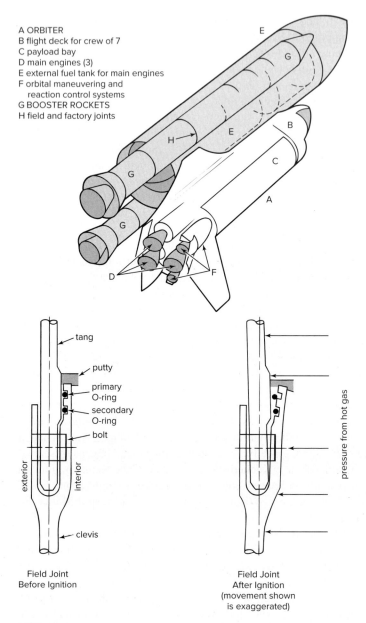

A ORBITER
B flight deck for crew of 7
C payload bay
D main engines (3)
E external fuel tank for main engines
F orbital maneuvering and
   reaction control systems
G BOOSTER ROCKETS
H field and factory joints

tang

putty

primary
O-ring

secondary
O-ring

bolt

exterior

interior

clevis

Field Joint
Before Ignition

pressure from hot gas

Field Joint
After Ignition
(movement shown
is exaggerated)

FIGURE 4-2
Space shuttle *Challenger*.

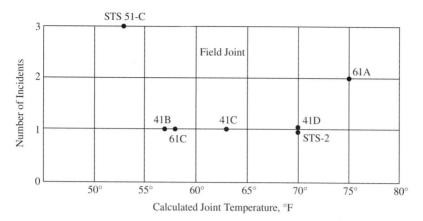

FIGURE 4-3
Plot of flights *with* incidents of O-ring thermal distress: Number of O-rings affected plotted against temperature. (*Source: Rogers Commission Report,* Report of the Presidential Commission on the Space Shuttle *Challenger* Accident *[Washington, DC: U.S. Government Printing Office, 1986].*)

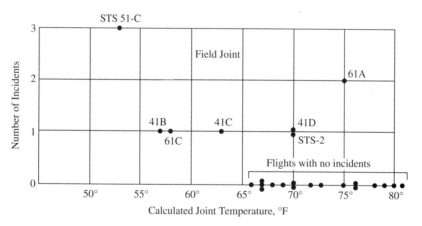

FIGURE 4-4
Plot of flights *with and without* incidents of O-ring thermal distress. (*Source: Rogers Commission Report,* Report of the Presidential Commission on the Space Shuttle *Challenger* Accident *[Washington, DC: U.S. Government Printing Office, 1986].*)

flights. In cold weather the problem is aggravated because the rings and the putty packing are less pliable then. But

> only limited consideration was given to the past history of O-ring damage in terms of temperature. The managers compared as a function of temperature the flights for which thermal distress of O-rings had been observed *[figure 4-3]*—not the frequency of occurrence based on all flights. When the entire history of flight experience is considered, including "normal" flights with no erosion or blow-by, the comparison is substantially different *[figure 4-4]*. Consideration of the entire launch

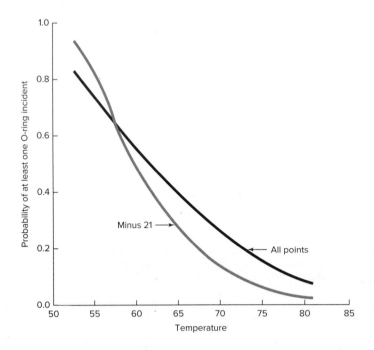

FIGURE 4-5
Binomial-logit model based on all data remaining after removing flight #61A in figure 4-4, and also *with* flight 61A (data point #21). The ordinate gives the probability of distress of at least one O-ring.

temperature history indicates that the probability of O-ring distress is increased to almost a certainty if the temperature of the joint is less than 65.[20]

If the graph depicted in figure 4-4 and other supporting renditions (such as figure 4-5 of later date) had been available at the meeting, upper management might just have seen the problem of cold O-rings more clearly, but with launch time approaching, deliberations were cut short.

The engineering managers, Bob Lund (vice president of engineering) and Joe Kilminster (vice president for booster rockets), agreed that there was a problem with safety. The team from Marshall Space Flight Center was incredulous. Since the specifications called for an operating temperature of the solid fuel prior to combustion of 40 to 90 degrees Fahrenheit, one could surely allow lower or higher outdoor temperatures, notwithstanding Boisjoly's testimony and recommendation that no launch should occur at less than 53 degrees. They were clearly annoyed at facing yet another postponement.

Top executives of Morton-Thiokol were also sitting in on the teleconference. Their concern was the image of the company, which was in the process of negotiating a renewal of the booster rocket contract with NASA. During a recess senior vice president Jerry Mason turned to Bob Lund and told him "to take off your engineering hat and put on your management hat." It was a subsequent vote (of the managers only) that produced the company's official finding that the seals could not be shown to be unsafe. The engineers' judgment was not considered sufficiently weighty. At Cape Kennedy, Allan McDonald refused to sign the formal recommendation to launch; Joe Kilminster had to. Accounts of the *Challenger* disaster tell of the cold Tuesday morning, the high seas that forced the recovery ships to seek coastal shelter, the ice at the launch site, and the concern expressed by Rockwell engineers that the ice might shatter and hit the orbiter or rocket casings.[21] The inability of these engineers to *prove* that the liftoff would be unsafe was presented to NASA as an approval by Rockwell to launch.

The countdown ended at 11:38 A.M. The temperature had risen to 36 degrees. As the rockets carrying *Challenger* rose from the ground, cameras recorded puffs of smoke that emanated from one of the field joints on the right booster rocket. Soon these turned into a flame that hit the external fuel tank and a strut holding the booster rocket. The hydrogen in the tank caught fire, the booster rocket broke loose, smashed into *Challenger*'s wing, then into the external fuel tank. At 76 seconds into the flight, by the time *Challenger* and its rockets had reached 50,000 feet, it was totally engulfed in a fireball. The crew cabin separated and fell into the ocean, killing all aboard: mission commander Francis (Dick) Scobee; pilot Michael Smith; mission specialists Gregory Jarvis, Ronald McNair, Ellison Onizuka, Judith Resnick; and Christa MacAuliffe, who was a high school teacher and the first civilian selected to go into space.

4.3.1 Safety Issues

Unlike the three-stage rockets that carried astronauts to the moon, the space shuttle could be involved in a simultaneous (inadvertent) ignition of all fuel carried aloft. An explosion close to the ground can have catastrophic effects. The crew had no escape mechanism, although McDonnell Douglas, in a losing shuttle proposal, had designed an abort module with its own thruster. It would have allowed the separation of the orbiter, triggered (among other events) by a field-joint leak. But such a safety measure was rejected as too expensive because of an accompanying reduction in payload.

Working with such constraints, why was safe operation not stressed more? First of all, we must remember that the shuttle program was indeed still a truly experimental and research undertaking. Next, it is quite clear that the members of the crews knew that they were embarking on dangerous missions. But it has also been revealed that the *Challenger* astronauts were not informed of particular problems such as the field joints. They were not asked for their consent to be launched under circumstances that experienced engineers had claimed to be unsafe.

The reason for the rather cavalier attitude toward safety is revealed in the way NASA assessed the system's reliability. For instance, recovered booster rocket casings had indicated that the field-joint seals had been damaged in many of the earlier flights. The waivers necessary to proceed with launches had become mere gestures. Richard Feynman made the following observations as a member of the Presidential Commission on the Space Shuttle Challenger Accident (called the Rogers Commission after its chairman):

> I read all of these [NASA flight readiness] reviews and they agonize whether they can go even though they had some blow-by in the seal or they had a cracked blade in the pump of one of the engines . . . and they decide "yes." Then it flies and nothing happens. Then it is suggested . . . that the risk is no longer so high. For the next flight we can lower our standards a little bit because we got away with it last time. . . . It is a kind of Russian roulette.[22]

Since the early days of unmanned space flight, about 1 in every 25 solid-fuel rocket boosters had failed. Given improvements over the years, Feynman thought that 1 in every 50 to 100 might be a reasonable estimate now. Yet NASA counted on only one crash in every 100,000 launches. Queried about these figures, NASA Chief Engineer Milton Silveira answered: "We don't use that number as a management tool. We know that the probability of failure is always sitting there."[23] So where was this number used? In a risk analysis needed by the Department of Energy to assure everyone that it would be safe to use small atomic reactors as power sources on deep-space probes and to carry both aloft on a space shuttle. As luck would have it, *Challenger* was not to carry the 47.6 pounds of lethal plutonium-238 until its next mission with the Galileo probe on board.[24]

Another area of concern was NASA's unwillingness to wait out risky weather. When serving as weather observer, astronaut John Young was dismayed to find his recommendations to postpone launches disregarded several times. Things had not changed much by March 26, 1987, when NASA ignored its devices monitoring electric storm conditions, launched a Navy communications satellite atop an Atlas-Centaur rocket, and had to destroy the $160 million system when it veered off course after being hit by lightning. The monitors had been installed after a similar event involving an Apollo command module 18 years before had nearly aborted a trip to the moon. Weather, incidentally, could be held partially responsible for the shuttle disaster because a strong wind shear may have contributed to the rupturing of the weakened O-rings.[25]

Veteran astronauts were also dismayed at NASA management's decision to land at Cape Kennedy as often as possible despite its unfavorable landing conditions, including strong crosswinds and changeable weather. The alternative, Edwards Air Force Base in California, is a better landing place but necessitates a piggyback ride for the shuttle on a Boeing 747 home to Florida. This costs time and money.

In 1982 Albert Flores had conducted a study of safety concerns at the Johnson Space Center. He found its engineers to be strongly committed to safety in all aspects of design. When they were asked how managers might further improve safety awareness, there were few concrete suggestions but many comments on how safety concerns were ignored or negatively affected by management. One engineer was quoted as saying, "A small amount of professional safety effort and upper management support can cause a quantum safety improvement with little expense."[26] This points to the important role of management in building a strong sense of responsibility for safety first and schedules second.

The space shuttle's field joints are designated criticality 1, which means there is no backup. Therefore a leaky field joint will result in failure of the mission and loss of life. There are 700 items of criticality 1 on the shuttle. A problem with any one of them should have been cause enough to do more than just launch more shuttles without modification while working on a better system. Improved seal designs had already been developed, but the new rockets would not have been ready for some time. In the meantime, the old booster rockets should have been recalled.

In several respects the ethical issues in the *Challenger* case resemble those of other such cases. Concern for safety gave way to institutional posturing. Danger signals did not go beyond Morton-Thiokol and Marshall Space Flight Center in the *Challenger* case. No effective recall was instituted. There were concerned engineers who spoke out, but ultimately they felt it only proper to submit to management decisions.

One notable aspect of the *Challenger* case is the late-hour teleconference that Allan McDonald had arranged from the *Challenger* launch site to get knowledgeable engineers to discuss the seal problem from a technical viewpoint. This tense conference did not involve lengthy discussions of ethics, but it revealed the virtues (or lack thereof) that allow us to distinguish between the "right stuff" and the "wrong stuff." This is well described by one aerospace engineer as arrogance, specifically "The arrogance that prompts higher-level decision makers to pretend that factors other than engineering judgement should influence flight safety decisions and, more important, the arrogance that rationalizes overruling the engineering judgement of engineers close to the problem by those whose expertise is naive and superficial by comparison."[27] Included, surely, is the arrogance of those who reversed NASA's (paraphrased) motto "Don't fly if it cannot be shown to be safe" to "Fly unless it can be shown not to be safe."

At Morton-Thiokol, some of the vice presidents in the space division have been demoted. The engineers who were outspoken at the prelaunch teleconference and again before the Rogers Commission kept their jobs at the company because of congressional pressure, but their jobs are of a pro forma nature. In a speech to engineering students at the Massachusetts Institute of Technology a year after the *Challenger* disaster, Roger Boisjoly said: "I have been asked by some if I would testify again if I knew in advance of the potential consequences to me and my career. My answer is always an immediate yes. I couldn't live with any self-respect if I tailored my actions based upon potential personal consequences as a result of my honorable actions."[28]

Today NASA has a policy that allows aerospace workers with concerns to report them anonymously to the Batelle Memorial Institute in Columbus, Ohio, but open disagreement still invited harassment for a number of years.

DISCUSSION QUESTIONS

1. Chairman Rogers asked Bob Lund: "Why did you change your decision [that the seals would not hold up] when you changed hats?" What might motivate you, as a midlevel manager, to go along with top management when told to "take off your engineering hat

and put on your management hat"? Applying the engineering-as-experimentation model, what might responsible experimenters have done in response to the question?

2. Under what conditions would you say it is safe to launch a shuttle without an escape mechanism for the crew?

3. Discuss the role of the astronauts in shuttle safety. To what extent should they (or at least the orbiter commanders) have involved themselves more actively in looking for safety defects in design or operation?

4. Consider the following actions or recommendations and suggest a plan of action to bring about safer designs and operations in a complex organization.

    a. Lawrence Mulloy represented Marshall Space Flight Center at Cape Kennedy. He did not tell Arnold Aldrich from the National Space Transportation Program at Johnson Space Center about the discussions regarding the field-joint seals even though Aldrich had the responsibility of clearing *Challenger* for launch. Why? Because the seals were "a Level III issue," and Mulloy was at Level III, while Aldrich was at a higher level (Level II) which ought not to be bothered with such details.

    b. The Rogers Commission recommended that an independent safety organization directly responsible to the NASA administrator be established. An anonymous reporting scheme now exists for aerospace industry employees working on NASA projects.

    c. Tom Peters advises managers to "involve everyone in everything. . . . Boldly assert that there is no limit to what the average person can accomplish if thoroughly involved."[29]

5. On October 4, 1930, the British airship *R 101* crashed about eight hours into its maiden voyage to India. Of the 54 persons aboard, only 6 survived. Throughout the craft's design and construction, Air Ministry officials and their engineers had been driven by strong political and competitive forces. Nevil Shute, who had worked on the rival, commercial *R 100,* wrote in his memoir, *Slide Rule,* that "if just one of [the men at the Air Ministry] had stood up [at a conference with Lord Thomson] and had said, 'This thing won't work, and I'll be no party to it. I'm sorry, gentlemen, but if you do this, I'm resigning,' . . . the disaster would almost certainly have been averted. It was not said, because the men in question put their jobs before their duty."[30] Examine the *R 101* case and compare it with the *Challenger* case, including the pressures not to delay the flight.

6. During *Columbia's* last launch, chunks of foam were falling from the fixture that initially keeps the shuttle attached to the external fueltank. The foam prevents ice from forming and, when breaking off during launch, damaging at high acceleration the fragile thermal insulation of the shuttle's wings, especially at the critical forward edges. Such hits were generally not considered serious, only a maintenance problem. But not by Allen J. Richardson. In the 1980's had developed an analysis tool that showed otherwise. This view was shared by engineers from three NASA research centers and from three aerospace companies. A scenario that resembles the warnings issued prior to the *Challenger* unfolded. Recommendations that NASA ask the *Columbia* to be photographed during its orbits by the Defense Department or intelligence agencies using their specialty cameras were sidetracked and never carried out. In the meantime Rodney Rocha of NASA's debris analysis team felt powerless to get is superiors' attention.

    Examine news reports (e.g., www.cnn.com/SPECIALS/2003/shuttle) and more recent official reports to determine to what extent the shuttle disasters can be ascribed to technical and/or management deficiencies. Was there a failure to learn from the earlier event? Note that during the 17 year period between the two events many engineers and managers had retired. How did experience get passed on—or not? Regarding the *Challenger* case, you may find material in addition to this Chapter 6 in the books by Rosa L. Pincus et al.* and Diane Vaughan.**

KEY CONCEPTS

—**Engineering as social experimentation:** Engineering projects can be viewed as social experiments in that (1) they are carried out in partial ignorance, (2) have uncertain outcomes, (3) require monitoring and feedback, and (4) mandate obtaining informed consent from those affected.

—**Informed consent to the use or effects of products:** consent that (1) is given voluntarily—without coercion, manipulation, or deception, (2) is based on having information that a rational person would want and other information requested, or is widely disseminated in an understandable form, (3) is given by a competent person or by a proxy group that represents the person's interests, concerns, and exposure to risk.

—**Engineers as responsible experimenters:** (1) conscientiously accept a primary obligation to protect the safety of human subjects and respect their right of consent; (2) maintain awareness of the experimental nature of any project, imaginatively foresee its possible side effects, and make a reasonable effort to monitor them; (3) have autonomous, personal involvement in engineering projects; (4) accept accountability for the results of projects.

—**Safe exits:** design and procedures ensuring that if a product fails it will fail safely and the user can safely avoid harm from the failed product.

—**Balanced outlook on law:** Reasonable laws and sanctions are appropriate components of engineering, but laws set the rules for minimal compliance rather than providing the full substance of engineering ethics.

REFERENCES

1. Walter Lord, *A Night to Remember* (New York: Holt, 1976); Wynn C. Wade, *The Titanic: End of a Dream* (New York: Penguin, 1980); Michael Davie, *The Titanic* (London: The Bodley Head, 1986).
2. Wade, *The Titanic,* p. 417.
3. "Yarrow Bridge," editorial, *The Engineer* 210 (October 1970), p. 415.
4. Robert Sugarman, "Nuclear Power and the Public Risk," *IEEE Spectrum* 16 (November 1979), p. 72.
5. Peter Borrelli, Mahlon Easterling, Burton H. Klein et al., *People, Power and Pollution* (Pasadena, CA: Environmental Quality Lab, California Institute of Technology, 1971), pp. 36–39.
6. Charles M. Culver and Bernard Gert, "Valid Consent," in *Conceptual and Ethical Problems in Medicine and Psychiatry,* ed. Charles M. Culver and Bernard Gert (New York: Oxford University Press, 1982).
7. Taft H. Broome, Jr., "Engineering Responsibility for Hazardous Technologies," *Journal of Professional Issues in Engineering* 113 (April 1987), pp. 139–49.
8. Gaylord Shaw, "Bureau of Reclamation Harshly Criticized in New Report on Teton Dam Collapse," *Los Angeles Times,* June 4, 1977, Part I, p. 3; Philip M. Boffey, "Teton Dam Verdict: Foul-up by the Engineers," *Science* 195 (January 1977), pp. 270–72.
9. Graham Haydon, "On Being Responsible," *The Philosophical Quarterly* 28 (1978), pp. 46–57.

---

*Rosa L. B. Pincus, Larry J. Shuman, Norman P. Hummon, Harvey Wolfe, *Engineering Ethics— Balancing Cost, Schedule, and Risk—Lessons Learned from the Space Shuttle* (Cambridge, UK: Cambridge University Press).

**Diane Vaughan, *The Challenger Launch Decision—Risky Technology, Culture, and Deviance at NASA* (Chicago: University of Chicago Press, 1998).

10. John Lachs, "'I Only Work Here': Mediation and Irresponsibility," in *Ethics, Free Enterprise, and Public Policy,* ed. Richard T. DeGeorge and Joseph A. Pichler (New York: Oxford University Press, 1978), pp. 201–13; Elizabeth Wolgast, *Ethics of an Artificial Person: Lost Responsibility in Professions and Organizations* (Stanford, CA: Stanford University Press, 1992).

11. Charles E. Harris, "The Good Engineer: Giving Virtue its Due in Engineering Ethics," *Science and Engineering Ethics* 14, no. 2 (2008): 153–264.

12. Stephen H. Unger, *Controlling Technology: Ethics and the Responsible Engineer,* 2nd ed. (New York: John Wiley & Sons, 1994).

13. Stanley Milgram, *Obedience to Authority* (New York: Harper & Row, 1974).

14. Christelle Didier, "Engineering Ethics," in *A Companion to the Philosophy of Technology,* ed. Jan Kyrre Berg Olsen, Stig Andur Pedersen and Vincent F. Hendricks (New York: John Wiley & Sons, 2009), pp. 426–32.

15. Hammurabi, *The Code of Hammurabi,* trans. R. F. Harper (University of Chicago Press, 1904).

16. William W. Lowrance, *Of Acceptable Risk* (Los Altos, CA: William Kaufmann, 1976), p. 134.

17. Stephen H. Unger, *Controlling Technology: Ethics and the Responsible Engineer,* 2nd ed. (New York: John Wiley & Sons, 1994), pp. 210–15. See also Paula Wells, Hardy Jones, and Michael Davis, *Conflicts of Interest in Engineering* (Dubuque, IA: Kendall/Hunt, 1986).

18. Alex Roland, "The Shuttle, Triumph or Turkey?" *Discover,* November 1985, pp. 29–49.

19. Wade Robison, Roger Boisjoly, David Hoeker, and Stefan Young, "Representation and Misrepresentation: Tufte and the Morton Thiokol Engineers on the *Challenger*," *Science and Engineering Ethics* 8, no. 1 (2002): 59–81.

20. Rogers Commission Report, *Report of the Presidential Commission on the Space Shuttle Challenger Accident* (Washington, DC: U.S. Government Printing Office, 1986).

21. Malcolm McConnell, *Challenger, a Major Malfunction* (Garden City, NY: Doubleday, 1987); Rosa Lynn B. Pinkus, Larry J. Shuman, Norman P. Hummon, and Harvey Wolfe, *Engineering Ethics: Balancing Cost, Schedule, and Risk—Lessons Learned from the Space Shuttle* (Cambridge: Cambridge University Press, 1997); Diane Vaughan, *The Challenger Launch Decision: Risky Technology, Culture, and Deviance at NASA* (Chicago: University of Chicago Press, 1996).

22. *Rogers Commission Report,* Report of the Presidential Commission.

23. Eliot Marshall, "Feynman Issues His Own Shuttle Report, Attacking NASA Risk Estimates," *Science* 232 (June 1986): 1596. See also Richard P. Feynman, *What Do You Care What Other People Think?* (New York: W.W. Norton & Co., 1988).

24. Karl Grossman, "Red Tape and Radioactivity," *Common Cause,* July-August 1986, pp. 24–27.

25. Trudy E. Bell, "Wind Shear Cited as Likely Factor in Shuttle Disaster," *The Institute* (IEEE) 11 (May 1987): 1. For effects of lightning, see Eliot Marshall, "Lightning Strikes Twice at NASA," *Science* 236 (May 1987): 903.

26. Albert Flores (ed.), *Designing for Safety: Engineering Ethics in Organizational Contexts* (Troy, NY: Rensselaer Polytechnic Institute, 1982), p. 79.

27. Calvin E. Moeller, "*Challenger* Catastrophe," *Los Angeles Times,* Letters to the Editor, March 11, 1986.

28. Roger M. Boisjoly, Speech on shuttle disaster delivered to MIT students, January 7, 1987. Printed in *Books and Religion* 15 (March-April 1987), p. 3. Also see Wade Robison, Roger Boisjoly, David Hoeker, Stefan Young, "Representation and Misrepresentation: Tufte and the Morton-Thiokol Engineers on the *Challenger*," *Science and Engineering Ethics* 8, no. 1 (2002): 59–81.

29. Tom Peters, *Thriving on Chaos* (New York: Alfred A. Knopf, 1987).

30. Nevil Shute, *Slide Rule* (New York: William Morrow, 1954), p. 140. Also see Henry Cord Meyer, "Politics, Personality, and Technology: Airships in the Manipulations of Dr. Hugo Eckener and Lord Thomson, 1919–1930," *Aerospace Historian* (September 1981): 165–72; Arthur M. Squires, *The Tender Ship: Governmental Management of Technological Change* (Boston: Birkhauser, 1986).

CHAPTER
5

SAFETY, RISK, AND DESIGN

Pilot Dan Gellert was flying an Eastern Airlines Lockheed L-1011, cruising at an altitude of 10,000 feet, when he inadvertently dropped his flight plan.[1] Being on autopilot control, he casually leaned down to pick it up. In doing so, he bumped the control stick. This should not have mattered, but immediately the plane went into a steep dive, terrifying the 230 passengers. Badly shaken himself, Gellert was nevertheless able to grab the control stick and ease the plane back on course. Though much altitude had been lost, the altimeter still read 10,000 feet.

Not long before this incident, one of Gellert's colleagues had been in a flight trainer when the autopilot and the flight trainer disengaged, producing a crash on an automatic landing approach. Fortunately it all happened in simulation. But just a short time later, an Eastern Airlines L-1011 actually crashed on approach to Miami. On that flight there seemed to have been some problem with the landing gear, so the plane had been placed on autopilot at 2000 feet while the crew investigated the trouble. Four minutes later, after apparently losing altitude without warning while the crew was distracted, it crashed in the Everglades, killing 103 people.

A year later Gellert was again flying an L-1011 and the autopilot disengaged once more when it should not have done so. The plane was supposedly at 500 feet and on the proper glide slope to landing as it broke through a cloud cover. Suddenly realizing it was only at 200 feet and above a densely populated area, the crew had to engage the plane's full takeoff power to make the runway safely.

The L-1011 incidents point out how vulnerable our intricate machines and control systems can be, how they can malfunction because of unanticipated circumstances, and how important it is to design for proper human-machine interactions whenever human safety is involved. In this chapter we discuss the role of safety as seen by the public and the engineer.

Typically, several groups of people are involved in safety issues, each with its own interests at stake. If we now consider that within each group there are differences of opinion regarding what is safe and what is not, it becomes obvious that "safety" can be an elusive term, as can "risk." Following a look at these basic concepts, we will then turn to safety and risk assessment and methods of reducing risk. Finally, in examining the nuclear power plant accidents at Three Mile Island and Chernobyl, we will consider the implications of an ever-growing complexity in engineered systems and the ultimate need for safe exits.

5.1   SAFETY AND RISK

We demand safe products and services because we do not wish to be threatened by potential harm, but we also realize that we may have to pay for this safety. To complicate matters, what may be safe enough for one person may not be for someone else—a power saw in the hands of a child will never be as safe as it can be in the hands of an adult. And an adult who is sick is more prone to suffer ill effects from air pollution than is a healthy adult.

Absolute safety, in the senses of (a) entirely risk-free activities and products, or (b) a degree of safety that satisfies all individuals or groups under all conditions, is neither attainable nor affordable. Yet it is important that we come to some understanding of what we mean by safety.

5.1.1   The Concept of Safety

One approach to defining safety would be to render the notion thoroughly subjective by defining it in terms of whatever risks a person judges to be acceptable. Such a definition was given by William W. Lowrance: "*A thing is safe if its risks are judged to be acceptable.*"[2] This approach helps underscore the notion that judgments about safety are tacitly value judgments about what is acceptable risk to a given person or group. Differences in appraisals of safety are thus correctly seen as reflecting differences in values.

Lowrance's definition, however, needs to be modified, for it departs too far from our common understanding of safety. This can be shown if we consider three types of situations. Imagine, first, a case where we seriously underestimate the risks of something, say of using a toaster we see at a garage sale. On the basis of that mistaken view, we judge it to be very safe and buy it. On taking it home and trying to make toast with it, however, it sends us to the hospital with a severe electric shock or burn. Using the ordinary notion of safety, we conclude we were wrong in our earlier judgment: The toaster was not safe at all! Given our values and our needs, its risks should not have been judged acceptable earlier. Yet, by Lowrance's definition, we would be forced to say that prior to the accident the toaster was entirely safe since, after all, at that time we had judged the risks to be acceptable.

Consider, second, the case where we grossly overestimate the risks of something. For example, we irrationally think fluoride in drinking water will kill a fifth of the populace. According to Lowrance's definition, the fluoridated water is unsafe, since we judge its risks to be unacceptable. It would, moreover, be impossible for someone to reason with us to prove that the water is actually safe. For again, according to his definition, the water became unsafe the moment we judged the risks of using it to be unacceptable for us.

Third, there is the situation in which a group makes no judgment at all about whether the risks of a thing are acceptable or not—they simply do not think about it. By Lowrance's definition, this means the thing is neither safe nor unsafe with respect to that group. Yet this goes against our ordinary ways of thinking about safety. For example, we normally say that some cars are safe and others unsafe, even though many people may never even think about the safety of the cars they drive.

There must be at least some objective point of reference outside ourselves that allows us to decide whether our judgments about safety are correct once we have settled on what constitutes to us an acceptable risk. An expanded definition could capture this element, without omitting the insight already noted that safety judgments are relative to people's value perspectives.[3] One option is simply to equate safety with the absence of risk. Because little in life, and nothing in engineering, is risk-free, we prefer to adopt a modified version of Lowrance's definition:

A thing is safe if, were its risks fully known, those risks would be judged acceptable by a reasonable person in light of their settled value principles.

In our view, then, safety is a matter of how people would find risks acceptable or unacceptable if they knew the risks and were basing their judgments on their most settled value perspectives. To this extent safety is an *objective* matter. It is a *subjective* matter to the extent that value perspectives differ. In what follows we will usually speak of safety simply as acceptable risk. But this is merely for convenience, and it should be interpreted as an endorsement of Lowrance's definition only as we have qualified it.

Safety is often thought of in terms of degrees and comparisons. We speak of something as "fairly safe" or "relatively safe" (compared with similar things). Using our definition, this translates as the degree to which a person or group, judging on the basis of their settled values, would decide that the risks of something are more or less acceptable in comparison with the risks of some other thing. For example, when we say that airplane travel is safer than automobile travel, we mean that for each mile traveled it leads to fewer deaths and injuries—the risky elements that our settled values lead us to avoid. Finally, we interpret "things" to include products as well as services, institutional processes, and disaster protection.

5.1.2 Risks

We say a thing is not safe if it exposes us to unacceptable risk, but what is meant by "risk"? *A risk is the potential that something unwanted and harmful may occur.* We take a risk when we undertake something or use a product or substance that has some probability of being not safe. William D. Rowe refers to the "potential for the realization of unwanted consequences from impending events."[4] Thus a future, possible occurrence of harm is postulated.

Risk, like harm, is a broad concept covering many different types of unwanted occurrences. In regard to technology, it can equally well include dangers of bodily harm, of economic loss, or of environmental degradation. These in turn can be caused by delayed job completion, faulty products or systems, or economically or environmentally injurious solutions to technological problems.

Good engineering practice has always been concerned with safety. But as technology's influence on society has grown, so has public concern about technological risks increased. For instance, public concerns about national security can increase due to the growing availability of new biotech tools to terrorist groups.[5] In addition to measurable and identifiable hazards arising from the use of consumer products and from production processes in factories, some of the less obvious effects of technology are now also making their way to public consciousness. While the latter are often referred to as new risks, many of them have existed for some time. They are new only in the sense that (1) they are now identifiable—because of changes in the magnitude of the risks they present, because they have passed a certain threshold of accumulation in our environment, or because of a change in measuring techniques, or (2) the public's perception of them has changed—because of education, experience, media attention, or a reduction in other hitherto dominant and masking risks.

Meanwhile, natural hazards continue to threaten human populations. Technology has greatly reduced the scope of some of these, such as floods, but at the same time it has increased our vulnerability to other natural hazards, such as earthquakes, as they affect our ever-greater concentrations of population and cause greater damage to our finely tuned technological networks of long lifelines for water, energy, and food. Of equal concern are our disposal services (sewers, landfills, recovery and neutralizing of toxic wastes) and public notification of potential hazards they present.

5.1.3   Acceptability of Risk

Having adopted a modified version of Lowrance's definition of safety as acceptable risk, we need to examine the idea of acceptability more closely. William D. Rowe says that "a risk is acceptable when those affected are generally no longer (or not) apprehensive about it."[6] Apprehensiveness depends to a large extent on how the risk is perceived. This is influenced by such factors as (1) whether the risk is accepted voluntarily; (2) the effects of knowledge on how the probabilities of harm (or benefit) are known or perceived; (3) if the risks are job-related or other pressures exist that cause people to be aware of or to overlook risks; (4) whether the effects of a risky activity or situation are immediately noticeable or are close at hand; and whether the potential victims are identifiable beforehand; (5) and the cultural tradition of a community or society. Let us illustrate these elements of risk perception by means of some examples.

(1) VOLUNTARISM AND CONTROL. John and Ann Smith and their children enjoy riding dirt bikes over rough terrain for amusement. They take voluntary risks, part of being engaged in such a potentially dangerous sport. They do not expect the manufacturer of their dirt bikes to adhere to the same standards of safety as they would the makers of a passenger car used for daily commuting. The bikes should be sturdy, but guards covering exposed parts of the engine, padded instrument panels, collapsible steering mechanisms, or emergency brakes are clearly unnecessary, if not inappropriate.

In discussing dirt bikes and the like we do not include all-terrain three-wheel vehicles. Those represent hazards of greater magnitude because of the false sense of security they give the rider. They tip over easily. During the five years before they were forbidden in the United States, they were responsible for nearly 900 deaths and 300,000 injuries. About half of the casualties were children under 16.

John and Ann live near a chemical plant. It is the only area in which they can afford to live, and it is near the shipyard where they both work. At home they suffer from some air pollution, and there are some toxic wastes in the ground. Official inspectors tell them not to worry. Nevertheless they do, and they think they have reason to complain—they do not care to be exposed to risks from a chemical plant with which they have no relationship except on an involuntary basis. Any beneficial link to the plant through consumer products or other possible connections is very remote and, moreover, subject to choice.

John and Ann behave as most of us would under the circumstances: We are much less apprehensive about the risks to which we expose ourselves voluntarily than about those to which we are exposed involuntarily. In terms of our "engineering as social experimentation" paradigm, people are more willing to be the subjects of their own experiments (social or not) than of someone else's.

Intimately connected with this notion of voluntarism is the matter of control. The Smiths choose where and when they will ride their bikes. They have selected their machines and they are proud of how well they can control them, or think they can. They are aware of accident figures, but they tell themselves those apply to other riders, not to them. In this manner they may well display the characteristically unrealistic confidence of most people when they believe hazards to be under their control.[7] But still, riding motorbikes, skiing, hang gliding, bungee jumping, horseback riding, boxing, and other hazardous sports are usually carried out under the assumed control of the participants. Enthusiasts worry less about their risks than the dangers of, say, air pollution or airline safety. Another reason for not worrying so much about the consequences of these sports is that rarely does any one accident injure any appreciable number of innocent bystanders.

(2) EFFECT OF INFORMATION ON RISK ASSESSMENTS. The manner in which information necessary for decision making is presented can greatly influence how risks are perceived. The Smiths are careless about using seat belts in their car. They know that the probability of their having an accident on any one trip is infinitesimally small. Had they been told, however, that in the course of

50 years of driving, at 800 trips per year, there is a probability of 1 in 3 that they will receive at least one disabling injury, then their seat belt habits, and their attitude about seat belt laws, would likely be different.[8] Studies have verified that a change in the manner in which information about a danger is presented can lead to a striking reversal of preferences about how to deal with that danger. Consider, for example, an experiment in which two groups of 150 people were told about the strategies available for combating a disease (that in some ways foreshadowed the SARS epidemic in 2003). The first group was given the following description:

> Imagine that the U.S. is preparing for the outbreak of an unusual disease originating from Asia, which is expected to kill 600 people. Two alternative programs to combat the disease have been proposed. Assume that the exact scientific estimate of the consequences of the programs are as follows:
> If Program A is adopted, 200 people will be saved.
> If Program B is adopted, there is 1/3 probability that 600 people will be saved, and 2/3 probability that no people will be saved.
> Which of the two programs would you favor?[9]

The researchers reported that 72 percent of the respondents selected program A, and only 28 percent selected program B. Evidently the vivid prospect of saving 200 people led many of them to feel averse to taking a risk on possibly saving all 600 lives.

The second group was given the same problem and the same two options, but the options were worded differently:

> If Program C is adopted 400 people will die.
> If Program D is adopted there is 1/3 probability that nobody will die, and 2/3 probability that 600 people will die.
> Which of the two programs would you favor?

This time only 22 percent chose program C, which is the same as program A. Seventy-eight percent chose program D, which is identical to program B.

One conclusion that we draw from the experiment is that options perceived as yielding firm gains will tend to be preferred over those from which gains are perceived as risky or only probable. A second conclusion is that options emphasizing firm losses will tend to be avoided in favor of those whose chances of success are perceived as probable. In short, people tend to be more willing to take risks in order to avoid perceived firm losses than they are to win only possible gains.

(3) JOB-RELATED RISKS. John Smith's work in the shipyard has in the past exposed him to asbestos. He is aware now of the high percentage of asbestosis cases among his coworkers, and after consulting his own physician finds that he is slightly affected himself. Even Ann, who works in a clerical position at the shipyard, has shown symptoms of asbestosis. John figured that he was being paid to do a job; he felt the masks that were occasionally handed out gave him sufficient protection, and he thought the company physician was giving him a clean bill of health.

In this regard, John's thinking is similar to that of many workers who take risks on their jobs in stride, and sometimes even approach them with bravado. Employees may have little choice other than to stick with their current job regardless of the health risks, especially in areas where jobs are scarce. What they are often not told about is their exposure to toxic substances and other dangers that cannot readily be seen, smelled, heard, or otherwise sensed.

Unions and occupational health and safety regulations (such as right-to-know rules regarding toxics) can correct the worst situations, but standards regulating conditions in the workplace (its air quality, for instance) are generally still far below those that regulate conditions in our general (public) environment. It may be argued that the "public" encompasses many people of only marginal health whose low thresholds for pollution demand a fairly clean environment. On the other hand, factory workers are seldom carefully screened for their work.

Engineers who design and equip workstations must take into account the cavalier attitude toward safety shown by many employers, especially when they pay their workers on a piecework basis creating an environment that rewards productivity above all else, including safety. And when one worker complains about unsafe conditions but others do not, the complaint should be investigated and taken seriously, not dismissed. Or consider professionals sitting at keyboards who may have carpal tunnel syndrome. Musculoskeletal disorders (MSD) like carpal tunnel syndrome are still considered by some industrial groups in the United States as not meriting worker's compensation. Nevertheless, all reports from the workplace regarding unsafe or health impairing conditions of any kind merit serious attention by engineers, whether specific rules are in place or not.

(4) MAGNITUDE AND PROXIMITY. Our reaction to risk is affected by the dread of a possible mishap, both in terms of its magnitude and of the personal identification or relationship we may have with the potential victims. A single major airplane crash in a remote country, the specter of a child we know or observe on the television screen trapped in a cave-in—these affect us more acutely than the ongoing but anonymous carnage on the highways, at least until someone close to us is involved in a car accident.

In terms of numbers alone we feel much more keenly about a potential risk if one of us out of a group of 20 intimate friends is likely to be subjected to great harm than if it might affect, say, 50 strangers out of a proportionally larger group of 1000. This proximity effect arises in perceptions of risk over time as well. A future risk is easily dismissed by various rationalizations including (1) the attitude of "out of sight, out of mind," (2) the assumption that predictions for the future must be discounted by using lower probabilities, or (3) the belief that a countermeasure will be found in time.

Misperceptions of numbers can easily make us overlook losses that are far greater than the numbers reveal by themselves. Consider the 75 men who died when the unfinished Quebec Bridge collapsed in 1907. As William Starna relates,

> Of those 75 men, no fewer than 35 were Mohawk Indians from the Caughnawaga Reserve in Quebec. Their deaths had a devastating effect on the Caughnawaga community, altering drastically its demographic profile, its economic base, and its social fabric. Mohawk steelworkers would never again work in such large crews, opting instead to work in small groups on several jobs.[10]

Engineers face two problems with public conceptions of safety. On the one hand, there is the overly optimistic attitude that things that are familiar, that have not hurt us before, and over which we have some control, present no real risks. On the other hand, there is the dread people feel when an accident kills or injures in large numbers, or harms those we know, even though statistically speaking such accidents might occur infrequently.

Leaders of industry are sometimes heard to proclaim that those who fear the effects of air pollution, toxic wastes, or nuclear power are emotional and irrational, or politically motivated. This in our view is a misperception of legitimate concerns expressed publicly by thoughtful citizens. Studies have shown that public often perceive emerging technologies such as nanotechnology as having greater risks and lesser benefits than experts.[11] It is important that engineers recognize as part of their work such widely held perceptions of risk and take them into account in their designs.

(5) CROSS-CULTURAL RISK PERCEPTION. Studies have shown that cultural factors including race, gender, socioeconomic class, and nationality can affect the ways in which risks are perceived. Paul Slovic and his colleagues compared public perceptions of risk in France and the United States. They found that French participants tended to rate some technological issues (e.g., genetically engineered bacteria, nuclear waste) as having higher risk than their American counterparts. They also found that in the two countries women tended to rate most of the technological issues as higher in risk than men.[12]

A more recent study has shown that the Chinese public held an overwhelmingly positive attitude toward nanotechnology compared to people from Western countries such as Canada, Switzerland, the UK, and the United States. 84.35 percent of the sample in the study indicated that benefits should outweigh or at least be equal to risks in the development of nanotechnology.[13] Such a strong support toward nanotechnology in China may be due to a variety of historical and cultural reasons including the scientism in Chinese thought,[14] China's ambitious national agenda for global competitiveness, and the penetrating influence of mass media in the Chinese society. Researchers of this study also note that the strong support of Chinese public toward nanotechnology was derived mainly from the public beliefs in nanotechnology rather than their knowledge of nanotechnology. Therefore, the strong support of nanotechnology among Chinese public may decline if more negative consequences of nanotechnology are disclosed publicly.

DISCUSSION QUESTIONS

1. Describe a real or imagined traffic problem in your neighborhood involving people who find it difficult to cross a busy street. Put yourself in the position of (a) a commuter traveling to work on that street; (b) the relative or spouse of someone who has to cross that street on occasion; (c) a police officer assigned to keep the traffic moving on that street; and (d) the town's traffic engineer working under a tight budget.

Describe how in these various roles you might react to (e) complaints about conditions dangerous to pedestrians at that crossing and (f) requests for a pedestrian crossing protected by traffic or warning lights.

2. In some technologically advanced nations, a number of industries restricted by safety regulations have resorted to dumping their waste on—or moving their production processes to—less-developed countries where higher risks are tolerated. Examples are the dumping of unsafe or ineffective drugs by pharmaceutical companies from highly industrialized countries, and in the past the transfer of asbestos processing from the United States to Mexico.[15] More recently, toxic wastes—from lead-acid batteries to nuclear wastes—have been added to the list of "exports." To what extent do differences in perception of risk justify the transfer of such hazards and production processes to other countries? Is this an activity that can or should be regulated?

3. Grain dust is pound for pound more explosive than coal dust or gunpowder. Ignited by an electrostatic discharge or other cause, it has ripped apart grain silos and killed or wounded many workers over the years. When 54 people were killed during Christmas week 1977, grain handlers and the U.S. government finally decided to combat dust accumulation.[16] Ten years, 59 deaths, and 317 serious injuries later, a compromise standard was agreed on that designates dust accumulation of 1/8 inch or more as dangerous and impermissible in silos in the United States. Nevertheless, on Monday, June 8th, 1998, a series of explosions killed seven workers performing routine maintenance at one of the largest grain elevators in the world, demolishing one of the 246 concrete, 120 feet high silos which stretch over a length of one-half mile in Haysville, Kansas. Use grain facility explosions for a case study of workplace safety and rule making.

5.2   ASSESSING AND REDUCING RISK

Any improvement in safety as it relates to an engineered product is often accompanied by an increase in the cost of that product. On the other hand, products that are not safe incur secondary costs to the manufacturer beyond the primary (production) costs that must also be taken into account—costs associated with warranty expenses, loss of customer goodwill and even loss of customers because of injuries sustained from the use of the product, litigation, possible downtime in the manufacturing process, and so forth (see figure 5-1). It is therefore important for manufacturers and users alike to reach some understanding of the risks connected with any given product and know what it might cost to reduce those risks (or not reduce them).

5.2.1   Uncertainties in Design

One would think that experience and historical data would provide good information about the safety of standard products. Much has been collected and published. Gaps remain, however, because (1) there are some industries where

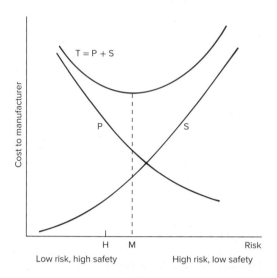

FIGURE 5-1

Why both low-risk and high-risk products are costly. P = primary cost of product, including cost of safety measures involved; S = secondary costs, including warranties, loss of customer goodwill, litigation costs, costs of downtime, and other secondary costs. T = total cost. Minimum total cost occurs at M, where incremental savings in primary cost (slope of P) are offset by an equal incremental increase in secondary cost (slope of S). Highest acceptable risk (H) may fall below risk at least cost (M), in which case H and its higher cost must be selected as the design or operating point.

information is not freely shared—for instance, when the cost of failure is less than the cost of fixing the problem, (2) problems and their causes are often not revealed after a legal settlement has been reached with a condition of nondisclosure, and (3) there are always new applications of old technology, or substitutions of materials and components, that render the available information less useful.

Risk is seldom intentionally designed into a product. It arises because of the many uncertainties faced by the design engineer, the manufacturing engineer, and even the sales and applications engineer.

To start with, there is the purpose of a design. Consider an airliner. Is it meant to maximize profits for the airline, or is it intended to give the highest possible return on investment? Investing $100 million in a jet to bring in maximum profits of, say, $20 million during a given time involves a lower return on investment than spending $48 million on a smaller jet to bring in a return of $12 million in that same period. Poorly designed artifacts can generate potential risks for even the most well-trained individuals. These artifacts, which Wade Robison calls "error-provocative designs," can permit or encourage errors even in the most commonly used circumstances.[17] For instance, a poorly designed autopilot system may provoke even the most competent pilot to make catastrophic mistakes.

Regarding applications, designs that do quite well under static loads may fail under dynamic loading. An historical example is the wooden bridge that collapsed when a contingent of Napoleon's army crossed it marching in step.

Such vibrations even affected one of Robert Stephenson's steel bridges, which shook violently under a contingent of marching British troops. Ever since then, soldiers are under orders to fall out of step when crossing a bridge. Wind can also cause destructive vibrations. Two examples are (1) "Galloping Gertie," the Tacoma Narrows Bridge that collapsed in 1940,[18] and (2) a high-voltage power line across the Bosporus in Turkey. When aerial cables of this power line oscillated during a strong wind, arcing melted them where they touched causing them to fall on houses and people below.

Apart from uncertainties about the applications of a product, there are uncertainties regarding the materials of which it is made and the level of skill that goes into designing and manufacturing it. For example, changing economic realities or hitherto unfamiliar environmental conditions such as extremely low temperatures may affect how a product is to be designed. A typical "handbook engineer" who extrapolates tabulated values without regard to their implied limits under different conditions will not fare well under such circumstances.

Caution is required even with standard materials specified for normal use. In 1981, a new bridge that had just replaced an old and trusted ferry service across the Mississippi at Prairie du Chien, Wisconsin, had to be closed because 11 of the 16 flange sections in both tie girders were found to have been fabricated from excessively brittle steel.[19] While strength tests are (supposedly) routinely carried out on concrete, the strength of steel is all too often taken for granted.

Such drastic variations from the standard quality of a given grade of steel are exceptional; more typically the variations are small. Nevertheless the design engineer should realize that the supplier's data on items like steel, resistors, insulation, optical glass, and so forth apply to statistical averages only. Individual components can vary considerably from the mean.

Engineers traditionally have coped with such uncertainties about materials or components, as well as incomplete knowledge about the actual operating conditions of their products, by introducing a comfortable "factor of safety." That factor is intended to protect against problems that arise when the stresses due to anticipated loads (duty) and the stresses the product as designed is supposed to withstand (strength or capability) depart from their expected values. Stresses can be of a mechanical or other nature—for example, an electric field gradient to which an insulator is exposed, or the traffic density at an intersection.

A product may be said to be safe if its capability exceeds its duty. But this presupposes exact knowledge of actual capability and actual duty. In reality, the stress calculated by the engineer for a given condition of loading and the stress that ultimately materializes at that loading may vary quite a bit. This is because each component in an assembly has been allowed certain tolerances in its physical dimensions and properties—otherwise the production cost would be prohibitive. The result is that the assembly's capability as a whole cannot be given by a single numerical value but must be expressed as a probability density that can be graphically depicted as a "capability" curve (figure 5-2). For a given point on a capability curve, the value along the vertical axis gives the probability that the capability, or strength, is equal to the corresponding value along the horizontal axis.

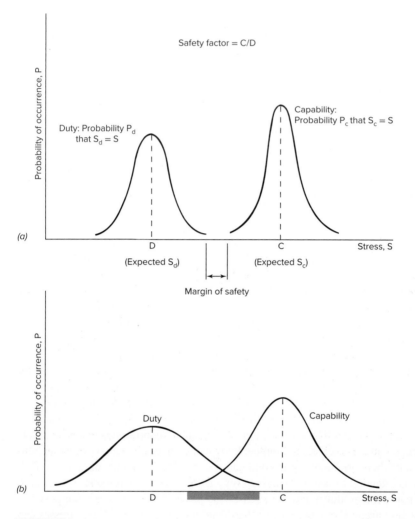

FIGURE 5-2
Probability density curves for stress in an engineered system. (*a*) Variability of stresses in a relatively safe case. (*b*) Lower safety due to overlap in stress distributions.

A similar curve can be constructed for the duty that the assembly will actually experience. The stress exposure varies because of differences in loads, environmental conditions, or the manner in which the product is used. Associated with the capability and duty curves are nominal or, statistically speaking, expected values C and D. We often think and act only in terms of nominal or expected values. And with such a deterministic frame of mind, we may find it difficult to conceive of engineering as involving experimentation. The "safety factor" C/D rests comfortably in our consciences. But how sure can we be sure that our materials are truly close to their specified nominal properties, or that the loads will not vary too widely from their anticipated values or occur in environments hostile to the proper functioning of the materials?

At times the probability density curves of capability and, or, duty will take on flatter and broader shapes because of larger than expected variances as indicated in figure 5-2*b*. If the respective values of D and C (shown on the horizontal axis for stress, S) remain the same, then so does the safety factor C/D. Now, however, there can be a pronounced overlap in the shaded region of the curves at worrisome values of probability. Edward B. Haugen has warned that "the safety factor concept completely ignores the facts of variability that result in different reliabilities for the same safety factor."[20]

A more appropriate measure of safety would be the "margin of safety," which is shown in figure 5-2*a*. If it is difficult to compute such a margin of safety for ordinary loads used every day, imagine the added difficulties that arise when repeatedly changing loads have to be considered.

5.2.2 Risk-Benefit Analyses

Many large projects, especially public works, are justified on the basis of a risk-benefit analysis. The questions answered by such a study are the following: Is the product worth the risks connected with its use? What are the benefits? Do they outweigh the risks?

We are willing to take on certain levels of risk as long as the project (the product, the system, or the activity that is risky) promises sufficient benefit or gain. If risk and benefit can both be readily expressed in a common set of units (say, lives or dollars), it is relatively easy to carry out a risk-benefit analysis and to determine whether we can expect to come out on the benefit side. For example, from an utilitarian perspective, an inoculation program may produce some deaths, but it is worth the risk if many more lives are saved by suppressing an imminent epidemic.

A closer examination of risk-benefit analyses reveals some conceptual difficulties.[21] Both risks and benefits lie in the future. Since there is some uncertainty associated with them, we should address their expected values (provided such a model fits the situation); in other words, we should multiply the magnitude of the potential loss by the probability of its occurrence, and similarly with the gain. But who establishes these values, and how? If the benefits are about to be realized in the near future but the risks are far off, how is the future to be discounted in terms of, say, an interest rate so we can compare present values? What if the benefits accrue to one party and the risks are incurred by another party?

The matter of delayed effects presents particular difficulties when an analysis is carried out during a period of high interest rates. Under such circumstances, the future is discounted too heavily because the very low present values of cost or benefit do not give a true picture of what a future generation will face.

How should one proceed when risks or benefits are composites of ingredients that cannot be added in a common set of units, as for instance in assessing effects on health plus aesthetics plus reliability? A similar challenge regularly encountering engineers is how to conduct risk-benefit analyses on projects that have impacts on the environment or historical sites.[22] At most, one can compare designs that satisfy some constraints in the form of "dollars not to exceed X, health not to drop below Y" and try to compare aesthetic values with those constraints. Or when the risks can be expressed and measured in one set of units (say, deaths on the highway) and benefits in another (speed of travel), we can employ the ratio of risks to benefits for different designs when comparing the designs.

It should be noted that risk-benefit analysis, like cost-benefit analysis, is concerned with the advisability of undertaking a *project*. When we judge the relative merits of different *designs*, however, we move away from this concern. Instead, we are dealing with something similar to cost-effectiveness analysis, which asks what design has the greater merit, given that the project is actually to be carried out. Sometimes the shift from one type of consideration to the other is so subtle that it passes unnoticed. Nevertheless, engineers should be aware of the differences so that they do not unknowingly carry the assumptions behind one kind of concern into their deliberations over the other.

These difficulties notwithstanding, there is a need in today's technological society for some commonly agreed-on process—or at least a process open to scrutiny and open to modification as needed—for judging the acceptability of potentially risky projects. What we must keep in mind is the following ethical question: "Under what conditions, if any, is someone in society entitled to impose a risk on someone else on behalf of a supposed benefit to yet others?"[23] Here we must not restrict our thoughts to average risks and benefits, but we should also consider those worst-case scenarios of persons exposed to maximum risks while they are also reaping only minimum benefits. Are their rights violated? Are they provided safer alternatives? In examining this problem further, we should also trace our steps back to an observation on risk perception made earlier: A risk to a known person (or to identifiable individuals) is perceived differently from statistical risks merely read or heard about. What this amounts to is that engineers do not affect just an amorphous public; their decisions have a direct impact on people who feel the impact acutely, and that fact should be taken into account just as seriously as are studies of statistical risk.

5.2.3 Personal Risk

Given sufficient information, an individual can decide whether to participate in (or consent to exposure to) a risky activity (an experiment). Chauncey Starr has prepared some widely used figures that indicate that individuals are more ready to assume voluntary risks than they are involuntary risks, or activities over which they have no control, even when the voluntary risks are 1000 times more likely to produce a fatality than the involuntary ones (figure 5-3).

The difficulty in assessing personal risks is magnified when we consider involuntary risks. Take John and Ann Smith and their discomfort over living near a chemical plant. Assume the general public was all in favor of building a new plant at that location, and assume the Smiths already lived in the area. Would they and others in their situation have been justified in trying to veto its construction? Would they have been entitled to compensation if the plant was built over their objections anyway? If so, how much compensation would have been adequate? These questions arise in many cases. Nuclear power plant siting is another example. Indeed, figure 5-3 was produced in the context of nuclear safety studies.

The problem of quantification alone raises innumerable problems in assessing personal safety and risk. How, for instance, is one to assess the dollar value of an individual's life? This question is as difficult as deciding whose life is worth saving, should such a choice ever have to be made. Some people might further argue that it is unethical to assign a monetary value to a human life.

Some would advocate that the assessment of personal risk needs to consider the role of marketplace in shaping social perceptions of values and risks. Nor are even more mundane gains and losses easily priced. If the market is being manipulated, or if there is a wide difference between "product" cost and sales price, it matters under what conditions the buying or selling takes place. For example, if one buys a loaf of bread, it can matter whether it is just one additional daily loaf among others; it is different when it is the first loaf available in weeks. Or, if you are compensated for a risk by an amount based on the exposure tolerance of the average person, yet your tolerance of a condition or your propensity to be harmed is much greater than average, the compensation is apt to be inadequate.

The result of these difficulties in assessing personal risk is that analysts employ whatever quantitative measures are ready at hand. In regard to voluntary

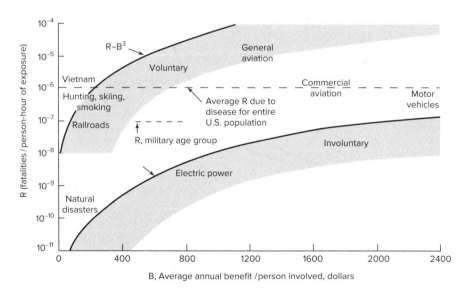

FIGURE 5-3

Willingness to assume voluntary risks as opposed to involuntary ones correlated to benefits those risks produce. (Adapted from C. Starr, "Social Benefit versus Technological Risk," *Science* 165: 1232–38.)

Reprinted with permission from "Social Benefit versus Technological Risk," *Science* 165: 1232–38. Copyright 1969 American Association for the Advancement of Science.

activities, one could possibly make judgments on the basis of the amount of home insurance a person buys. Is that individual going to offer the same amount to a home owner affected by flood to be freed? Or is there likely to be a difference between future events (requiring insurance) and present events (demand for natural disaster/flood relief)? In assessing a hazardous job, one might look at the increased wages a worker demands to carry out the task. Faced with the wide range of variables possible in such assessments, one can only suggest that an open procedure, overseen by trained arbiters, be employed in each case as it arises. On the other hand, for people taken in a population-at-large context, it is much easier to use statistical averages without giving offense to anyone in particular.

5.2.4 Public Risk and Public Acceptance

Risks and benefits to the public at large are more easily determined because individual differences tend to even out as larger numbers of people are considered. The contrast between the costs of a disability viewed from the standpoint of a private value system and from that of a societal value system, for example, is vividly illustrated in figure 5-4. Also, assessment studies relating to technological safety can be conducted more readily in the detached manner of a macroscopic view as statistical parameters take on greater significance. In that context, the

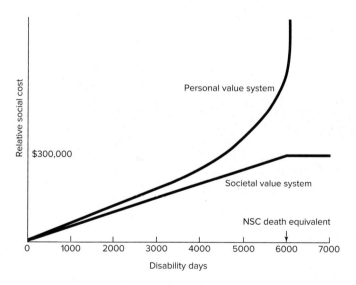

FIGURE 5-4

Value systems for social costs of disability. Using the National Safety Council equivalent of 6000 disability days for death and L. A. Sagan's 1972 assumed rate of $50 per day of disability (Sagan, "Human Cost of Nuclear Power," *Science* 177, pp. 487–93) yields a "death equivalent" of $300,000—valid for societal value analysis only. (Adapted from Starr, Rundman, and Whipple, "Philosophical Basis for Risk Analysis," *Annual Review of Energy* 1, pp. 629–62).

National Highway Traffic Safety Administration (NHTSA) has proposed a value for human life based on loss of future income and other costs associated with an accident. Intended for study purposes only, NHTSA's "blue-book value" amounted to $200,725 in 1972 dollars. This is certainly a more convenient measure than sorting out the latest figures from court cases. In one recent case, a mechanic killed in an automobile accident was awarded $6 million for the loss of the breadwinner. Punitive damages amounted to an additional $20 million.

Shulamit Kahn finds a labor market value of life in the amount of $8 million to be an acceptable average value to many people who were questioned by various investigators. Interestingly, people place higher values on *other* persons' lives, on the order of 115 to 230 percent more. Kahn says, "Yet even the $8 million figure is higher than is typically used in policy analysis. The unavoidable implication . . . is therefore that policy analysts do not evaluate the risk of their subjects' lives as highly as people evaluate risks to their own (and others') lives. Consequently, too many risks are taken."[24]

NHTSA, incidentally, emphasized that "placing a value on a human life can be nothing more than a play with figures. We have provided an estimate of some of the quantifiable losses in social welfare resulting from a fatality and can only hope that this estimate is not construed as some type of basis for determining the 'optimal' (or even worse, the 'maximum') amount of expenditure to be allocated to saving lives."[25]

5.2.5 Examples of Improved Safety

This is not a treatise on design; therefore, only a few simple examples will be given to show that safety need not rest on elaborate contingency features.

The first example is the magnetic door catch introduced on refrigerators to prevent death by asphyxiation of children accidentally trapped in them. The catch in use today permits the door to be opened from the inside without major effort. It also happens to be cheaper than the older types of latches.

The second example is the enabling handle used by the engineer (engine driver) to control a train's speed. The train is powered only as long as some pressure is exerted on the handle. If the engineer becomes incapacitated and lets go of the handle, the train stops automatically. Self-driving cars such as Tesla will sound warnings if drivers take their hands off the wheel for one minute at speeds above 45 miles per hour. If drivers ignore three warnings in an hour, the system will temporarily shut off until the car is parked.

Railroads provide the third example as well. Over a hundred years ago, to signal to a train that it could proceed, a ball was raised to the top of a mast, to signal a stop the ball was lowered. Later, semaphores used a mechanical arm, but both methods required a cable to be pulled, and they incorporated a fail-safe approach in that an accidentally cut cable would let the ball or the arm drop to the STOP position all by itself.

The motor-reversing system shown in figure 5-5 gives still another example of a situation in which the introduction of a safety feature involves merely the

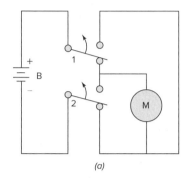
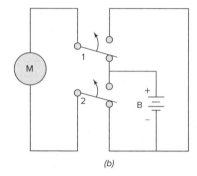

(a)   (b)

FIGURE 5-5
Reversing switch for a permanent magnet motor. (*a*) Arms 1 and 2 of the switch are both raised by a solenoid (not shown). If either one does not move—say, a contact sticks—while the other does, there is a short across the battery. The battery will discharge and be useless even after the trouble is detected. (*b*) By exchanging the positions of battery and motor, a stuck switch will cause no harm to the battery. (The motor can be shorted without harm.)

proper arrangement of functions at no additional expense. As the mechanism is designed in figure 5-5*a*, sticky contacts could cause battery B to be shorted, thus making it unavailable for further use even after the contacts are coaxed loose. A simple reconnection of wires as shown in figure 5-5*b* removes that problem altogether.

In the rush to bring a product onto the market, safety considerations are often slighted. This would not be so much the case if the venture were regarded as an experiment—an experiment that is about to enter its active phase as the product comes into the hands of the user. Space flights were carried out with such an attitude, but more mundane ventures involve less obvious dangers, and therefore less attention is usually paid to safety. If moral concerns alone do not sway engineers and their employers to be more heedful of potential risks, then recent trends in product liability law should certainly do so.

DISCUSSION QUESTIONS

1. A worker accepts a dangerous job after being offered an annual bonus of $2000. The probability that the worker may be killed in any one year is 1 in 10,000. This is known to the worker. The bonus may therefore be interpreted as a self-assessment of life with a value equal to $2000 divided by 1/10,000, or $20 million. Is the worker more or less likely to accept the job if presented with the statistically nearly identical figures of a $100,000 bonus over 50 years (neglecting interest) and a 1/200 probability of a fatal accident during that period?

2. "Airless" paint spray guns do not need an external source of compressed air connected to the gun by a heavy hose (although they do need a cord to attach them to a power source) because they have incorporated a small electric motor and pump. One common design uses an induction motor that does not cause sparking because it does not require a commutator and brushes (which are sources of sparking). Nevertheless the gun carries a label

warning users that electrical devices operated in paint spray environments pose special dangers. Another type of gun that, like the first, also requires only a power cord is designed to weigh less by using a high-speed universal motor and a disk-type pump. The universal motor does require a commutator and brushes, which cause sparking. This second kind of spray gun carries a warning similar to that attached to the first, but it states in addition that the gun should never be used with paints that employ highly volatile and flammable thinners such as naphtha. The instruction booklet is quite detailed in its warnings.

A painter had been lent one of the latter types of spray guns. In order to clean the apparatus, they partially filled it with paint thinner and operated it. It caught fire, and the painter was severely burned as the fire spread. The instruction booklet was in the cardboard box where the gun was kept, but the painter had not read it. They had, however, used the first type of airless paint spray gun in a similar manner without mishap. The warning messages on both guns looked pretty much the same. Do you see any ethical problems in continuing over-the-counter sales of this second type of spray gun? What should the manufacturer of this novel, lightweight device do?

In answering these questions, consider the fact that courts have ruled that hidden design defects are not excused by warnings attached to the defective products or posted in salesrooms. Informed consent must rest on a more thorough understanding than can be transmitted to buyers by warning labels.

5.3 THREE MILE ISLAND, CHERNOBYL, AND SAFE EXITS

As our engineered systems grow more complex, it becomes more difficult to operate them. As Charles Perrow argues, our traditional systems tended to incorporate sufficient slack, which allowed system aberrations to be corrected in a timely manner.[26] Nowadays, he points out, subsystems are so tightly coupled within more complex total systems that it is not possible to alter a course safely unless it can be done quickly and correctly. Often the supposedly corrective action taken by operators can make matters worse because they do not know what the problem is. For instance, during the emergency at Three Mile Island, to be described next, so many alarms had to be recorded by a printer that it fell behind by as much as $2\frac{1}{2}$ hours in displaying the events.

Designers hope to ensure greater safety during emergencies by removing the need of human operators and mechanizing their functions. The control policy would be based on predetermined rules. This in itself creates problems because (1) not all eventualities are foreseeable, and (2) even those that can be predicted will be programmed by an error-prone human designer. In addition, another problem arises when the mechanized system fails and a human operator has to replace the computer in an operation that demands many rapid decisions.

Operator errors were the main cause of the nuclear reactor accidents at Three Mile Island and Chernobyl. Beyond these errors a major deficiency surfaced at both installations: inadequate provisions for evacuation of nearby populations. This lack of safe exit is found in too many of our amazingly complex systems.

5.3.1   Three Mile Island

Walter Creitz, president of Metropolitan Edison, the power company in the Susquehanna Basin, was obviously annoyed by a series of articles in the *Record,* a local daily newspaper of York, Pennsylvania. The *Record* had cited unsafe conditions at Metropolitan Edison's Three Mile Island nuclear power plant Unit 2. Creitz dismissed the stories as "something less than a patriotic act—comparable in recklessness . . . to shouting 'Fire!' in a crowded theater." A few days later a minor malfunction in the plant set off a series of events that made Three Mile Island (TMI) a household name across the world.[27]

Briefly, this is what happened.[28] At 4:00 A.M. on March 28, 1979, Unit TMI-2 was operating under full automatic control at 97 percent of its rated power output. For 11 hours a maintenance crew had been working on a recurring minor problem. Resin beads were used in several demineralizers (labeled 14 in figure 5-6) to clean or "polish" the water on its way from the steam condenser (12) back to the steam generator (3). Some beads clogged the resin pipe from a demineralizer to a tank in which the resin was regenerated. In flushing the pipe with water, perhaps a cupful of water backed up into an air line that provided air for fluffing the resin in its regeneration tank. But that air line was connected to the air system that also served the control mechanisms of the large valves at the outlet of the demineralizers. Thus it happened that these valves closed unexpectedly.

With water flow interrupted in the secondary loop (26), all but one of the condensate booster pumps turned off. That caused the main feedwater pumps (23) and the turbine (10) to shut down as well. In turn, an automatic emergency system started up the auxiliary feedwater pumps (25). But with the turbines inoperative, there was little outlet for the heat generated by the fission process in the reactor core. The pressure in the reactor rose to over 2200 pounds per square inch, opening a pressure-relief valve (7) and signaling a SCRAM, in which control rods are lowered into the reactor core to stop the main fission process.

The open valve succeeded in lowering the pressure, and the valve was readied to be closed. Its solenoid was de-energized and the operators were so informed by their control-panel lights. But something went wrong: The valve remained open, contrary to what the control panel indicated. Apart from this failure everything else had proceeded automatically as it was supposed to. Everything, that is, except for one other serious omission: The auxiliary pumps (25) that had been started automatically could not supply the auxiliary feedwater because block valves (24) had inadvertently been left closed after maintenance work done on them two days earlier. Without feedwater in the loop (26), the steam generator (3) boiled dry. Now there was practically no heat removal from the reactor, except through the relief valve. Water was pouring out through it at the rate of 220 gallons per minute. The reactor had not yet cooled down, and even with the control rods shutting off the main fission reaction there would still be considerable heat produced by the continuing radioactive decay of waste products.

Loss of water in the reactor caused one of a group of pumps, positioned at 15, to start automatically; another one of these pumps was started by the operators

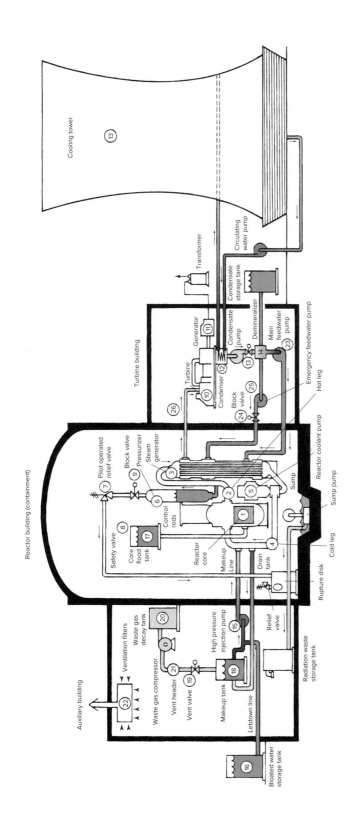

FIGURE 5-6

Schematic diagram of Three Mile Island nuclear power plant Unit 2. Pressurized water reactor system: Heat from reactor core (1) is carried away by water in a primary loop (1, 2, 3, 5, 4). In steam generator (3) the heat is transferred to water in a secondary loop (26) at lower pressure. The secondary-loop water turns to steam in the steam generator or boiler (3), drives the turbine (10), turns into water in the condenser (12), and is circulated back to (3) by means of pumps (13, and 23 and 25). (Adapted from John F. Mason, "The Technical Blow-by-Blow: An Account of the Three Mile Island Accident," *IEEE Spectrum*, 16 [November 1979], copyright © 1979 by the Institute of Electrical and Electronics Engineers, Inc., and from Mitchell Rogovin and George T. Frampton, Jr., *Three Mile Island: A Report to the Commissioners and the Public*, vol. 1, Nuclear Regulatory Commission Special Inquiry Group, NUREG/CR=1250, Washington, DC [January 1980], 1980.)

to rapidly replenish the water supply for the reactor core. Soon thereafter, the full emergency core-cooling system went into operation in response to low reactor pressure. Low reactor pressure can promote the formation of steam bubbles that reduce the effectiveness of heat transfer from the nuclear fuel to the water. There is a pressurizer that is designed to keep the reactor water under pressure. (The relief valve sits atop this pressurizer.) The fluid level in the pressurizer was also used as an indirect—and the only—means of measuring the water level in the reactor.

The steam in the reactor vessel caused the fluid level in the pressurizer to rise. The operators, thinking they had resolved the problem and that they now had too much water in the reactor, shut down the emergency core-cooling system and all but one of the emergency pumps. Then they proceeded to drain water at a rate of 160 gallons per minute from the reactor, causing the pressure to drop. At this point they were still unaware of the water escaping through the open relief valve. Actually, they assumed some leakage, which occurred because of poor valve seating even under normal circumstances. It was this that made them disregard the high-temperature readings in the pipes (beyond location 7).

The steam bubbles in the reactor water covered much of the fuel, and the tops of the fuel rods began to crumble. The chemical reaction between the steam and zircaloy covering the fuel elements produced hydrogen, some of which was released into the containment structure, where it exploded.

The situation was becoming dire when, two hours after the initial event, the next shift arrived for duty. With some fresh insights into the situation, the relief valve was deduced to be open. Blocking valve 9 in the relief line was then closed by the crew. Soon thereafter, with radiation levels in the containment building rising, a general alarm was sounded. While there had been telephone contact with the Nuclear Regulatory Commission (NRC), as well as with Babcock & Wilcox (B&W), who had built the reactor facility, no one answered at NRC's regional office and a message had to be left with an answering service. The fire chief of nearby Middletown was to hear about the emergency on the evening news.

In the meantime, a pump was transferring the drained water from the main containment building to the adjacent auxiliary building, but not into a holding tank as intended; because of a blown rupture disk, the water landed on the floor. When there was indication of sufficient airborne radiation in the control room to force evacuation, all but essential personnel wearing respirators stayed behind. The respirators made communication difficult.

Eventually the operators decided to turn the high-pressure injection pumps on again, as the automatic system had been set to do all along. The core was covered once more with water, though there were still some steam and hydrogen bubbles on the loose. Thirteen and one-half hours after the start of the episode, there was finally hope of getting the reactor under control. Confusion over the actual state of affairs, however, continued for several days.

Nationwide, the public watched television coverage in disbelief as responsible agencies displayed their lack of emergency preparedness at both the reactor site and evacuation-planning centers. Years later one still reads about the steadily accumulating costs of decommissioning (defueling, decontaminating, entombing) Unit 2 at TMI, $1 billion so far, of which one-third is passed on to ratepayers—all this for what cost $700 million to build. Three Mile Island was a financial disaster and a blow to the reputation of the industry, but fortunately radioactive release was low, and cancer rates downwind are reported to be only slightly higher than normal. The reactor cleanup started in August 1979 and ended in December 1993 at a cost of $975 million. The other reactor (TMI-1) was restarted in 1985 and subsequently changed ownership.

5.3.2   Chernobyl

The nuclear power plant complex at Chernobyl, near Kiev (Ukraine, then of the U.S.S.R.) had four reactors in place by 1986.[29] With the planned addition of Units 5 and 6, for which foundation work was under way, the site would be the world's second-largest electric power plant park, with an output of 6000 megawatts (electrical). The reactors were of a type called RBMK; they are graphite-moderated and use boiling-water pressure tubes. Chernobyl and the Soviet nuclear power program were prominently featured in 1985 issues of the English-language periodical *Soviet Life*. The articles featured the safety of atomic energy and the low risk of accidents and radiation exposure. Since that time "Chernobyl" has become a household word because of a terrible reactor fire that occurred there.

On April 25, 1986, a test was under way on Reactor 4 to determine how long the mechanical inertia of the turbine-generator's rotating mass could keep the generator turning and producing electric power after the steam supply was shut off (figure 5-7). This was of interest because reactor coolant pumps and other vital electric machinery have to continue functioning although the generators may have had to be disconnected suddenly from a malfunctioning power grid. Special diesel generators will eventually start to provide emergency power for the plant, but diesel units cannot always be relied on to come up promptly. This test was undertaken as part of a scheduled plant shutdown for general maintenance purposes.

It requires 3600 megawatts of thermal power in the RBMK reactor to produce 1200 megawatts at the generator output. The output of Unit 4 had been gradually throttled from 3200 megawatts (thermal) to 1600 megawatts and was to be slowly taken down to between 1000 and 700 megawatts, but at 2:00 P.M. the power dispatch controller at Kiev requested that output be maintained to satisfy an unexpected demand. This meant a postponement of the test. In preparation for the test, the reactor operators had disconnected the emergency core-cooling system so its power consumption would not affect the test results. This was to be the first of many safety violations.

Another error occurred when a control device was not properly reprogrammed to maintain power at the 700- to 1000-megawatt level. When, at 11:10 P.M., the plant was authorized to reduce power, its output dropped all the way to 30 megawatts, where the reactor is difficult to control. Instead of shutting down the reactor, the operators tried to keep the test going by raising the control rods to

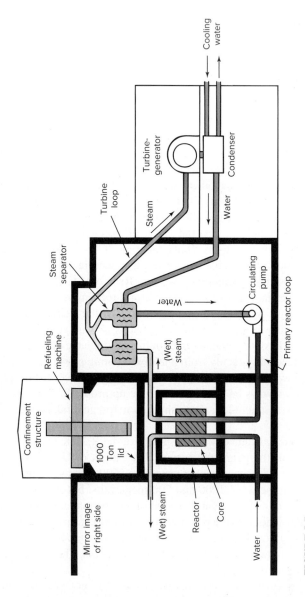

FIGURE 5-7

Schematic diagram of Reactor 4 at Chernobyl. This RBMK-type reactor produces steam for two 500-megawatt steam turbine generators, only one of which is shown. (Adapted from Ahearne, "Nuclear Power after Chernobyl," *Science* 236: 674.) Reprinted with permission from "Nuclear Power after Chernobyl," *Science* 236 (May 1987), p. 674. Copyright 1987 American Association for the Advancement of Science.

increase power. Instead of leaving 15 controls inserted as required, the operators raised almost all control rods because at the low power level the fuel had become poisoned by a buildup of xenon-135, which absorbs neutrons.

The power output stayed steady at 200 megawatts (thermal)—still below what the test called for—but the test continued. In accordance with the test protocol, two additional circulating pumps were turned on to join the six already in operation. Under normal levels of power output this would have contributed to the safety of the reactor, but at 200 megawatts it required many manual adjustments to maintain the balance of steam and water. In addition, "the operators at this point recognized that because of the instabilities in this reactor and the way xenon poisoning builds up, once the reactor is shut down, they would have to wait a long time before starting it up again."[30] So, deciding to proceed with the test, the operators blocked the emergency signals and automatic shutdown controls because they would have been activated upon removal of the electrical load.

This left the reactor in a precarious position: "The reactor was now running free, isolated from the outside world, its control rods out, and its safety system disconnected." As Valery Legasov, then U.S.S.R. representative to the International Atomic Energy Commission (IAEC), reported at a review of the accident: "The reactor was free to do as it wished."[31]

At 1:23 A.M. the test began. When the steam valves were closed and its load was effectively removed, the reactor's power and temperature rose sharply. Unlike water-moderated reactors, the graphite-moderated RBMK reactor uses water only as a heat-transfer medium, not as a moderator. As the core becomes hotter it allows fission to increase. This positive feedback effect produced a surge of power in Chernobyl's Reactor 4, from 7 percent to hundreds of times its rated thermal output: "The effect was the equivalent of a ton of TNT exploding in the core. . . . The fuel did not have time to melt . . . it simply shattered into fragments."[32] The fuel, bereft of its cladding, came in contact with the water. A second explosion occurred (very likely a steam explosion). It lifted and shifted a 1000-ton concrete floor pad separating the reactor from the refueling area above it. The zirconium cladding of the fuel rods interacted with the circulating water to form hydrogen. This produced a spectacular display of fireworks. A shower of glowing graphite and fuel spewed over the compound while a radioactive plume was driven sky-high by the heat.

What followed was as inexcusable as what had caused the accident. While valiant firefighters lost their lives extinguishing the blaze, it took hours to warn the surrounding communities. Only when alert nuclear plant operators in Sweden detected an increase in radioactivity did Moscow learn that something was amiss. The Soviet republics and the rest of Europe were not prepared to handle such a grave event, especially the radioactive fallout. Many countries blamed Moscow for not notifying them, but these countries had no monitoring devices of their own, not even to check on their local nuclear plants. Subsequent instructions on what to do about drinking milk, eating vegetables, letting children play outside, and other concerns of the populations of Europe depended more on the political leanings and the pronuclear or antinuclear stance of their respective health departments.

Acute radiation sickness, combined with burns, severely affected about 200 Chernobyl plant workers, of whom 31 died very quickly. By 1992 the number of excess-radiation death cases attributable to Chernobyl were reported to have exceeded 6000 within the Ukraine alone, and in Belarus so many children had enlarged thyroids that 10,000 cancer cases above the usual were expected.[33] The 1000 families living in a workers' settlement one mile from the plant were evacuated 12 hours after the explosion, but the plant had no responsibility for, nor direct link with, the communities beyond a 1.5-mile radius. The evacuation of nearby Pripyat and 71 villages within 18 miles of the plant started the next day. About 135,000 people had to be moved by buses and trucks. Numerous new villages were constructed to house the displaced. By arbitrarily announcing a new, "safe" radiation dose pegged at 10 times the former level, the Politburo saved itself the burden of evacuating another 1.25 million people from surrounding areas and having to give medical care to countless more exposed to radiation. High government officials, however, were quietly moving their families away from as far away as Kiev while the masses were asked to turn out for open-air May Day celebrations.

The near- and long-term effects of radiation on the people and fauna of Europe will be widely discussed for many years. Seven years after the accident it had become clear that more radiation escaped than had been estimated earlier because the red glow that was targeted by the pilots with their loads of sand, clay, and dolomite came not from the reactor as thought but from a small, ejected core element 50 feet from the reactor. Contamination was also spread by agricultural products from the affected areas (milk and meats), which were exported to other parts of the Soviet Union after they were mixed into clean products.

After the accident the reactor was encased in a concrete sarcophagus, but not an airtight one—so it had to be replaced. Eventually some of the other reactors at the plant experienced difficulties as well, leading the government (of what had by then become the Republic of the Ukraine) to decommission the entire plant by the end of the year 2000 with international financial support.

5.3.3 Safe Exits

It is almost impossible to build a completely safe product or one that will never fail. The best one can do is to assure that when a product fails, (1) it will fail safely, (2) the product can be abandoned safely, or—at least—(3) the user can safely escape the product. Let us refer to these three conditions as *safe exit*. It is not obvious who should take the responsibility for providing safe exit. But apart from questions of who will build, install, maintain, and pay for a safe exit system there remains the crucial question of who will recognize the need for a safe exit.

It is our position that providing for a safe exit is an integral part of the experimental procedure and sound engineering. The experiment is to be carried out without causing bodily or financial harm. If safety is threatened, the experiment must be terminated. The full responsibility cannot fall on the shoulders of a lone engineer, but one can expect the engineer to issue warnings when a safe exit does not exist or the experiment must be terminated.

Here are some examples of what this might involve. Ships need lifeboats with enough spaces for all passengers and crew members. Buildings need usable fire escapes. The operation of nuclear power plants calls for realistic ways to evacuate nearby communities. The foregoing are examples of safe exits for people. Provisions are also needed for safe disposal of dangerous products and materials: Altogether too many truck accidents and train derailments have exposed communities to toxic gases, and too many dumps have let toxic wastes get in to the groundwater or into the hands of children. Finally, avoiding system failure may require redundant or alternative means of continuing a process when the original process fails. Examples would be backup systems for computer-based data banks, air traffic control systems, automated medical treatment systems, or sources of water for fighting fires.

Apart from a safety conscious design and thorough testing of any potentially dangerous product before it is delivered for use, it is of course necessary that its user have in place procedures for regular maintenance and safety checks. Beyond such measures there should also be in place (a) avenues for employees to freely report hazardous conditions regarding the design or the operation of the product without having to resort to whistle-blowing, and (b) emergency procedures based on human factors engineering which takes into account how people react and interact under conditions of stress as occurred during the TMI accident.

DISCUSSION QUESTIONS

1. Discuss what you see as the main similarities and differences between Three Mile Island and Chernobyl.
2. It has been said that Three Mile Island showed us the risks of nuclear power and the Arab oil embargo the risk of having no energy. Forcing hazardous products or services from the market has been criticized as closing out the options of those individuals or countries with rising aspirations who can now afford them and who may all along have borne more than their share of the risks without any of the benefits. Finally, pioneers have always exposed themselves to risk. Without risk there would be no progress. Discuss this problem of "the risk of no risk."[34]
3. Discuss the notion of safe exit, using evacuation plans for communities near nuclear power plants or chemical process plants.
4. Valery Legasov, nuclear engineer and later U.S.S.R. representative to the IAEC, played an important role in containing the reactor fire at Chernobyl. He said that human errors brought on the accident, and if earlier we looked at safety technology as a means of protecting us from machines, now technology must be protected from us.[35] Discuss the change in Legasov's viewpoint and in what ways you agree or disagree with his statement.
5. The toxic waste cases known as the Love Canal and Woburn Cancer Cluster episodes have received wide attention in the United States and are well documented in the technical and popular literature. The latter is the topic of a movie, *A Civil Action*, based on

a book by Jonathan Harr that describes lawyer Jan Schlichtmann's efforts to see justice done on behalf of some of Woburn's families. Compare these two toxic waste cases and discuss the bottlenecks that prevent the release of pertinent information to public bodies and litigants in such court cases.

6. Gather information on accidents that plagued Japan's atomic energy industry during the past decade. Reluctant to admit to problems apt to have serious political repercussions, electric utility managers on several occasions decided not to report problems that had been encountered. For employees to go public with such information would have been considered treachery. So it was left to an American engineer working for the subcontracting General Electric Company to report the deficiencies. He was able to do so under a new law providing "legal protection to third parties informing regulators of improprieties by operators of nuclear facilities." (You may search for Tokyo Electric Power Co. or TEPCO at website www.platt.com and for a case study by Hiroshi Iino at www.onlineethics.org.)

KEY CONCEPTS

—**Safety** (as acceptable risk): the risks about the technology, were they fully known, would be judged acceptable by a reasonable person in light of their settled value principles.

—**Risk:** the potential that something unwanted and harmful may occur.

—**Risk perception factors:** whether the risk is assumed voluntarily, how the probabilities of harm or benefit are presented, job-related or other pressures, magnitude, and proximity.

—**Risk-benefit analysis:** studies determining the risks, the benefits, and whether the project or product is worth the risks connected with its use.

—**Cost-effectiveness analysis:** studies of which design has the greatest merit.

—**Safe exits:** design and procedures ensuring that if a product fails it will fail safely and the user can safely escape the product.

REFERENCES

1. Dan Gellert, "Whistle-Blower: Dan Gellert, Airline Pilot," *The Civil Liberties Review,* September 1978.
2. William W. Lowrance, *Of Acceptable Risk* (Los Altos, CA: William Kaufmann, 1976), p. 8.
3. Council for Science and Society, *The Acceptability of Risks* (England: Barry Rose, Ringwood, Hants, 1977), p. 13.
4. William D. Rowe, *An Anatomy of Risk* (New York: John Wiley & Sons, 1977), p. 24.
5. Camino Kavanagh, *New Tech, New Threats, and New Governance Challenges: An Opportunity to Craft Smarter Responses?* (Washington, DC: Carnegie Endowment for International Peace, 2019).
6. William D. Rowe, "What Is an Acceptable Risk and How Can It Be Determined?" in *Energy Risk Management,* ed. G. T. Goodman and W. D. Rowe (New York: Academic Press, 1979), p. 328.
7. Paul Slovic, Baruch Fischhoff, and Sarah Lichtenstein, "Weighing the Risks: Which Rights Are Acceptable?" *Environment* 21 (April 1979): 14–20 and (May 1979): 17–20, 32–38; Paul Slovic, Baruch Fischhoff, and Sarah Lichtenstein, "Risky Assumptions," *Psychology Today* 14 (June 1980): 44–48.
8. Richard J. Arnould and Henry Grabowski, "Auto Safety Regulation: An Analysis of Market Failure," *The Bell Journal of Economics* 12 (Spring 1981): 35.

9. Amos Tversky and Daniel Kahneman, "The Framing of Decisions and the Psychology of Choice," *Science* 211 (January 30, 1981): 453. Kahneman received the 2002 Nobel Prize in Economics, and had Tversky not died earlier he would undoubtedly have shared the prize.

10. William A. Starna, "A Disaster's Toll," letter to the editor, *American Heritage of Invention and Technology,* Summer 1986, commenting on "A Disaster in the Making" in the Spring 1986 issue.

11. Shirley S. Ho, Dietram A. Scheufele, and Elizabeth A. Corley, "Value Predispositions, Mass Media, and Attitudes Toward Nanotechnology: The Interplay of Public and Experts," *Science Communication* 33 (June 2011): 167–200.

12. Paul Slovic, James Flynn, C. K. Mertz, Claire Mays, and Mark Poumadere, "Nuclear Power and the Public: A Comparative Study of Risk Perception in France and the United States," in *Cross-Cultural Risk Perception: A Survey of Empirical Studies,* ed. O. Renn and B. Rohrmann (Dordrecht, Netherlands: Springer Science+Business Media, 2000), pp. 55–102.

13. Jing Zhang, Guoyu Wang, and Deming Lin, "High Support for Nanotechnology in China: A Case Study in Dalian," *Science and Public Policy* 43 (2016): 115–27.

14. Daniel W. Y. Kwok, *Scientism in Chinese Thought, 1900–1950* (New Haven, CT: Yale University Press, 1965).

15. Milton Silverman, Philip Lee, and Mia Lydecker, *Prescription for Death: The Drugging of the Third World* (San Francisco: University of California Press, 1981); and Henry Shue, "Exporting Hazards," *Ethics* 91 (July 1981): 586.

16. Eliot Marshall, "Deadlock Over Explosive Dust," *Science* 222 (November 4, 1983): 485–87; discussion, p. 1183.

17. Wade L. Robison, *Ethics within Engineering: An Introduction* (London, UK: Bloomsbury Academic, 2017).

18. Henry Petroski, *To Engineer Is Human: The Role of Failure in Successful Design* (New York: St. Martin's Press, 1985); M. Levy and M. Salvadori, *Why Buildings Fall* (New York: C.C. Norton & Co., 1992); Frederic D. Schwarz, "Why Theories Fall Down," *American Heritage of Invention & Technology,* Winter 1993, pp. 6–7.

19. "Suit Claims Faulty Bridge Steel," *ENR (Engineering News Record),* March 12, 1981, p. 14; see also March 26, 1981, p. 20; April 23, 1981, pp. 15–16; November 19, 1981, p. 28.

20. Edward B. Haugen, *Probabilistic Approaches to Design* (New York: John Wiley & Sons, 1968), p. 5. Also see Haugen's *Probabilistic Mechanical Design* (N.Y.: Wiley, 1980), p. 357.

21. See especially Matthew D. Adler and Eric A. Posner (eds.), *Cost-Benefit Analysis: Legal, Economic, and Philosophical Perspectives* (Chicago: University of Chicago Press, 2001).

22. Martin Peterson, *Ethics for Engineers* (New York, NY: Oxford University Press, 2020).

23. Council for Science and Society, *The Acceptability of Risks,* p. 37.

24. Shulamit Kahn, "Economic Estimates of the Value of Life," *IEEE Technology and Society Magazine,* June 1986, pp. 24–29. Reprinted in Albert Flores, *Ethics and Risk Management in Engineering* (Boulder, CO: Westview Press, 1988).

25. Brian O'Neill and A. B. Kelley, "Costs, Benefits, Effectiveness, and Safety: Setting the Record Straight," *Professional Safety,* August 1975, p. 30.

26. Charles Perrow, *Normal Accidents: Living with High-Risk Technologies* (Princeton, NJ: Princeton University Press, 1999).

27. Mitchell Rogovin and George T. Frampton Jr., *Three Mile Island: A Report to the Commissioners and the Public,* vol. 1, Nuclear Regulatory Commission Special Inquiry Group, NUREG/CR-1250, Washington, DC (January 1980), p. 3. Diagram in text used with permission of Mitchell Rogovin.

28. Kemeny Commission Report, *Report of the President's Commission on the Accident at Three Mile Island* (New York: Pergamon Press, 1979); Daniel F. Ford, *Three Mile Island* (New York: Viking, 1982); John F. Mason, "The Technical Blow-by-Blow: An Account of the Three Mile Island Accident," *IEEE Spectrum* 16 (November 1979): 33–42; Thomas H. Moss and David L. Sills (eds.), *The Three Mile Island Nuclear Accident: Lessons and Implications,* vol. 365 (New York: Annals of the New York Academy of Sciences, 1981); Bill Keisling, *Three Mile Island* (Seattle, WA: Veritas, 1980); Daniel Martin, *Three Mile Island: Prologue or Epilogue?* (Cambridge, MA: Ballinger, 1980).

29. David R. Marples, *Chernobyl and Nuclear Power in the USSR* (London: MacMillan Press, 1986); Mike Edwards, "Chernobyl—One Year After," *National Geographic* 171 (May 1987): 632–53; Grigori Medvedev, *The Truth about Chernobyl* (New York: Basic Books, 1991); Alla Yaroshinskaya, *Chernobyl, the Forbidden Fruit* (Oxford: Jon Carpenter, 1994).

30. John F. Ahearne, "Nuclear Power after Chernobyl," *Science* 236 (May 1987): 673–79; discussion by E. G. Silver and J. F. Ahearne, *Science* 238 (October 1987): 144–45.

31. Nigel Hawkes, Geoffrey Lean, David Leigh, et al., *Chernobyl—The End of the Nuclear Dream* (New York: Vintage, 1986), p. 102.

32. Ibid.

33. Testimony by Murray Feshbach before U.S. Senate Hearing in 1992; S.Hrg.102-765.

34. Aaron Wildavsky, "No Risk Is the Highest Risk of All," in *Ethical Problems in Engineering*, 2nd ed., ed. Albert Flores (Troy, NY: Rensselaer Polytechnic Institute 1980), pp. 221–26.

35. Mike Moore, editorial, *Bulletin of Atomic Science,* May-June 1996.

CHAPTER

6

WORKPLACE CULTURES, RESPONSIBILITIES AND RIGHTS

Data General Corporation grew spectacularly during its first decade of operation, quickly becoming a Fortune 500 company that was ranked third in overall sales of small computers. However, it began to fall behind the competition and desperately needed a powerful new microcomputer to sustain its share of the market. The development of that computer is chronicled by Tracy Kidder in his Pulitzer Prize–winning book *The Soul of a New Machine*.

Tom West, one of Data General's most trusted engineers, convinced management that he could build the new computer within one year—an unprecedented time for a project of its importance. West assembled a team of 15 exceptionally motivated though relatively inexperienced young engineers, many of whom were just out of school. Within six months they designed the central processing unit, and they delivered the complete computer ahead of schedule. Named the Eclipse MV/8000, the computer immediately became a major marketing success.

The remarkable success was possible because the engineers came to identify themselves with the project and the product: "Ninety-eight percent of the thrill comes from knowing that the thing you designed works, and works almost the way you expected it would. If that happens, part of you is in that machine."[1] The "soul" of the new machine was not any one person. Instead, it was the team of engineers who invested themselves in the product through their personal commitment to work together creatively with colleagues as part of a design group. As might be expected, personality clashes occurred during the sometimes frenzied work schedule, but conflicts were minimized by a commitment to teamwork, collegiality, shared commitment, and identification with the group's project. More worrisome, there were times when the engineers pushed themselves to their limits, imposing burdens on their families and their health, but for the most part those times remained limited. (The computer appears on www.simulogics.com.)

Kidder ends his book by quoting a regional sales manager speaking to the sales representatives preparing to market the new computer:

> " 'What motivates people?' he asked.
> He answered his own question, saying, 'Ego and the money to buy things that they and their families want.' "[2]

The engineers, of course, cared about money and ego, but Kidder makes it clear that those motives could not explain how it was possible for them to accomplish what they did. Professionalism involves much more, including both a sense of fun and excitement and personal commitments that have moral dimensions.

The kind of commitment shown by the engineers understandably ranks high on the list of expectations that employers have of the engineers they employ or of the engineers they engage as consultants. Engineers in turn should see top performance at a professional level as their main responsibility, accompanied by others such as maintaining confidentiality and avoiding conflicts of interest. But engineers also need the opportunity to perform responsibly, and this means that their professional rights must be observed.

In this chapter we first discuss the moral aspects of teamwork. Next we address the need to maintain confidentiality and to avoid conflicts of interest as issues related to an engineer's loyal service to employers. Then we turn to an engineer's rights as a professional and as an employee. We end with the issue of whistleblowing, which can pit loyalty to the employer against the engineer's responsibilities to the public.

6.1 TEAMWORK

Loyalty to corporations, respect for authority, collegiality, and other teamwork virtues are enormously important in engineering. Yet they are virtues only within the context of corporations that maintain an ethical climate, rather than pursuing ends and using means that are morally objectionable.

6.1.1 An Ethical Corporate Climate

An ethical climate is a working environment that is conducive to morally responsible conduct. Within corporations it is produced by a combination of formal organization and policies, informal traditions and practices, and personal attitudes and commitments. Engineers can make a vital contribution to such a climate, especially as they move into technical management and then more general management positions.

Professionalism in engineering would be threatened at every turn in a corporation devoted primarily to powerful egos. Sociologist Robert Jackall describes several such corporations in his book *Moral Mazes* as organizations that reduce (and distort) corporate values to merely following orders: "What is right in the corporation is what the guy above you wants from you. That's what morality is in the corporation."[3] Jackall describes a world in which professional standards are disregarded by top-level managers preoccupied with maintaining self-promoting images and forming power alliances with other managers. Hard work, commitment to worthwhile and safe products, and even profit-making take a back seat to personal survival in the tumultuous world of corporate takeovers and layoffs. It is noteworthy that Jackall's book is based primarily on his study of several large chemical and textile companies during the 1980s, companies notorious for indifference to worker safety (including cotton dust poisoning) and environmental degradation (especially chemical pollution).

In contrast, most corporations seek to establish an ethical climate, which may include going to great expense to establish ethics programs. For example, in 1985 Martin Marietta Corporation (now Lockheed Martin), the large aerospace and defense contractor, began an ethics program emphasizing basic values like honesty and fairness, which later expanded to include responsibilities for the environment and for high product quality.[4] Part of the stimulus for the program was public scrutiny of the defense industry, and indeed Martin Marietta was being investigated at that time for improper billings to the government for travel expenses. Lockheed also had a troubled past. In 1972 and 1973 it had engaged in paying bribes to facilitate the sale of its TriStar jumbo jets to Japan Airlines. Earlier, well-placed connections had also enabled the company to license, manufacture, and sell components of its F-104 Starfighter to airplane builders in Germany (where eventually 175 crashed, killing 85 pilots) and in Japan (where 54 jets were lost).

Nevertheless, the ethics program developed was not merely a reaction designed to avoid legal penalties, but also a concerted effort to institutionalize ethical commitments throughout the newly merged corporation. Specifically, the ethics team drafted a code of conduct, conducted ethics workshops for managers, and created effective procedures for employees to express their ethical concerns. An ethics network links a central ethics office with ethics representatives appointed at each major facility. In 1991, when the company had about 60,000 employees, some 9000 confidential employee inquiries or complaints entered the network, and during the following year 684 investigations were conducted. More recently, in 2019, 442 ethics investigations were conducted in the company. In the meantime, the company also produced an ethics training game featuring Dilbert, a popular comic-strip character.

Not all attempts to establish corporate ethics programs are successful. During the late 1980s another large defense contractor established a program that included top-level ethics planning, the appointment of division-level ethics directors, the establishment of new channels for handling complaints, and training programs for employees.[5] Higher management viewed the program as a success because the company avoided scandals faced by competitors, but a group of professional employees assessed the program as a sham intended for public relations and window dressing. The primary difficulty seemed to be a gap between the intentions of top management and the unchanged conduct of senior line managers, a gap that engendered employee cynicism. The company also emphasized a negative approach by requiring employees to sign cards stating that they understood the new requirements and by widely publicizing sanctions for specific violations.

How did this company appear to its clients and the government? Probably quite acceptable; after all, there was an ethics compliance program that made sure all relevant laws were made known. Indeed, the U.S. government would treat transgressions more leniently because of the compliance programs. Nevertheless, such inauthentic efforts are bound to harm the company in the long run.

What are the defining features of an ethical corporate climate? The preceding examples suggest at least four features. First, ethical values in their full complexity are widely acknowledged and appreciated by managers and employees alike. Responsibilities to all constituencies of the corporation are affirmed—not only to stockholders, but also to customers, employees, and all other stakeholders in the corporation. That does not mean that profits are neglected, nor does it downplay the special obligations that employees of corporations have to promote the interests of the corporation. For the most part, the public good is promoted through serving the interests of the corporation. Nevertheless, the moral limits on profit-seeking go beyond simply obeying the law and avoiding fraud. Precisely what those limits are is a matter of ongoing discussion in democracies; they concern, for example, such things as tobacco, weapons, and dangerous recreational vehicles.

Second, in an ethical corporate climate, the use of ethical language is honestly applied and recognized as a legitimate part of corporate dialogue. One way to emphasize this legitimacy is to make prominent a corporate code of ethics. Another way is to explicitly include a statement of ethical responsibilities in the job descriptions of all layers of management.

Third, top management sets a moral tone in words, in policies, and by personal example. Official pronouncements asserting the importance of professional conduct in all areas of the corporation must be backed by support for professionals who work according to the guidelines outlined in codes of ethics. Whether or not there are periodic workshops on ethics or formal brochures on social responsibility distributed to all employees, what is most important is fostering confidence that management is serious about ethics. Sometimes the real test arises in connection with mergers and acquisitions: some ethics programs quietly vanish, others are strengthened, and some appear where none had existed before.

Fourth, there are procedures for conflict resolution. One avenue, exemplified by Martin Marietta, is to create ombudspersons or designated executives with whom employees can have confidential discussions about moral concerns. More recently, Lockheed Martin created a pamphlet "How the Ethics Process Works" that introduces the process and avenues for their employees to report suspected misconduct. Equally important is educating managers about conflict resolution, about which we say more later.

In building an ethical corporate climate it is particularly important not to fall into the trap of relying solely on conveniently *legalistic* compliance strategies. These appear to be much favored by lawyers and executives who can then lay sole blame for organizational failures on individuals who have supposedly acted contrary to the organization's rules, whether these actually promote ethical behavior or not. Specific compliance rules are suitable only in very structured settings, such as purchasing and contracting. What needs to be encouraged is responsible ethical decision making that is understood to be a part of loyalty to the corporation.

6.1.2 Loyalty and Collegiality

Loyalty to an employer can mean two things.[6] *Agency-loyalty* is acting to fulfill one's contractual duties to an employer. These duties are specified in terms of the particular tasks for which one is paid, as well as the more general activities of cooperating with colleagues and following legitimate authority within the corporation.

Attitude-loyalty, by contrast, has as much to do with attitudes, emotions, and a sense of personal identity as it does with actions. It can be understood as agency-loyalty that is motivated by a positive identification with the group to which one is loyal. It implies seeking to meet one's moral duties to a group or organization willingly, with personal attachment and affirmation, and with a reasonable degree of trust. People who do their work grudgingly or spitefully are not loyal in this sense, even though they may adequately perform all their work responsibilities and hence manifest agency-loyalty.

When codes of ethics assert that engineers ought to be loyal (or faithful) to employers, is agency-loyalty or attitude-loyalty meant? Within proper limits, agency-loyalty to employers is an obligation, or rather it comprises the sum total of obligations to employers to serve the corporation in return for the contractual benefits from the corporation. But it is not the sole or paramount obligation of engineers. To think otherwise would be to lapse into a type of "corporate egoism," the attitude that the corporation is all that matters or is more important than human life itself. According to the NSPE Code of Ethics, the overriding obligation of engineers remains to "hold paramount the safety, health and welfare of the public."

What about attitude-loyalty: Is it obligatory? In our view, attitude-loyalty is often a virtue but not strictly an obligation. It is good when it contributes to a sense of corporate community and, thereby, increases the prospects for corporations to meet their desirable goals of productivity. We might say that loyalty is a "dependent virtue": its desirability depends on the value of the projects and communities to which it contributes.[7]

When engineering codes of ethics mention collegiality, they generally cite acts that constitute *dis*loyalty. The National Society of Professional Engineers (NSPE) code, for example, states that "Engineers shall not attempt to injure, maliciously or falsely, directly or indirectly, the professional reputation, prospects, practice or employment of other engineers. Engineers who believe others are guilty of unethical or illegal practice shall present such information to the proper authority for action" (Sec. III-7).

These injunctions not to defame colleagues unjustly and not to condone unethical practice are important, but collegiality also has a more positive dimension. Craig Ihara suggests that "collegiality is a kind of connectedness grounded in respect for professional expertise and in a commitment to the goals and values of the profession, and . . . as such, collegiality includes a disposition to support and cooperate with one's colleagues."[8] In other words, the central elements of collegiality are: (1) respect for colleagues, valuing their professional expertise and their devotion to the social goods promoted by the profession; (2) commitment, in the sense of sharing a devotion to the moral ideals inherent in one's profession, and (3) connectedness, or awareness of participating in cooperative projects based on shared commitments and mutual support. As such, collegiality is a virtue defining the teamwork essential for pursuing shared goods.

In teaching professional ethics to engineering students, philosopher William J. Frey argues that four ethical values are fundamental for creating a responsible culture of teamwork: (1) justice (in the distribution of work); (2) responsibility (in specifying tasks, assigning blame, and awarding credit); (3) reasonableness (ensuring participation, resolving conflict, and reaching consensus); and (4) honesty (avoiding deception, corruption, and impropriety).[9] It is worth noting that there may be some potential obstacles to ethical and effective teamwork such as free riders (those who attempt to "ride for free" on the work done by others in the group) and groupthink (strong leaders disregard and defend against information that is different from their plans and beliefs).

6.1.3 Managers and Engineers

Respect for authority is important in meeting organizational goals. Decisions must be made in situations where allowing everyone to exercise unrestrained individual discretion would create chaos. Moreover, clear lines of authority provide a means for identifying areas of personal responsibility and accountability.

The relevant kind of authority has been called *executive authority:* the corporate or institutional right given to a person to exercise power based on the resources of an organization.[10] It is distinguishable from *power* (or influence) in getting the job done. It is distinguishable, too, from *expert authority:* the possession of special knowledge, skill, or competence to perform some task or to give sound advice. Employees *respect authority* when they accept the guidance and obey the directives issued by the employer having to do with the areas of activity covered by the employer's institutional authority, assuming the directives are legal and do not violate norms of moral decency.

Within this general framework of authority, however, there are wide variations in how engineers and managers relate to each other. At one extreme, there is the rigid, top-down control described by Jackall. At the other extreme, there is something more like how professors and administrators interact within universities. Several such corporations were studied by Michael Davis and his colleagues, working under a grant from the Hitachi Foundation of America.

Davis and his colleagues found subtle variations among the corporations they studied, which they grouped into three categories.[11] "Engineer-oriented companies" focus primarily on the quality of products. Engineers' judgments about safety and quality were given great weight, and they were overridden rarely, when considerations such as cost and scheduling became especially important. "Customer-oriented companies" make their priority the satisfaction of customers. In these companies safety considerations were also given high priority, but engineers were expected to be more assertive in speaking as advocates for safety, so that it received a fair hearing amidst managers' preoccupation with satisfying the needs of customers. Because of this sharper differentiation of managers' and engineers' points of view, communication problems tended to arise more frequently. Finally, "finance-oriented companies" make profit the primary focus (which is what Milton Friedman argued all corporations should do, unless they are nonprofit corporations). The group did not happen to encounter such companies, and they introduced this third category as a contrast to the first two.

Davis reports that in addition to having different roles and authority, managers and engineers typically have different attitudes and approaches. Managers tend to be more distanced from the technical details of jobs; they focus more on jobs in their entirety, from wider perspectives; and they are more focused on people than things.

6.1.4 Managing Conflict

Effectively dealing with conflicts, including value disagreements, is an essential managerial task in guiding and integrating employees' work, but conflict resolution aimed at maintaining teamwork is also a responsibility of all engineers. Managers have authority and the responsibility to resolve or prevent damaging conflicts that threaten corporate efficiency. Their ultimate weapon is force, but today overt reliance on force is generally regarded as a self-defeating, authoritarian abuse of authority. In general, it is a misconception to equate managing people with issuing orders and then demanding unquestioning obedience, an approach that is generally ineffective in maintaining long-term productive relationships among professionals.[12] Certainly within technological corporations, successful management means evoking the fullest contribution of employees, and that sometimes means tolerating and even inviting some forms of conflict. The shared task is to create climates in which conflicts are addressed constructively, and that requires the contributions of all engineers.

The types and relative intensity of conflicts among persons differ according to the level of management, as well as the corporate setting. One study ranked the seven most common conflicts confronted by engineering project managers, in order of priority of overall intensity (as perceived by managers), as follows: (1) conflicts over schedules, especially where managers must rely on support departments over which the manager has little control, (2) conflicts over which projects and departments are most important to the organization at a given time, (3) conflicts over personnel resources made available for projects, (4) conflicts over technical issues, in particular over alternative ways to solve a technical problem within cost, schedule, and performance objectives, (5) conflicts over administrative procedures, such as the extent of the manager's authority, accountability procedures and reviews, and administrative support, (6) personality conflicts, and (7) conflicts over costs.[13] All of these areas can involve explicit or tacit value disagreements.

The study noted that while personality conflicts ranked relatively low in intensity, they tended to be the most difficult to resolve. In part the difficulty was that they were sometimes difficult to pinpoint since they were interwoven with other conflicts, such as disagreements over technical issues and communication problems. Perhaps an additional explanation is that project managers have less training in dealing with personality conflicts.

When properly managed, both ethical and technical disagreements are usually fruitful rather than harmful. By highlighting options, different points of view provide an occasion for increased creativity. They remind us that the full truth is often a blend of complementary insights, rather than the exclusive possession of one party to a disagreement. Accordingly, rather than hastily quieting disagreement, it is important to sustain an environment that encourages the expression of a variety of viewpoints about technological problems and ethical issues and facilitates a discussion of their rationales.[14]

All types of conflicts among persons, not just personality conflicts, have been explored by the Harvard Negotiation Project in recent decades. The project has sought ways to avoid both the "win-lose" style of managing conflict in which one adversary wins and the other is humiliated, and the "being nice" style that is too eagerly yielding to others or that tries to avoid conflict altogether, even when conflict is creative. Among the central ideas generated by that project are the following four widely applicable principles for conflict resolution.[15]

1. "People: Separate the people from the problem."

   This does not mean that only the problem is important. The personal aspect of conflicts is distinguished from the problem in order to be able to better deal with both. Of course, sometimes the people are the problem, as with personality clashes. Even there the focus should be on the problem arising from behavior, not on blaming people for their character. Moreover, all conflict involves persons who must be respected. That implies understanding the problem from their point of view, even though one need not share their point of view. It also implies communicating with them as clearly and honestly as possible. Above all, it means including them in the decision-making process as fully as possible, so that even when their view of the problem is rejected they do not feel rejected.

2. "Interests: Focus on interests, not positions."

   This principle applies most clearly to personnel matters and ethical perspectives, rather than technical disputes (although often they are intertwined). "Positions" refers to stated views, not only those serving as bargaining ploys but those the person may think (incorrectly) accurately express their best interests. Consider the case of an engineer nearing retirement who insists on a 3 percent royalty for the innovative high-speed stamping tool he developed.[16] The company believes that 1.5 percent is a reasonable amount. After months of fruitless haggling, and amid threats of lawsuits, a mediator discovers that the reason the engineer wanted the higher percentage was to protect himself from lawsuits in the case of injuries from workers using the tool. The mediator also discovered that the company could include the engineer under its liability policy at a modest cost to the company. The result led to an agreement in which the engineer was happy to accept a royalty around 1 percent.

3. "Options: Generate a variety of possibilities before deciding what to do."

As the previous example also illustrates, often the best solutions are not compromises that split the differences between stated positions, but instead creative options that have not been brought into focus. Especially in conflicts over technological solutions and ethical priorities, it is crucial to consider a wide range of options in order to overcome the effects of tunnel vision.

4. "Criteria: Insist that the result [of conflict resolution] be based on some objective standard."

Within corporate settings it is usually clear what general standards are to be used in evaluating results. But beyond the goals of efficiency, quality, and customer satisfaction, it is important to develop a sense of fair process in how the goals are met. Otherwise, disagreements easily degenerate into contests of will.

DISCUSSION QUESTIONS

1. You are a production engineer and technical manager for a corporation that manufactures medical equipment. Injuries involving cuts and lacerations are rare, but they do occur. Coworkers learn that one of the production specialists under your supervision is HIV-positive and may have developed AIDS. The coworkers come to you asking that either he or they be transferred. You indicate that no transfers to comparable positions are available, but the workers insist. How might you best resolve or help resolve this conflict?

2. You are a project leader for developing a testing mechanism for use on an experimental laser that was contracted for by the Navy, but which also has a potential civilian market. The project proves far more complex than originally anticipated, and you are already behind schedule and over budget (due to overtime salary). You have several meetings with the project director, who is your supervisor. He tells you to complete the project and send it to the Navy as scheduled, even if not all the glitches have been worked out. Should you do so, as told? If not, what steps might you take to best resolve the disagreement?

3. You are a middle-level manager who writes evaluations on 10 project engineers. Your supervisor wants to promote one of those engineers to a higher management position. You gave that engineer an overall grade of 4 on a scale of 5, whereas company policy normally requires a 5 for promotion. Your supervisor tells you to rewrite the evaluation to justify a 5. What should you do?

4. You decide to rewrite the report in question 3. Two days later, another project engineer, who also wanted the promotion, learns of the change and comes to you for an explanation. What should you say?

5. Jim Serra, vice president of engineering, must decide who to recommend for a new director-level position that was formed by merging the product (regulatory) compliance group with the environmental testing group.[17] The top inside candidate is Diane Bryant, senior engineering group manager in charge of the environmental testing group. Bryant is 36, exceptionally intelligent and highly motivated, and a well-respected leader. She is also five months pregnant and is expected to take an eight-week maternity leave two months before the first customer ship deadline (six months away) for a new product. Bryant applies for the job and in a discussion with Serra assures him that she will be available at all crucial stages of the project. Your colleague David Moss, who is vice president of product engineering, strongly urges you to find an outside person, insisting that there is no guarantee that Bryant will be available when needed. Much is at stake. A schedule delay could cost several million dollars in revenues lost to competitors. At the same time, offending Bryant could lead her and perhaps other valuable engineers whom she supervises to leave the company. What procedure would you recommend in reaching a solution? Is there an ethical issue involved here?

6.2 CONFIDENTIALITY AND CONFLICTS OF INTEREST

Maintaining confidentiality and avoiding harmful conflicts of interest are especially important aspects of teamwork and trustworthiness. Let us begin with confidentiality and then turn to conflicts of interest, in each case seeking a clearer understanding of what is at stake morally.

6.2.1 Confidentiality: Definition

The duty of confidentiality is the duty to keep secret all information deemed desirable to keep secret. Deemed by whom? Basically, it is any information that the employer or client would like to have kept secret in order to compete effectively against business rivals. Often this is understood to be any data concerning the company's business or technical processes that are not already public knowledge. While this criterion is somewhat vague, it clearly points to the employer or client as the main source of the decision as to what information is to be treated as confidential.

"Keep secret" is a relational expression. It always makes sense to ask, "Secret with respect to whom?" In the case of some government organizations, such as the FBI and CIA, highly elaborate systems for classifying information have been developed that identify which individuals and groups may have access to what information. Within other governmental agencies and private companies, engineers and other employees are usually expected to withhold information labeled "confidential" from unauthorized people both inside and outside the organization.

Several related terms should be distinguished. *Privileged information* literally means "available only on the basis of special privilege," such as the privilege accorded an employee working on a special assignment. *Proprietary information* is information that a company owns or is the proprietor of, and hence is a term carefully defined by property law. A rough synonym for "proprietary information" is *trade secret,* which can be virtually any type of information that has not become public, which an employer has taken steps to keep secret, and which is thereby given limited legal protection in common law (law generated by previous court rulings) that forbids employees from divulging it. *Patents* legally protect specific products from being manufactured and sold by competitors without the express permission of the patent holder. In the United States, the patent system was initially created to protect the benefits of inventors and encourage them to further improve and innovate technologies. Trade secrets have no such protection, and a corporation may learn about a competitor's trade secrets through legal means—for instance, "reverse engineering," in which an unknown design or process can be traced out by analyzing the final product. But patents do have the drawback of being public and thus allowing competitors an easy means of working around them by finding alternative designs. Copyrights, patents, trademarks, and trade secrets are often considered as four types of intellectual properties for businesses.

6.2.2   Confidentiality and Changing Jobs

The obligation to protect confidential information does not cease when employees change jobs. If it did, it would be impossible to protect such information. Former employees would quickly divulge it to their new employers, or perhaps for a price sell it to competitors of their former employers. Thus, the relationship of trust between employer and employee in regard to confidentiality continues beyond the formal period of employment. Unless the employer gives consent, former employees are barred indefinitely from revealing trade secrets. This provides a clear illustration of the way in which the professional integrity of engineers involves much more than mere loyalty to one's present employer.

Yet thorny problems arise in this area. Many engineers value professional advancement more than long-term ties with any one company, and so they change jobs frequently. Engineers in research and development are especially likely to have high rates of turnover. They are also the people most likely to be exposed to important new trade secrets. Moreover, when they transfer into new companies they often do the same kind of work as before—precisely the type of situation in which trade secrets of their old companies may have relevance, a fact that could have strongly contributed to their having readily found new employment.

A high-profile case of trade secret violations was settled in January 1997 (without coming to trial) when Volkswagen AG (VW) agreed to pay General Motors Corporation (GM) and its German subsidiary Adam Opel $100 million in cash and to buy $1 billion in parts from GM over the next seven years. Why? Because in March 1993, Jose Ignacio Lopez, GM's highly effective manufacturing expert, left GM to join VW, a fierce competitor in Europe, and took with him not only three colleagues and know-how, but also copies of confidential GM documents.

A more legally important case concerned Donald Wohlgemuth, a chemical engineer who at one time was manager of B.F. Goodrich's space suit division.[18] Technology for space suits was undergoing rapid development, with several companies competing for government contracts. Dissatisfied with his salary and the research facilities at B.F. Goodrich, Wohlgemuth negotiated a new job with International Latex Corporation as manager of engineering for industrial products. International Latex had just received a large government subcontract for developing the Apollo astronauts' space suits, and that was one of the programs Wohlgemuth would manage.

The confidentiality obligation required that Wohlgemuth not reveal any trade secrets of Goodrich to his new employer. This was easier said than done. Of course it is possible for employees in his situation to refrain from explicitly stating processes, formulas, and material specifications. Yet in exercising their general skills and knowledge, it is virtually inevitable that some unintended "leaks" will occur. An engineer's knowledge base generates an intuitive sense of what designs will or will not work, and trade secrets form part of this knowledge base. To fully protect the secrets of an old employer on a new job would thus virtually require that part of the engineer's brain be removed.

Is it perhaps unethical, then, for employees to change jobs in cases where unintentional revelations of confidential information are a possibility? Some companies have contended that it is. Goodrich, for example, charged Wohlgemuth with being unethical in taking the job with International Latex. Goodrich also went to court seeking a restraining order to prevent him from working for International Latex or any other company that developed space suits. The Ohio Court of Appeals refused to issue such an order, although it did issue an injunction prohibiting Wohlgemuth from revealing any Goodrich trade secrets. Their reasoning was that while Goodrich had a right to have trade secrets kept confidential, it had to be balanced against Wohlgemuth's personal right to seek career advancement. And this would seem to be the correct moral verdict as well.

6.2.3 Confidentiality and Management Policies

What might be done to recognize the legitimate personal interests and rights of engineers and other employees while also recognizing the rights of employers in this area?[19] One approach is to use employment contracts that place special restrictions on future employment. Traditionally, those restrictions centered on the geographical location of future employers, the length of time after leaving the present employer before one can engage in certain kinds of work, and the type of work it is permissible to do for future employers. Thus, Goodrich might have required as a condition of employment that Wohlgemuth sign an agreement that if he sought work elsewhere he would not work on space suit projects for a competitor in the United States for five years after leaving Goodrich.

Yet such contracts are hardly agreements between equals, and they threaten the right of individuals to pursue their careers freely. For this reason the courts have tended not to recognize such contracts as binding, although they do uphold contractual agreements forbidding the disclosure of trade secrets.

A different type of employment contract is perhaps not so threatening to employee rights in that it offers positive benefits in exchange for the restrictions it places on future employment. Consider a company that normally does not have a portable pension plan. It might offer such a plan to an engineer in exchange for an agreement not to work for a competitor on certain kinds of projects for a certain number of years after leaving the company. Or another clause might offer an employee a special post-employment annual consulting fee for several years on the condition that he or she not work for a direct competitor during that period.

Other tactics aside from employment contract provisions have been attempted by various companies. One is to place tighter controls on the internal flow of information by restricting access to trade secrets except where absolutely essential. The drawback to this approach is that it may create an atmosphere of distrust in the workplace. It might also stifle creativity by lessening the knowledge base of engineers involved in research and development.

One potential solution is for employers to help generate a sense of professional responsibility among their staff that reaches beyond merely obeying the directives of current employers. Engineers can then develop a real sensitivity to the moral conflicts they may be exposed to by making certain job changes. They can arrive at a greater appreciation of why trade secrets are important in a competitive system, and they can learn to take the steps necessary to protect them. In this way, professional concerns and employee loyalty can become intertwined and reinforce each other.

6.2.4 Confidentiality: Justification

Upon what moral basis does the confidentiality obligation rest, with its wide scope and obvious importance? The primary justification is to respect the autonomy (freedom, self-determination) of individuals and corporations and to recognize their legitimate control over some private information concerning themselves.[20] Without that control, they could not maintain their privacy and protect their self-interest insofar as it involves privacy. Just as patients should be allowed to maintain substantial control over personal information, so employers should have some control over the private information about their companies. All the major ethical theories recognize the importance of autonomy, whether it is understood in terms of rights to autonomy, duties to respect autonomy, the utility of protecting autonomy, or the virtue of respect for others.

Additional justifications include trustworthiness: Once practices of maintaining confidentiality are established socially, trust and trustworthiness can grow. Thus, when clients go to attorneys or tax accountants they expect them to maintain confidentiality, and the professional indicates that confidentiality will be maintained. Similarly, employees often make promises (in the form of signing contracts) not to divulge certain information considered sensitive by the employer.

In addition, there are public benefits in recognizing confidentiality relationships within professional contexts. For example, if patients are to have the best chances of being cured, they must feel completely free to reveal the most personal information about themselves to physicians, and that requires trust that the physician will not divulge private information. Likewise, the economic benefits of competitiveness within a free market are promoted when companies can maintain some degree of confidentiality about their products. Developing new products often requires investing large resources in acquiring new knowledge. The motivation to make those investments might diminish if that knowledge were immediately dispersed to competitors who could then quickly make better products at lesser cost, since they did not have to make comparable investments in research and development.

Confidentiality has its limits, particularly when it is invoked to hide misdeeds. Investigations into a wide variety of white-collar crimes covered up by management in industry or public agencies have been thwarted by invoking confidentiality or false claims of secrecy based on national interest.

6.2.5 Conflicts of Interest: Definition and Examples

We turn now to some equally thorny issues concerning conflicts of interest. Professional conflicts of interest are situations where professionals have an interest that, if pursued, might keep them from meeting their obligations to their employers or clients. Sometimes such an interest involves serving in some other professional role, say, as a consultant for a competitor's company. Other times it is a more personal interest, such as making substantial private investments in a competitor's company.

Concern about conflicts of interest largely centers on their potential to distort good judgment in faithfully serving an employer or client.[21] Exercising good judgment means arriving at beliefs on the basis of expertise and experience, as opposed to merely following simple rules. Thus, we can refine our definition of conflicts of interest by saying that they typically arise when *two* conditions are met: (1) the professional is in a relationship or role that requires exercising good judgment on behalf of the interests of an employer or client, and (2) the professional has some additional or side interest that could threaten good judgment in serving the interests of the employer or client—either the good judgment of that professional or the judgment of a typical professional in that situation. Why the reference to "a typical professional"? There might be conclusive evidence that the actual persons involved would never allow a side interest to affect their judgment, yet they are still in a conflict of interest.

"Conflict of interest" and "conflicting interests" are not synonyms.[22] A student, for example, may have interests in excelling on four final exams. She knows, however, that there is time to study adequately for only three of them, and so she must choose which interest not to pursue. In this case "conflicting interests" means a person has two or more desires that cannot all be satisfied given the circumstances. But there is no suggestion that it is morally wrong or problematic to try pursuing them all. Similarly, professionals such as engineers may encounter situations involving "conflicts of commitment" in which they hold commitments to different people or entities. For instance, a researcher needs to make decisions about how to divide their time between research and other responsibilities (e.g., how to serve her scientific disciplines or professional fields, how to respect her employer's values, how to represent science to the society).[23] By contrast, in professional conflicts of interest it is often physically or economically possible to pursue all of the conflicting interests, but doing so would be morally problematic.

Because of the great variety of possible outside interests, conflicts of interest can arise in innumerable ways, and with many degrees of subtlety. We will sample only a few of the more common situations involving (1) gifts, bribes, and kickbacks, (2) interests in other companies, and (3) insider information.

(1) GIFTS, BRIBES, AND KICKBACKS. A bribe is a substantial amount of money or goods offered beyond a stated business contract with the aim of winning an advantage in gaining or keeping the contract, and where the advantage is unfair or otherwise unethical.[24] *Substantial* is a vague term, but it alludes to amounts, beyond acceptable gratuities, that are sufficient to distort the judgment of a typical person. Typically, though not always, bribes are made in secret. Gifts are not bribes as long as they are small gratuities offered in the normal conduct of business. Prearranged payments made by contractors to companies or their representatives in exchange for contracts actually granted are called kickbacks. When suggested by the granting party to the party bidding on the contract, the latter often defends its participation in such an arrangement as having been subjected to "extortion."

Often, companies give gifts to selected employees of government agencies or partners in trade. Many such gifts are unobjectionable, some are intended as bribes, and still others create conflicts of interest that do not, strictly speaking, involve bribes. What are the differences? In theory, these distinctions may seem clear, but in practice they become blurry. Bribes are illegal or immoral because they are substantial enough to threaten fairness in competitive situations, while gratuities are of smaller amounts. Some gratuities play a legitimate role in the normal conduct of business, while others can bias judgment like a bribe does. Much depends on the context, and there are many gray areas, which is why companies often develop elaborate guidelines for their employees.

Philosopher Charles E. Harris and colleagues have created the method "line drawing" to address practical ethical issues in engineering practice such as bribes.[25] Line drawing starts with creating a paradigm case that is clear-cut, unproblematic case of a bribe. Such a pragmatic case also includes several aspects of the situation (or "features") that are relevant in making the situation a paradigmatic bribe. These defining aspects or features include gift size, timing (when the gift is provided), reason (why the gift is provided), responsibility (who is responsible for decision-making or whether the gift recipient is the sole decision-maker), product quality, and product cost. Then, based on the same group of defining features (e.g., gift size, etc.), an ideal or ethical paradigmatic case at the other extreme will be created. Such a case depicts a situation that is clearly not a bribe. Engineers who are unsure about whether a gift-giving activity constitutes a bribe or not then compare the actual case they are countering with the two paradigmatic cases (one is clearly a bride and the other is clearly not a bribe) and make some intuitive judgment on their actual situation. Engineers will have to analyze whether their actual case is closer to the clearly unethical case or the clearly unproblematic case in terms of the defining features of a bribe.

(2) INTERESTS IN OTHER COMPANIES. Some conflicts of interest consist in having an interest in a competitor's or a subcontractor's business. One blatant example is actually working for the competitor or subcontractor as an employee or consultant. Another example is partial ownership or substantial stockholdings in the competitor's business. Does holding a few shares of stock in a company one has occasional dealings with constitute a conflict of interest? Usually not, but as the number of shares of stock increases, the issue becomes blurry. Again, is there a conflict of interest if one's spouse works for a subcontractor to one's company? Usually not, but a conflict of interest arises if one's job involves granting contracts to that subcontractor.

Should there be a general prohibition on *moonlighting,* that is, working in one's spare time for another company? That would violate the rights to pursue one's legitimate self-interest. Moonlighting usually creates conflicts of interest only in special circumstances, such as working for competitors, suppliers, or customers. Even then, in rare situations, an employer sometimes gives permission for exceptions, as for example when the experience gained would greatly promote business interests. A special kind of conflict of interest arises, however, when moonlighting leaves one exhausted and thereby harms job performance.[26]

Conflicts of interest arise in academic settings as well. For example, a professor of electrical engineering at a west coast university was found to have used $144,000 in grant funds to purchase electronic equipment from a company he owned in part. He had not revealed his ownership to the university; he had priced the equipment much higher than market value, and some of the purchased items were never received. The Supplier Information Form and Sole Source Justification Statements had been submitted as required, but with falsified content. In addition, the professor had hired a brother and two sisters for several years, concealing their relationship to him in violation of anti-nepotism rules and paying them for research work they did not perform. All told, he had defrauded the university of at least $500,000 in research funds. Needless to say, the professor lost his university position and had to stand trial in civil court when an internal audit and subsequent hearings revealed these irregularities.

(3) INSIDER INFORMATION. An especially sensitive conflict of interest consists in using "inside" information to gain an advantage or set up a business opportunity for oneself, one's family, or one's friends. The information might concern one's own company or another company with which one does business.

For example, engineers might tell their friends about the impending announcement of a revolutionary invention, which they have been perfecting, or of their corporation's plans for a merger that will greatly improve the worth of another company's stock. In doing so, they give those friends an edge on an investment promising high returns. Owning stock in the company for which one works is of course not objectionable, and this is often encouraged by employers. But that ownership should be based on the same information available to the general public.

6.2.6 Moral Status of Conflicts of Interest

What is wrong with employees having conflicts of interest? Most of the answer is obvious from our definition: Employee conflicts of interest occur when employees have interests that if pursued could keep them from meeting their obligations to serve the interests of the employer or client for whom they work. Such conflicts of interest should be avoided because they threaten to prevent one from fully meeting those obligations.

More needs to be said, however. Why should mere threats of possible harm always be condemned? Suppose that substantial good might sometimes result from pursuing a conflict of interest?

In fact, it is not always unethical to pursue conflicts of interest. In practice, some conflicts are thought to be unavoidable, or even acceptable. One illustration of this is that the government allows employees of aircraft manufacturers, like Boeing or McDonnell Douglas, to serve as government inspectors for the Federal Aviation Agency (FAA). The FAA is charged with regulating airplane manufacturers and making objective safety and quality inspections of the airplanes they build. Naturally the dual roles—government inspector and employee of the manufacturer being inspected—could bias judgments. Yet with careful screening of inspectors, the likelihood of such bias is said to be outweighed by the practical necessities of airplane inspection. The options would be to greatly increase the number of nonindustry government workers (at great expense to taxpayers) or to do without government inspection altogether (putting public safety at risk).

Even when conflicts of interest are unavoidable or reasonable, employees are still obligated to inform their employers and obtain approval. This suggests a fuller answer to why conflicts of interest are generally prohibited: (1) The professional obligation to employers is very important in that it overrides in the vast majority of cases any appeal to self-interest on the job, and (2) the professional obligation to employers is easily threatened by self-interest (given human nature) in a way that warrants especially strong safeguards to ensure that it is fulfilled by employees.

Many conflicts of interest violate trust, in addition to undermining specific obligations. Employed professionals are in fiduciary (trust) relationships with their employers and clients. Allowing side interests to distort one's judgment violates that trust. And additional types of harm can arise as well. Many conflicts of interest are especially objectionable in business affairs precisely because they pose risks to free competition. In particular, bribes and large gifts are objectionable because they lead to awarding contracts for reasons other than the best work for the best price.

As a final point, we should note that even the appearance of conflicts of interest, especially appearances of seeking a personal profit at the expense of one's employer, is considered unethical because the appearance of wrongdoing can harm a corporation as much as any actual bias that might result from such practices.

DISCUSSION QUESTIONS

1. Consider the following example:

"Who Owns Your Knowledge? Ken is a process engineer for Stardust Chemical Corp., and he has signed a secrecy agreement with the firm that prohibits his divulging information that the company considers proprietary.

"Stardust has developed an adaptation of a standard piece of equipment that makes it highly efficient for cooling a viscous plastics slurry. (Stardust decides not to patent the idea but to keep it as a trade secret.) Eventually, Ken leaves Stardust and goes to work for a candy-processing company that is not in any way in competition. He soon realizes that a modification similar to Stardust's trade secret could be applied to a different machine used for cooling fudge, and at once has the change made."[27]

Has Ken acted unethically?

2. In the following case, are the actions of Client A morally permissible?

Client A solicits competitive quotations on the design and construction of a chemical plant facility. All the bidders are required to furnish as a part of their proposals the processing scheme planned to produce the specified final products.

The process generally is one which has been in common use for several years. All of the quotations are generally similar in most respects from the standpoint of technology.

Contractor X submits the highest-price quotation. He includes in his proposals, however, a unique approach to a portion of the processing scheme. Yields are indicated to be better than current practice, and quality improvement is apparent. A quick laboratory check indicates that the innovation is practicable.

Client A then calls on Contractor Z, the low bidder, and asks him to evaluate and bid on an alternate scheme conceived by Contractor X. Contractor Z is not told the source of alternative design. Client A makes no representation in his quotation request that replies will be held in confidence.[28]

3. American Potash and Chemical Corporation advertised for a chemical engineer having industrial experience with titanium oxide. It succeeded in hiring an engineer who had formerly supervised E. I. Du Pont de Nemours and Company's production of titanium oxide. Du Pont went to court and succeeded in obtaining an injunction prohibiting the engineer from working on American Potash's titanium oxide projects. The reason given for the injunction was that it would be inevitable that the engineer would disclose some of du Pont's trade secrets.[29] Defend your view as to whether the court injunction was morally warranted or not.

4. "Facts: Engineer Doe is employed on a full-time basis by a radio broadcast equipment manufacturer as a sales representative. In addition, Doe performs consulting

engineering services to organizations in the radio broadcast field, including analysis of their technical problems and, when required, recommendation of certain radio broadcast equipment as may be needed. Doe's engineering reports to his clients are prepared in form for filing with the appropriate governmental body having jurisdiction over radio broadcast facilities. In some cases Doe's engineering reports recommend the use of broadcast equipment manufactured by his employer."

"Question: May Doe ethically provide consulting services as described?"[30]

5. "Henry is in a position to influence the selection of suppliers for the large volume of equipment that his firm purchases each year.

"At Christmas time, he usually receives small tokens from several salesmen, ranging from inexpensive ballpoint pens to a bottle of liquor. This year, however, one salesman sends an expensive briefcase stamped with Henry's initials."[31]

Should Henry accept the gift? Should he take any further course of action?

6. Consider the following case:

"Scott Bennett is the engineer assigned to deal with vendors who supply needed parts to the Upscale Company. Larry Newman, sales representative from one of Upscale's regular vendors, plays in the same golf league as Scott. One evening they go off in the same foursome. Sometime during the round Scott mentions that he is really looking forward to vacationing in Florida next month. Larry says his uncle owns a condo in Florida that he rents out during the months he and his family are up north. Larry offers to see if the condo is available next month—assuring Scott that the rental cost would be quite moderate.

"What should Scott say?"[32]

Does your answer turn on whether Scott's company policy indicates a clear answer to this question?

6.3   RIGHTS OF ENGINEERS

Engineers have several types of moral rights, which fall into the sometimes overlapping categories of human, employee, contractual, and professional rights. As *humans,* engineers have fundamental rights to live and freely pursue their legitimate interests, which implies, for example, rights not to be unfairly discriminated against in employment on the basis of sex, race, or age. As *employees,* engineers have special rights, including the right to receive one's salary in return for performing one's duties and the right to engage in the nonwork political activities of one's choosing without reprisal or coercion from employers. As *professionals,* engineers have special rights that arise from their professional role and the obligations it involves. We begin with professional rights, most of which can be viewed as aspects of a fundamental right of professional conscience. We will then move to a discussion of employee rights.

6.3.1   Professional Rights

Three professional rights have special importance: (1) the basic right of professional conscience, (2) the right of conscientious refusal, and (3) the right of professional recognition.

(1) RIGHT OF PROFESSIONAL CONSCIENCE. The right of professional conscience is the moral right to exercise professional judgment in pursuing professional responsibilities. Pursuing those responsibilities involves exercising both technical judgment and reasoned moral convictions. This right has limits, of course, and must be balanced against responsibilities to employers and colleagues of the sort discussed earlier.

If the duties of engineers were so clear that it was obvious to every sane person what was morally proper in every situation, there would be little point in speaking of "conscience" in specifying this basic right. Instead, we could simply say it is the right to do what everyone agrees it is obligatory for the professional engineer to do. But engineering, like other professions, calls for morally complex decisions. It requires autonomous moral judgment in trying to uncover the most morally reasonable courses of action, and the correct courses of action are not always obvious.

As with most moral rights, the basic professional right is an entitlement giving one the moral authority to act without interference from others. It is a "liberty right" that places an obligation on others not to interfere with its proper exercise. Yet occasionally, special resources may be required by engineers seeking to exercise the right of professional conscience in the course of meeting their professional obligations. For example, conducting an adequate safety inspection may require that special equipment be made available by employers. Or, more generally, in order to feel comfortable about making certain kinds of decisions on a project, the engineers involved may need an environment conducive to trust and support, which management may be obligated to help create and sustain. In this way the basic right is also in some respects a "positive right" placing on others an obligation to do more than merely not interfere.

There are two general ways to justify the basic right of professional conscience. One is to proceed piecemeal by reiterating the justifications given for the specific professional *duties*. Whatever justification there is for the specific duties will also provide justification for allowing engineers the *right* to pursue those duties. Fulfilling duties, in turn, requires the exercise of moral reflection and conscience, rather than rote application of simplistic rules. Hence the justification of each duty ultimately yields a justification of the right of conscience with respect to that duty.

The second way is to justify the right of professional conscience directly, which involves grounding it more directly in the ethical theories. Thus, duty ethics regards professional rights as implied by general duties to respect persons, and rule-utilitarianism would accent the public good of allowing engineers to pursue their professional duties. Rights ethics would justify the right of professional conscience by reference to the rights of the public not to be harmed and the right to be warned of dangers from the "social experiments" of technological innovation.

The right of professional conscience implies more specific rights, corresponding to specific professional obligations. In section 6.4 we will discuss the right to whistleblow in some situations where the public good is severely threatened. Here we cite two further examples: the right of conscientious refusal and the right to recognition.

(2) RIGHT OF CONSCIENTIOUS REFUSAL. The right of conscientious refusal is the right to refuse to engage in unethical behavior and to refuse to do so solely because one views it as unethical. This is a kind of second-order right. It arises because other rights to honor moral obligations within the authority-based relationships of employment sometimes come into conflict.

There are two situations to be considered: (1) where there is widely shared agreement in the profession as to whether an act is unethical and (2) where there is room for disagreement among reasonable people over whether an act is unethical.

It seems clear enough that engineers and other professionals have a moral right to refuse to participate in activities that are illegal and clearly unethical (for example, forging documents, altering test results, lying, giving or taking bribes, or padding payrolls). And coercing employees into acting by means of threats (to their jobs) plainly constitutes a violation of this right of theirs.

The troublesome cases concern situations where there is no shared agreement about whether a project or procedure is unethical. Do engineers have any rights to exercise their personal consciences in these more cloudy areas? Just as prolife physicians and nurses have a right not to participate in abortions, engineers should be recognized as having a *limited* right to turn down assignments that violate their personal consciences in matters of great importance, such as threats to human life, even where there is room for moral disagreement among reasonable people about the situation in question. We emphasize the word *limited* because the right is contingent on the organization's ability to reassign the engineer to alternative projects without serious economic hardship to itself. The right of professional conscience does not extend to the right to be paid for not working.

(3) RIGHT TO RECOGNITION. Engineers have a right to professional recognition for their work and accomplishments. Part of this involves fair monetary remuneration, and part nonmonetary forms of recognition. The right to recognition, and especially fair remuneration, may seem to be purely a matter of self-interest rather than morality, but it is both. Without a fair remuneration, engineers cannot concentrate their energies where they properly belong—on carrying out the immediate duties of their jobs and on maintaining up-to-date skills through formal and informal continuing education. Their time will be taken up by money worries, or even by moonlighting in order to maintain a decent standard of living.

The right to reasonable remuneration is clear enough to serve as a moral basis for arguments against corporations that make excessive profits while engineers are paid below the pay scales of blue-collar workers. It can also serve as the basis for criticizing the unfairness of patent arrangements that fail to give more than nominal rewards to the creative engineers who make the discoveries leading to the patents. If a patent leads to millions of dollars of revenue for a company, it is unfair to give the discoverer no more than a nominal bonus and a thank-you letter.

But the right to professional recognition is not sufficiently precise to pinpoint just what a reasonable salary is or what a fair remuneration for patent discoveries is. Such detailed matters must be worked out cooperatively between employers and employees, for they depend on both the resources of a company and the bargaining position of engineers. Professional societies can be of help by providing general guidelines.

6.3.2 Employee Rights

Employee rights are any rights, moral or legal, that involve the status of being an employee. They overlap with some professional rights, of the sort just discussed, and they also include institutional rights created by organizational policies or employment agreements, such as the right to be paid the salary specified in one's contract. However, here we will focus on human rights that exist even if unrecognized by specific contract arrangements.

Many of these human rights are discussed more fully in *Freedom Inside the Organization* by David Ewing (editor of *The Harvard Business Review* from 1949 to 1985).[33] Ewing refers to employee rights as the "black hole in American rights." The Bill of Rights in the Constitution was written, to apply to government, not to business. But when the Constitution was written, no one envisaged the giant corporations that have emerged in our century. Corporations wield enormous power politically and socially, often in multinational settings; they operate much as mini-governments, and they are often comparable in size to those governments the authors of the Constitution had in mind. For instance, American Telephone & Telegraph in the 1970s employed twice the number of people that inhabited the largest of the original 13 colonies when the Constitution was written.

Ewing proposes that large corporations ought to recognize a basic set of employee rights. As examples we will discuss rights to privacy and to equal opportunity.

(1) **PRIVACY.** The right to pursue outside activities can be thought of as a right to personal privacy in the sense that it means the right to have a private life off the job. In speaking of the right to privacy here, however, we mean the right to control the access to and the use of information about oneself. As with the right to outside activities, this right is limited in certain instances by employers' rights, but even then who among employers has access to confidential information is restricted. For example, the personnel division needs medical and life insurance information about employees, but immediate supervisors usually do not.

Consider a few examples of situations in which the functions of employers conflict with the right employees have to privacy:

1. Job applicants at the sales division of an electronics firm are required to take personality tests that include personal questions about alcohol use and sexual conduct. The rationale given for asking those questions is a sociological study showing correlations between sales ability and certain data obtained from

answers to the questionnaire. (That study has been criticized by other sociologists.)

2. A supervisor unlocks and searches the desk of an engineer who is away on vacation without the permission of that engineer. The supervisor suspects the engineer of having leaked information about company plans to a competitor and is searching for evidence to prove those suspicions.

3. A large manufacturer of expensive pocket computers has suffered substantia losses from employee theft. It is believed that more than one employee is involved. Without notifying employees, hidden surveillance cameras are installed.

4. A rubber products firm has successfully resisted various attempts by a union to organize its workers. It is always one step ahead of the union's strategies, in part because it monitors the phone calls of employees who are union sympathizers. It also pays selected employees bonuses in exchange for their attending union meetings and reporting on information gathered. It considered, but rejected as imprudent, the possibility of bugging the rest areas where employees were likely to discuss proposals made by union organizers.

We may disagree about which of these examples involve abuse of employer prerogatives. Yet the examples remind us of the importance of privacy and of how easily rights of privacy are abused. Employers should be viewed as having the same trust relationship with their employees concerning confidentiality that doctors have with their patients and lawyers have with their clients.[34] In all of these cases, personal information is given in trust on the basis of a special professional relationship.

(2) EQUAL OPPORTUNITY: NONDISCRIMINATION. Perhaps nothing is more demeaning than to be discounted because of one's sex, race, skin color, age, or political or religious outlook. These aspects of biological makeup and basic conviction lie at the heart of self-identity and self-respect. Such discrimination—that is, morally unjustified treatment of people on arbitrary or irrelevant grounds—is especially pernicious within the work environment, for work is itself fundamental to a person's self-image. Accordingly, human rights to fair and decent treatment at the workplace and in job training are vitally important.

Consider the following examples:

1. An opening arises for a chemical plant manager. Normally such positions are filled by promotions from within the plant. The best qualified person in terms of training and years of experience is an African-American engineer. Management believes, however, that the majority of workers in the plant would be disgruntled by the appointment of a nonwhite manager. They fear lessened employee cooperation and efficiency. They decide to promote and transfer a white engineer from another plant to fill the position.

2. A farm equipment manufacturer has been hit hard by lowered sales caused by a flagging produce economy. Layoffs are inevitable. During several

clandestine management meetings, it is decided to use the occasion to "weed out" some of the engineers within 10 years of retirement in order to avoid payments of unvested pension funds.

These examples involve discrimination. They also involve the violation of antidiscrimination laws, in particular the Civil Rights Act of 1964: "It shall be an unlawful employment practice for an employer to fail or refuse to hire or to discharge any individual, or otherwise to discriminate against any individual with respect to his compensation, terms, conditions, or privileges of employment, because of such individual's race, color, religion, sex, or national origin" (Title VII–Equal Employment Opportunity). Age discrimination was added in the 1967 Age Discrimination in Employment Act, and discrimination based on disability was forbidden in the 1994 Americans With Disabilities Act. On June 15, 2020, the U.S. Supreme Court ruled that Title VII protects gay and transgender workers from workplace discrimination. An employer who fires an individual simply for sexual orientation or gender identity defies the federal law.

(3) EQUAL OPPORTUNITY: SEXUAL HARASSMENT. Beginning in 1991, several events focused national attention on sexual harassment. First, in October of that year, Anita Hill testified against confirming Supreme Court nominee Clarence Thomas, charging that he made lewd remarks and unwanted sexual provocations to her years earlier when she had worked for him at the Justice Department. Hill was a respected attorney and law professor, and at the time one-third of Americans were convinced she was telling the truth. The majority of the public sided with Clarence Thomas, however, and he was confirmed as Supreme Court justice amid controversies over what sexual harassment is, how it is to be proven, and how best to prevent it.

Second, a series of scandals followed in the military, first the Navy and then the Army. More recently, officers of the highest ranks in the U.S. military have been discharged for engaging in sexual relations with the wives of subordinates, even when conducted on a consensual basis.

Third, corporations and universities were caught up in numerous incidents of sexual harassment. These incidents continue today, even among individuals who are responsible for preventing it.

One definition of sexual harassment is: "the unwanted imposition of sexual requirements in the context of a relationship of unequal power."[35] It takes two main forms: *quid pro quo* and hostile work environment.

Quid pro quo includes cases where supervisors require sexual favors as a condition for some employment benefit (a job, promotion, or raise). It can take the form of a sexual threat (of harm) or sexual offer (of a benefit in return for a benefit). *Hostile work environment,* by contrast, is any sexually oriented aspect of the workplace that threatens employees' rights to equal opportunity. It includes unwanted sexual proposals, lewd remarks, sexual leering, posting nude photos, and inappropriate physical contact.

What is morally objectionable about sexual harassment? Sexual harassment is a particularly invidious form of sex discrimination, involving as it does not only the abuse of gender roles and authority relationships, but the abuse of sexual intimacy itself. Sexual harassment is a display of power and aggression through sexual means. Accordingly, it has appropriately been called "dominance eroticized."[36] Insofar as it involves coercion, sexual harassment constitutes an infringement of one's autonomy to make free decisions concerning one's body. But whether or not coercion and manipulation are used, it is an assault on the victim's dignity. In abusing sexuality, such harassment degrades people on the basis of a biological and social trait central to their sense of personhood.

Thus a duty ethicist would condemn it as violating the duty to treat people with respect, to treat them as having dignity and not merely as means to personal aggrandizement and gratification of one's sexual and power interests. A rights ethicist would see it as a serious violation of the human right to pursue one's work free from the pressures, fears, penalties, and insults that typically accompany sexual harassment. And a utilitarian would emphasize the impact it has on the victim's happiness and self-fulfillment, and on women in general. This also applies to men who experience sexual harassment.

According to a survey conducted by the *PE* magazine of NSPE in 2017, 34 percent of respondents had witnessed sexual harassment and 15 percent had personally experienced it in the workplace. Almost half of women (45 percent) mentioned that they had witnessed and more than half of them (52 percent) said they personally had experienced sexual harassment at work.[37] Sexual harassment can be a particular challenge in fields that have male-dominated, hierarchical working environments such as engineering. A majority of women who experienced sexual harassment in the workplace chose to stay silent mainly due to their fear of retaliation. Engineering educators and policymakers feel concerned that sexual harassment may significantly discourage women who earn engineering degrees leave the profession. Therefore, to attract more talented engineers from diverse backgrounds in engineering (many of them are women), it is critical to create an inclusive and just culture of the engineering profession.

(4) EQUAL OPPORTUNITY: AFFIRMATIVE ACTION. Affirmative action, as the expression is usually defined, is giving a preference or advantage to a member of a group that in the past was denied equal treatment, in particular, women and minorities. It is also called "reverse preferential treatment" in that it reverses the historical order of preferences, which favored white males and excluded women and minorities. The *weak form* of preferential treatment consists in hiring a woman or a member of a minority over an equally qualified white male. The *strong form,* by contrast, consists in giving preference to women or minorities over better-qualified white males.

Affirmative action began to be used during the 1960s. A major legal challenge to it came in 1978 in the *Regents of the University of California v. Bakke.* In that case, Allan Bakke, a white male engineer, was denied entrance to the medical school at the University of California, Davis (UC Davis), which reserved 16 of 100 openings for applicants who were either Black, Latino, Asian, or American Indians. He sued, arguing that his credentials were superior to many of the minority students accepted. The Supreme Court ruled that the UC Davis admissions program was unconstitutional because it used explicit numerical quotas for minorities, which prevented person-to-person comparisons among all applicants. Yet the court also ruled that using race as one of many factors in comparing applicants is permissible, as long as quotas are avoided and the intent is to ensure the goal of diversity among students—an important educational goal.

The same basic line of reasoning was reaffirmed in two Supreme Court rulings concerning the University of Michigan on June 23, 2003.[38] In *Grutter v. Bollinger,* the Court approved of the University of Michigan's law school, which took race into account as one of many factors in order to achieve a diverse student body, ensuring a "critical mass" of minority students who could feel comfortable in expressing their viewpoints without being narrowly stereotyped (which tends to occur when there is only a token representation of minorities). In *Gratz v. Bollinger,* by contrast, the court ruled that the University of Michigan's undergraduate admissions program was unconstitutional. That program gave an automatic 20 points to members of minorities, out of the 100 needed for entrance (and out of a total possible 150 points). Such a rigid point system, the Court ruled, functioned too much like a quota system.

The rulings in both *Bakke* and *Grutter* were close: 5 to 4. Furthermore, in *Grutter* the Court made it clear that eventually, certainly within the next 25 years, it was expected that there would no longer be a need for affirmative action programs. Ironically, by 2003, many businesses and the military, which had in the 1960s opposed affirmative action, joined most educational institutions in desiring affirmative action policies as a way to achieve diversity for their own needs. Yet, even within education, there is no consensus on the issue. For example, in 1996 California voted (in Proposition 209) to forbid the use of race in granting admission to public universities, and that ruling still stands. In short, affirmative action remains a lively and contentious issue because of the important and clashing moral values at stake.

Can such preferences, either in the weak or strong form, ever be justified morally (as distinct from legally)? There are compelling arguments on both sides of the issue.[39]

Arguments favoring preferential treatment take three main forms, which look to the past, present, and future. First, there is an argument based on compensatory justice: Past violations of rights must be compensated. Ideally such compensation should be given to those specific individuals who in the past were denied jobs. But the costs and practical difficulties of determining such discrimination on a case-by-case basis through the job-interviewing process permits giving preference on the basis of membership in a group that has been disadvantaged in the past. Second, sexism and racism still permeate our society today, and to counterbalance their insidious impact reverse preferential treatment is warranted in order to ensure equal opportunity for minorities and women. Third, reverse preferential treatment has many good consequences: integrating women and minorities into the economic and social mainstream (especially in male-dominated professions like engineering), providing role models for minorities that build self-esteem, and strengthening diversity.

Arguments against reverse preferential treatment condemn it as reverse discrimination. It violates the rights to equal opportunity of white males and others who are now not given a fair chance to compete on the basis of their qualifications. Granted, past violations of rights may call for compensation, but only compensation to specific individuals who are wronged and only in ways that do not violate the rights of others who did not personally wrong minorities. It is also permissible to provide special funding for educational programs for economically disadvantaged children, but not to use jobs as a compensatory device. Moreover, reverse preferential treatment has many negative effects: lowering economic productivity by using criteria other than qualifications in hiring, encouraging racism by generating intense resentment among white males and their families, encouraging traditional stereotypes that minorities and women cannot make it on their own without special help, and thereby adding to self-doubts of members of these groups.

Various attempts have been made to develop intermediate positions sensitive to all of the preceding arguments for and against strong preferential treatment. For example, one approach rejects blanket preferential treatment of special groups as inherently unjust, but it permits reverse preferential treatment within companies that can be shown to have a history of bias against minorities or women. Another approach is to permit weak reverse preferential treatment but to forbid strong forms.

DISCUSSION QUESTIONS

1. Present and defend your view as to whether affirmative action is morally permissible and desirable in (a) admissions to engineering schools, (b) hiring and promoting within engineering corporations.

2. The majority of employers have adopted mandatory random drug testing on their employees, arguing that the enormous damage caused by the pervasive use of drugs in our society carries over into the workplace. Typically the tests involve taking urine or blood samples under close observation, thereby raising questions about personal privacy as well as privacy issues about drug usage away from the workplace that is revealed by the tests. Present and defend your view concerning mandatory drug tests at the workplace.

   In your answer, take account of the argument set forth by Joseph R. DesJardins and Ronald Duska that, except where safety is a clear and present danger (as in the work of pilots, police, and the military), such tests are unjustified.[40] They contend that employers have a right to the level of performance for which they pay employees, a level typically specified in contracts and job descriptions. When a particular employee fails to meet that level of performance, then employers will take appropriate disciplinary action based on observable behavior. Either way, it is employee performance that is relevant in evaluating employees, not drug usage per se.

3. A company advertises for an engineer to fill a management position. Among the employees the new manager is to supervise is a woman engineer, Ms. X, who was told by her former boss that she would soon be assigned tasks with increased responsibility. The prime candidate for the manager's position is Mr. Y, a recent immigrant from a

country known for confining the roles for women. Ms. X was alerted by other women engineers to expect unchallenging, trivial assignments from a supervisor with Mr. Y's background. Is there anything she can and should do? Would it be ethical for her to try to forestall the appointment of Mr. Y?

4. Jim Serra, vice president of engineering, must decide who to recommend for a new director-level position that was formed by merging the product (regulatory) compliance group with the environmental testing group.[41] The top inside candidate is Diane Bryant, senior engineering group manager in charge of the environmental testing group. Bryant is 36, exceptionally intelligent and highly motivated, and a well-respected leader. She is also five months pregnant and is expected to take an eight-week maternity leave two months before the first customer ship deadline (six months away) for a new product. Bryant applies for the job and in a discussion with Serra assures him that she will be available at all crucial stages of the project. Your colleague David Moss, who is vice president of product engineering, strongly urges you to find an outside person, insisting that there is no guarantee that Bryant will be available when needed. Much is at stake. A schedule delay could cost several million dollars in revenues lost to competitors. At the same time, offending Bryant could lead her and perhaps other valuable engineers whom she supervises to leave the company. What procedure would you recommend in reaching a solution?

5. In the past, engineering societies have generally portrayed participation by engineers in unions and collective bargaining in engineering as unprofessional and disloyal to employers. Critics reply that such generalized prohibitions reflect the excessive degree to which engineering is still dominated by corporations' interests. Discuss this issue with regard to the following case. What options might be pursued, and would they still involve "collective coercive action"?

Managers at a mining and refinery operation have consistently kept wages below industry-wide levels. They have also sacrificed worker safety in order to save costs by not installing special structural reinforcements in the mines, and they have made no effort to control excessive pollution of the work environment. As a result, the operation has reaped larger-than-average profits. Management has been approached both by individuals and by representatives of employee groups about raising wages and taking the steps necessary to ensure worker safety, but to no avail. A nonviolent strike is called, and the metallurgical engineers support it for reasons of worker safety and public health.

6.4   WHISTLEBLOWING

No topic in engineering ethics is more controversial than whistleblowing. A host of issues are involved. When is whistleblowing morally permissible? Is it ever morally obligatory, or is it beyond the call of duty? To what extent, if any, do engineers have a right to whistleblow, and when is doing so immoral and imprudent? When is whistleblowing an act of disloyalty to an organization? What procedures ought to be followed in blowing the whistle? Before considering these questions we need to define *whistleblowing*.

6.4.1   Whistleblowing: Definition

Whistleblowing occurs when an employee or former employee conveys information about a significant moral problem to someone in a position to take action on the problem, and does so outside approved organizational channels (or against strong pressure). The definition has four main parts.

1. *Disclosure:* Information is intentionally conveyed outside approved organizational (workplace) channels or in situations where the person conveying it is under pressure from supervisors or others not to do so.
2. *Topic:* The information concerns what the person believes is a significant moral problem for the organization (or an organization with which the company does business). Examples of significant problems are serious threats to public or employee safety and well-being, criminal behavior, unethical policies or practices, and injustices to workers within the organization.
3. *Agent:* The person disclosing the information is an employee or former employee (or someone else closely associated with the organization).
4. *Recipient:* The information is conveyed to a person or organization that is in a position to act on the problem (as opposed, for example, to telling it to a relative or friend who is in no position to do anything).[42] The desired response or action might consist in remedying the problem or merely alerting affected parties. Typically, though not always, the information being revealed is new or not fully known to the person or group receiving it.

Using this definition, we will speak of *external whistleblowing* when the information is passed outside the organization. *Internal whistleblowing* occurs when the information is conveyed to someone within the organization (but outside approved channels or against pressures to remain silent).

The definition also allows us to distinguish between open and anonymous whistleblowing. In *open whistleblowing,* individuals openly reveal their identity as they convey the information. *Anonymous whistleblowing,* by contrast, involves concealing one's identity. There are also overlapping cases that are partly open and partly anonymous, such as when individuals acknowledge their identities to a journalist but insist their names be withheld from anyone else.

Notice that the definition does not mention the motives involved in the whistleblowing, and hence it avoids assumptions about whether those motives are good or bad. Nor does it assume that the whistleblower is correct in believing there is a serious moral problem. In general, it leaves open the question of whether whistleblowing is justified. In turning to issues about justification, let us begin with two classic case studies, one in which the whistle is blown and one in which it was not.

6.4.2 Two Cases

ERNEST FITZGERALD AND THE C-5A. One of the most publicized instances of open, external whistleblowing occurred on November 13, 1968. On that day Ernest Fitzgerald was one of several witnesses called to testify before Senator William Proxmire's Subcommittee on Economy in Government concerning the C-5A, a giant cargo plane being built by Lockheed Aircraft Corporation for the

Air Force. Fitzgerald, who had previously been an industrial engineer and management consultant, was then a deputy for management systems under the assistant secretary of the Air Force. During the preceding two years he had reported huge cost overruns in the C-5A project to his superiors, overruns that by 1968 had hit $2 billion. He had argued forcefully against similar overruns in other projects, so forcefully that he had become unpopular with his superiors. They pressured him not to discuss the extent of the C-5A overruns before Senator Proxmire's committee. Yet when Fitzgerald was directly asked to confirm Proxmire's own estimates of the overruns on that November 13, he told the truth.

Doing so turned his career into a costly nightmare for himself, his wife, and his three children.[43] He was immediately stripped of his duties and assigned trivial projects, such as examining cost overruns on a bowling alley in Thailand. He was shunned by his colleagues. Within 12 days he was notified that his promised civil service tenure was a computer error. And within four months the bureaucracy was restructured so as to abolish his job. It took four years of extensive court battles before federal courts ruled that he had been wrongfully fired and ordered the Air Force to rehire him. After years of further litigation, involving fees of around $900,000, he was reinstated in his former position in 1981.

DAN APPLEGATE AND THE DC-10. In 1974 the first crash of a fully loaded DC-10 jumbo jet occurred over the suburbs of Paris; 346 people were killed, a record for a single-plane crash. It was known in advance that such a crash was bound to occur because of the jet's defective design.[44]

The fuselage of the plane was developed by Convair, a subcontractor for McDonnell Douglas. Two years earlier, Convair's senior engineer directing the project, Dan Applegate, had written a memo to the vice president of the company itemizing the dangers that could result from the design. He accurately detailed several ways the cargo doors could burst open during flight, depressurize the cargo space, and thereby collapse the floor of the passenger cabin above. Since control lines ran along the cabin floor, this would mean a loss of full control over the plane. Applegate recommended redesigning the doors and strengthening the cabin floor. Without such changes, he stated, it was inevitable that some DC-10 cargo doors would open in midair, resulting in crashes.

In responding to this memo, top management at Convair disputed neither the technical facts cited by Applegate nor his predictions. Company officials maintained, however, that the possible financial liabilities Convair might incur prohibited them from passing on this information to McDonnell Douglas. These liabilities could be severe since the cost of redesign and the delay to make the necessary safety improvements would be very high and would occur at a time when McDonnell Douglas would be placed at a competitive disadvantage.

6.4.3 Moral Guidelines

Under what conditions are engineers justified in whistleblowing? This really involves two questions: When are they morally permitted, and when are they morally obligated, to do so? In our view, it is permissible to whistleblow when the following conditions have been met.[45] Under these conditions there is also an obligation to whistleblow, although the obligation is prima facie and in some situations can be overridden by other moral considerations.

1. The actual or potential harm reported is serious.
2. The harm has been adequately documented.
3. The concerns have been reported to immediate superiors.
4. After not getting satisfaction from immediate superiors, regular channels within the organization have been used to reach up to the highest levels of management.
5. There is reasonable hope that whistleblowing can help prevent or remedy the harm.

These conditions are not always necessary for permissible and obligatory whistleblowing, however.[46] Condition 2 might not be required in situations where cloaks of secrecy are imposed on evidence that, if revealed, could supposedly aid commercial competitors or a nation's adversaries. In such cases it might be very difficult to establish adequate documentation, and the whistleblowing would consist essentially of a request to the proper authorities to carry out an external investigation or to request a court to issue an order for the release of information.

Again, conditions 3 and 4 may be inappropriate in some situations, such as when one's supervisors are the main source of the problem or when extreme urgency leaves insufficient time to work through all regular organizational channels.

Finally, when whistleblowing demands great sacrifices, one cannot overlook that personal obligations to family, as well as rights to pursue one's career, militate against whistleblowing. Where blowing the whistle openly could result not only in the loss of one's job but also in being blacklisted within the profession, the sacrifice may become supererogatory—more than one's basic moral obligations, all things considered, require. Engineers share responsibilities with many others for the products they help create. It seems unfair to demand that one individual bear the harsh penalties for picking up the moral slack for other irresponsible persons involved. Most important, the public also shares some responsibilities for technological ventures and hence for passing reasonable laws protecting responsible whistleblowers. When those laws do not exist or are not enforced, the public has little basis for demanding that engineers risk their means of livelihood.[47]

Certainly Fitzgerald's action was morally permissible and admirable when he engaged in whistleblowing. His case seems to us clear-cut for several reasons: He had made every effort to first seek a remedy to the abuses he uncovered by working within accepted organizational channels; his views were well founded on hard evidence; the harm done to the Air Force by his disclosures was far outweighed by the benefits that accrued to the public; he was a public servant with especially strong obligations to the public, which his organization, the Air Force, is committed to serve; and to have withheld the information from Senator Prox-mire would have involved lying and participating in a cover-up.

Was Fitzgerald obligated to do what he did, all things considered? In his situation, as is often true, failure to blow the whistle would have amounted to complicity in wrongdoing. The Code of Ethics for the United States Government Service says that employees should "put loyalty to the highest moral principles and to country above loyalty to persons, party, or government department" and that they should expose "corruption wherever discovered." A cover-up of a $2 billion expenditure of taxpayers' money in contract overruns would seem to qual-ify as corruption. If we feel any hesitation in saying Fitzgerald was obligated to whistleblow, all things considered, it concerns whether it might be asking too much of someone in his position to do what he did. Is it not beyond the call of duty to require such an incredible degree of personal sacrifice in performing one's job?

How about Applegate? As a loyal employee Applegate had a responsibility to follow company directives, at least reasonable ones. Perhaps he also had family responsibilities that made it important for him not to jeopardize his job. Yet as an engineer he was obligated to protect the safety of those who would use or be affected by the products he designed. Given the great public hazard involved, few would question whether it would be morally *permissible* for him to blow the whistle, either to the FAA or to the newspapers. Was he also morally *obligated* to blow the whistle? We leave this as a study question.

Not all whistleblowing, of course, is admirable, obligatory, or even permis-sible. Certainly inaccurate whistleblowing can cause unjustified harm to compa-nies that unfairly receive bad publicity that hurts employees, stockholders, and sometimes the economy.[48] But is there a general presumption against whistleblow-ing that at most is overridden in extreme situations? The most common argument against whistleblowing is that it constitutes disloyalty to the corporation, although we will expand loyalty to include collegiality and respect for authority.

Taken together, loyalty, collegiality, and respect for authority do create a presumption against whistleblowing, but it is a presumption that can be overrid-den. Loyalty, collegiality, and respect for authority are not excuses or justification for shielding irresponsible conduct. To think otherwise would be to lapse into a form of *corporate egoism:* the view that the corporation is more important than the wider good of the public. In addition to corporate virtues, there are public-oriented virtues, especially respect for the public's safety.

6.4.4   Protecting Whistleblowers

Most whistleblowers have suffered unhappy and even tragic fates. In the words of one lawyer who defended a number of them: "Whistle-blowing is lonely, unre-warded, and fraught with peril. It entails a substantial risk of retaliation which is difficult and expensive to challenge. Furthermore, 'success' may mean no more than retirement to a job where the bridges are already burned, or monetary compensation that cannot undo damage to a reputation, career and personal relationships."[49]

Yet the vital service to the public provided by many whistleblowers has led increasingly to public awareness of a need to protect them against retaliation by employers. In particular, whistleblowers played a vital role in informing the public and investigators about recent corporate and government scandals, including whistleblowers Sherron Watkins at Enron, Cynthia Cooper at Worldcom, and Coleen Rowley at the FBI—three women who appeared on the cover of *Time Magazine* as "Persons of the Year" for 2002.[50]

Government employees have won important protections. Various federal laws related to environmental protection and safety and the Civil Service Reform Act of 1978 protect them against reprisals for lawful disclosures of information believed to show "a violation of any law, rule, or regulation, mismanagement, a gross waste of funds, an abuse of authority, or a substantial and specific danger to public health and safety."[51] The fact that few disclosures are made appears to be due mostly to a sense of futility—the feeling that no corrective action will be undertaken. In the private sector, employees are covered by statutes forbidding firing or harassing of whistleblowers who report to government regulatory agencies the violations of some 20 federal laws, including those covering coal mine safety, control of water and air pollution, disposal of toxic substances, and occupational safety and health. In a few instances, unions provide further protection. Overall, legal aid for whistleblowers was growing twenty years ago, but federal enforcement tends to fluctuate as Administrations change.[52]

Laws, when they are carefully formulated and enforced, provide two types of benefits for the public, in addition to protecting the responsible whistleblower: episodic and systemic. The *episodic* benefits are in helping to prevent harm to the public in particular situations. The *systemic* benefits are in sending a strong message to industry to act responsibly or be subject to public scrutiny once the whistle is blown.

Laws alone will usually not suffice. When officialdom is not ready to enforce existing laws—or introduce obviously necessary laws—engineering associations and employee groups need to act as watchdogs ready with advice and legal assistance. Successful examples are the Government Accountability Project (GAP) and some professional societies that report the names of corporations found to have taken unjust reprisals against whistleblowers. The Institute of Electrical and Electronics Engineers (IEEE) had taken an active role by assisting members with a help line, backing them when faced with retaliatory court action, helping unjustly discharged engineers find new jobs, and honoring courageous whistleblowers with public recognitions. Apparently fearing potential legal repercussions, the IEEE has recently become much less pro-active.

6.4.5   Commonsense Procedures

It is clear that a decision to whistleblow, whether within or outside an organization, is a serious matter that deserves careful reflection. And there are several rules of practical advice and common sense that should be heeded before taking this action.[53]

1. Except for extremely rare emergencies, always try working first through normal organizational channels. Get to know both the formal and informal (unwritten) rules for making appeals within the organization.
2. Be prompt in expressing objections. Waiting too long may create the appearance of plotting for your advantage and seeking to embarrass a supervisor.
3. Proceed in a tactful, low-key manner. Be considerate of the feelings of others involved. Always keep focused on the issues themselves, avoiding any personal criticisms that might create antagonism and deflect attention from solving those issues.
4. As much as possible, keep supervisors informed of your actions, both through informal discussion and formal memorandums.
5. Be accurate in your observations and claims, and keep formal records documenting relevant events.
6. Consult trusted colleagues for advice—avoid isolation.
7. Before going outside the organization, consult the ethics committee of your professional society.
8. Consult a lawyer concerning potential legal liabilities.

6.4.6 Beyond Whistleblowing

Sometimes whistleblowing is a practical moral necessity. But generally it holds little promise as the best possible method for remedying problems and should be viewed as a last resort.

The obvious way to remove the need for internal whistleblowing is for management to allow greater freedom and openness of communication within the organization. By making those channels more flexible and convenient, the need to violate them would be removed. But this means more than merely announcing formal "open-door" policies and appeals procedures that give direct access to higher levels of management. Those would be good first steps, and a further step would be the creation of an ombudsperson or an ethics review committee with genuine freedom to investigate complaints and make independent recommendations to top management. The crucial factor that must be involved in any structural change, however, is the creation of an atmosphere of positive affirmation of engineers' efforts to assert and defend their professional judgments in matters involving ethical considerations.

What about external whistleblowing? Much of it can also be avoided by the same sorts of intra-organizational modifications. Yet there will always remain troublesome cases where top management and engineers differ in their assessments of a situation even though both sides may be equally concerned to meet their professional obligations to safety. To date, the assumption has been that management has the final say in any such dispute. But our view is that engineers have a right to some further recourse in seeking to have their views heard, including confidential discussions with the ethics committees of their professional societies.

Conscientious engineers sometimes find the best solution to be to resign and engage in protest, as in the following example. David Parnas, a computer scientist, lost his initial enthusiasm for the Strategic Defense Initiative (SDI), also known as the Star Wars project.[54] He resigned from an advisory panel on computing after only the first meeting of the panel. When agency officials would not seriously listen to his doubts about the feasibility of the project, he gradually succeeded through journal articles, open debates, and public lectures to convince the profession that Star Wars did not differ much from conventional anti-ballistic-missile defense without overcoming earlier shortcomings. Indeed, the system's complexity made it practically impossible to write software as reliable as it ought to be in tight-trigger situations. For his efforts on behalf of the public interest, he was honored with the Norbert Wiener Award by the society of Computer Professionals for Social Responsibility (CPSR).

Sometimes, engineers need to be morally creative about if and how other alternative approaches to whistleblowing might be taken. For instance, in typical professional ethics cases such as "when your boss tells you to fake the tests, it is easy to think that either I agree and lose my self-respect or refuse and threaten my job,"[55] engineers are encouraged to explore some creative "win-win" resolutions that are less adversarial and more empathetic and communicative, such as asking the boss what pressures she is facing, how to help her respond to these pressures, and how to help her make sense of the ethical problems associated with the given situation and her associated request. Nevertheless, as suggested earlier, whether these creative resolutions can be effective or not largely depends on the culture of the organization (e.g., whether the organization allows openness of communication).

DISCUSSION QUESTIONS

1. According to Kenneth Kipnis, a professor of philosophy, Dan Applegate and his colleagues share the blame for the death of the passengers in the DC-10 crash. Kipnis contends that the engineers' overriding obligation was to obey the following principle: "Engineers shall not participate in projects that degrade ambient levels of public safety unless information concerning those degradations is made generally available."[56] Do you agree or disagree with Kipnis, and why? Was Applegate obligated to blow the whistle?

2. Present and defend your view as to whether in the following case the actions of Ms. Edgerton and her supervisor were morally permissible, obligatory, or admirable. Did Ms. Edgerton have a professional moral right to act as she did? Was hers a case of legitimate whistleblowing?

    In 1977, Virginia Edgerton was senior information scientist on a project for New York City's Criminal Justice Coordinating Council. The project was to develop a computer system for use by New York district attorneys in keeping track of data about court cases. It was to be added to another computer system, already in operation, which dispatched police cars in response to emergency calls. Ms. Edgerton, who had 13 years of data processing experience, judged that adding on the new system might result in overloading the existing system in such a way that the response time for dispatching emergency vehicles might increase. Because it might risk lives to test the system in operation, she recommended that a study be conducted ahead of time to estimate the likelihood of such overload.

She made this recommendation to her immediate supervisor, the project director, who refused to follow it. She then sought advice from the IEEE, of which she was a member. The Institute's Working Group on Ethics and Employment Practices referred her to the manager of systems programming at Columbia University's computer center, who verified that she was raising a legitimate issue.

Next she wrote a formal memo to her supervisor, again requesting the study. When her request was rejected, she sent a revised version of the memo to New York's Criminal Justice Steering Committee, a part of the organization for which she worked. In doing so she violated the project director's orders that all communications to the Steering Committee be approved by him in advance. The project director promptly fired her for insubordination. Later he stated: "It is . . . imperative that an employee who is in a highly professional capacity, and has the exposure that accompanies a position dealing with top level policy makers, follow expressly given orders and adhere to established policy."[57]

3. A controversial area of recent legislation allows whistleblowers to collect money. Federal tax legislation, for example, pays informers a percentage of the money recovered from tax violators. And the 1986 False Claims Amendment Act allows 15 to 25 percent of the recovered money to go to whistleblowers who report overcharging in federal government contracts to corporations. These sums can be substantial because lawsuits can involve double and triple damages as well as fines. A recent study revealed the following statistics: overall, about 23 percent of these lawsuits succeed; 80 percent succeed if the federal government joins the case as a plaintiff; 5 percent succeed when the government does not join the case.[58] Discuss the possible benefits and drawbacks of using this approach in engineering and specifically concerning safety matters. Is the added incentive to whistleblow worth the risk of encouraging self-interested motives in whistleblowing?

4. Do you see any special moral issues raised by anonymous whistleblowing?[59]

6.5 THE BART CASE

The Bay Area Rapid Transit system (BART) is a suburban rail system that links San Francisco with the cities across its bay. It was constructed during the late 1960s and early 1970s, and its construction led to a now-classic case of whistleblowing. The case is important because it remains controversial, because it involved a precedent-setting intervention by an engineering professional society, and especially because it became the subject of the first book-length scholarly study of an instance of whistleblowing, *Divided Loyalties* by Robert M. Anderson and his colleagues.[60]

6.5.1 Background

The example of the pioneering years of railroading indicates that technological experimentation is usually highly fruitful. For example, early fears about the effects of high-speed travel—of sparks showering the countryside, of animals being frightened by the noise and fast movement—proved to be unfounded. The benefits to agriculture, industry, and commerce, moreover, were immense. And society learned that to secure those benefits it could live with the loss of forests to railroad ties and fuel, or with the cycle of settlement building and abandonment entailed by the construction of new railroads.

As technological innovation in railroading accelerated, however, the trend to do the fashionable thing for its own sake increasingly predominated. For example, railroads took over in instances where common sense would have dictated the continued use of barges on canals. To some extent BART is a recent example of that trend. Developed to incorporate the latest "space age" technology in its design, it became (at least in its initial design) more expensive and less reliable than its traditional counterparts.

The BART system was built with tax funds, and its construction was characterized by tremendous cost overruns and numerous delays. Much of this can be ascribed to the introduction of innovative methods of communicating with individual trains and of controlling them automatically. In addition, plain fail-safe operation was replaced by complex redundancy schemes. (Fail-safe features simply cause a train to stop if something breaks down; redundancy features try to keep trains running by switching the faulted components to alternate ones.) The rationale given for this approach was that the system could be sold to the public only if it involved glamorous and exciting gadgetry.

6.5.2 Responsibility and Experimentation

The opportunity to build a rail system from scratch, unfettered by old technology, was a challenge that excited many engineers and engineering firms. Yet among the engineers who worked on it were some who came to feel that too much experimentation was going on without proper safeguards. Safety features were given insufficient attention, and quality control was poor, they thought.

Three engineers in particular, Holger Hjortsvang, Robert Bruder, and Max Blankenzee, identified dangers that were to be recognized by management only much later. They saw that the automatic train control was unsafely designed. Moreover, schedules for testing it and providing operator training prior to its public use were inadequate. Computer software problems continued to plague the system. Finally, there was insufficient monitoring of the work of the various contractors hired to design and construct the railroad. These inadequacies were to become the main causes of several early accidents.[61]

The three engineers wrote a number of memos and voiced their concerns to their employers and colleagues. Their initial efforts were directed through organizational channels to both their immediate supervisors and the two next higher levels of management, but to no avail. Yet they refused to wait passively for accidents to occur and resolved to do more.

Up to this point Hjortsvang, Bruder, and Blankenzee were conscientious in refusing to lose sight of their primary obligation to the public—that is, their obligation to what was, in effect, the "subject" of this particular engineering "experiment." They were imaginative in foreseeing dangers. They were personally and autonomously involved. And they were willing to accept moral accountability for their participation in the project.

Of special interest in the case is that for the most part the three engineers were not specifically assigned or authorized by the BART organization to check into the safety of the automatic control system. Hjortsvang, for example, first identified the dangers when he was sent to Westinghouse (a BART subcontractor) primarily to observe, not supervise, the development of the control system. Similarly, Robert Bruder worked for the construction department, not the operations department, which had responsibility for the train control. Thus, both engineers looked to the wider implications of the specific tasks assigned to them within the organization. They refused to have their moral responsibility confined within a narrow organizational bailiwick.

6.5.3 Controversy

The controversial events that followed as the engineers sought to pursue their concerns further are described and interpreted from the opposing viewpoints of the engineers and management (and others) in the book *Divided Loyalties*. An account of five of those events follows.

First, Hjortsvang wrote an anonymous memo summarizing the problems, and distributed copies of it to nearly all levels of management, including the project's general manager. The memo argued that a new systems engineering department was needed, a department that Hjortsvang had also requested in an earlier signed memo. Distribution of such an unsigned memo was regarded by management as suspicious and unprofessional since it was done outside the normal channels of accountability within the organization. Later, when its author was identified, management decided Hjortsvang was motivated by self-interest and a desire for power since it could be assumed that he wished to become the head of such a department.

Second, the three engineers contacted several members of BART's board of directors when their concerns were not taken seriously by lower levels of management. By doing so, they departed from approved organizational channels, since BART's general manager allowed only himself and his designates to deal directly with the board. Since BART was a publicly funded organization governed by the public board of directors, it could be argued that this was an instance of internal whistleblowing.

Third, in order to obtain an independent view, the engineers contacted a private engineering consultant who on his own wrote an evaluation of the automatic train control.

Fourth, one of the directors, Dan Helix, listened sympathetically and agreed to contact top management while keeping the engineers' names confidential. But to the shock of the three engineers, Helix released copies of their unsigned memos and the consultant's report to the local newspapers. It would be the engineers, not Helix, who would be penalized for this act of external whistleblowing.

Fifth, management immediately sought to locate the source of Helix's information. Fearing reprisals, the engineers at first lied to their supervisors and denied their involvement.

6.5.4 Aftermath

At Helix's request the engineers later agreed to reveal themselves by going before the full board of directors in order to seek a remedy for the safety problems. On that occasion they were unable to convince the board of those problems. One week later they were given the option of resigning or being fired. The grounds given for the dismissal were insubordination, incompetence, lying to their superiors, causing staff disruptions, and failing to follow understood organizational procedures.

These dismissals were damaging to the engineers. Robert Bruder could not find engineering work for eight months. He had to sell his house, go on welfare, and receive food stamps. Max Blankenzee was unable to find work for nearly five months, lost his house, and was separated from his wife for one and a half months. Holger Hjortsvang could not obtain full-time employment for 14 months, during which time he suffered from extreme nervousness and insomnia.

The impact on BART, by comparison, was minor. Subsequent studies proved that the safety judgments of the engineers were sound. Changes in the design of the automatic train control were made, but it is unclear whether those changes would have been made in any case. During its decade of development BART was plagued by many technical problems of the type the engineers drew attention to. And the inability of BART management to deal effectively with the engineers' concerns was typical of many other instances of poor management.

Two years later the engineers sued BART for damages in the sum of $875,000 on the grounds of breach of contract, harming their future work prospects, and depriving them of their constitutional rights under the First and Fourteenth Amendments. A few days before the trial began, however, they were advised by their attorney that they could not win the case because they had lied to their employers during the episode. They settled out of court for $75,000 minus 40 percent for lawyers' fees.

In the development of their case the engineers were assisted by an amicus curiae ("friend of the court") brief filed by the IEEE. This legal brief noted in their defense that it is part of each engineer's professional duty to promote the public welfare, as stated in IEEE's code of ethics. In 1978 IEEE presented each of them with its Award for Outstanding Service in the Public Interest for "courageously adhering to the letter and spirit of the IEEE code of ethics."

6.5.5 Comments

The discussion questions that follow ask about the extent to which the three engineers and BART's management acted responsibly. The complexities revealed in *Divided Loyalties* show the case is hardly a simple one. Here we wish to comment on two attitudes held by the authors of that book, attitudes germane to the topic of moral responsibility and deserving mention because of the frequency with which similar arguments are heard in other contexts.

The final verdict of the authors of *Divided Loyalties* is that the BART case "can be viewed as not really involving safety or ethics to any marked degree."[62] We disagree. The main basis for that verdict seems to be the claim that BART's complex organizational structure alone was to blame for the conflicts that helped precipitate the incidents. For example, the engineers were given considerable freedom to determine for themselves the specific tasks they were to pursue, but granted little authority to implement changes they felt were needed. Frustration on their part was therefore to be expected.

This argument, however, fails to show that ethical issues were not involved. On the contrary, it shows how ethical issues can arise out of problems associated with organizational structure. Indeed, the conflicts engendered by the social, political, and economic settings of an organization quite often form the background for the ethical problems engineers confront when they are concerned about how best to ensure the safety of their projects.

The authors' verdict may also have resulted from a lack of clarity about ethical issues. They emphasize that there were no villains in the BART episode. Those involved were basically good people trying in the main to do their jobs responsibly even if they were influenced to some degree by self-interest. This seems to imply that ethical situations must always involve bad people who are opposed by good people—a melodramatic view of morality. Yet surely the question of how best to assure safety in any engineering project is a moral issue, whatever the ultimate personal motivations of the people involved in it. Ethics can involve a decision between good and better just as much as a conflict between good and bad.

DISCUSSION QUESTIONS

1. Present and defend your view as to whether, and in what respects, the BART engineers and BART management acted responsibly. In doing so, discuss alternative courses of action that either or both groups might have pursued. Discuss and apply De George's criteria (from section 6.4.3) for when whistleblowing is morally permissible and obligatory, as well as any modifications of those criteria that you see as appropriate. Focus especially on (a) Hjortsvang's anonymous memo distributed within BART, (b) the act of contacting BART's board of directors, and (c) lying to the supervisors when questioned about their involvement.

2. The authors of *Divided Loyalties* suggest that

   management shares with the three engineers responsibility for the political naïveté which permitted them to carry their grievance as far as they did. It is clear that the engineers took a narrow and technical view of the issues which disturbed them, and failed to place them in the context of the whole BART development. At the same time, management fostered this naïveté by failing adequately to sensitize its professional employees to the political and economic climate surrounding and influencing the activities of the organization.[63]

Presumably this is a criticism of the act of contacting the board of directors of a public project for which a positive public image is needed to sustain support and continued funding. Do you agree with these authors that political considerations should have entered into the decisions of the three engineers? Or do you agree with IEEE that the engineers acted in a courageous way in trying to protect public safety?

3. The movie "Erin Brockovich," based on a screenplay by Susannah Grant, describes how toxic effluents from an electric power plant contaminated the nearby residents' drinking water, causing many cancer cases and deaths, especially among infants.[64] The power company disclaims any responsibility. Erin is a clerk without legal training, but she succeeds in gathering evidence of hazardous pollution and patients ready to seek damages, whereupon her law firm employer brings a winning suit for compensation. The defendant claims to have settled the suit in order to terminate the distraction. Some newspapers laud the settlement, others claim that bogus science was at play. Examine the case and give your opinion, drawing on similar but earlier cases.

KEY CONCEPTS

—**Ethical corporate climate:** (1) ethical values in their full complexity are widely acknowledged and appreciated by managers and employees alike; (2) ethical language is honestly applied and recognized as a legitimate part of corporate dialogue; (3) top management sets a moral tone; (4) there are procedures for conflict resolution.

—**Loyalty to a corporation** can mean either (1) agency-loyalty—acting to fulfill one's contractual duties, or (2) attitude-loyalty—agency-loyalty that is motivated by identification with the corporation.

—**Collegiality:** a virtue of teamwork that includes respect for colleagues, commitment to the moral ideals inherent in one's profession, and connectedness in the sense of awareness of participating in cooperative projects based on shared commitments and mutual support.

—**Executive authority:** the corporate or institutional right given to a person to exercise power based on the resources of an organization. It is distinct from mere power and from expert authority (knowledge or skill).

—**Managing conflict:** dealing with conflicts, including value disagreements, in order to maintain teamwork.

—**Confidentiality:** keeping secret the information specified by an employer or client in order to compete effectively against business rivals, especially proprietary information and trade secrets (owned by a company) but also privileged information concerning a project.

—**Conflicts of interest:** situations where professionals or other employees have an interest that if pursued, might keep them from meeting their obligations to their employers or clients. Examples include accepting gifts, bribes, kickbacks, having large interests in competing companies, and using insider information.

—**Professionals' rights:** the rights of professionals needed to meet their responsibilities. They include the right of professional conscience (to exercise professional judgment in pursuing responsibilities), the right of conscientious refusal (to refuse directives to engage in unethical behavior), and the right of recognition (to be fairly recognized for one's accomplishments).

—**Employee rights:** rights as an employee, including rights to privacy, nondiscrimination, equal opportunity.

—**Sexual discrimination:** unwanted imposition of sexual requirements, both *quid pro quo* (where sexual favors are made in exchange for a benefit) and *hostile work environment* (in which a sexually oriented aspect of the workplace threatens equal opportunity).

—**Affirmative action** (as preferential treatment): giving preferences, especially in hiring or promoting, on the basis of race or gender. The weak form occurs when a woman or minority is equally qualified with a white male, and the strong form occurs when the preference is over a more qualified white male.

—**Whistleblowing:** when an employee or former employee conveys information about a significant moral problem to someone in a position to take action on the problem, and does so outside approved organizational channels (or against strong pressure). With **external whistleblowing** the information is passed outside the organization, and with **internal whistleblowing** it is conveyed within the organization. With **open whistleblowing** individuals reveal their identities, and with **anonymous whistleblowing** they withhold their identities.

REFERENCES

1. Tracy Kidder, *The Soul of a New Machine* (New York: Avon Books, 1981), p. 273.
2. Ibid., p. 291.
3. Robert Jackall, *Moral Mazes: The World of Corporate Managers* (New York: Oxford University Press, 1988), pp. 6, 109 (italics removed).
4. Lynn Sharp Paine, "Managing for Organizational Integrity," *Harvard Business Review*, March-April 1994, pp. 106–17.
5. Ibid., pp. 136–42.
6. John Ladd, "Loyalty," in *The Encyclopedia of Philosophy,* vol. 4, ed. Paul Edwards (New York: Macmillan, 1967), pp. 97–98. Also see Andrew Oldenquist, "Loyalties," *Journal of Philosophy* 79 (1982): 173–93; George P. Fletcher, *Loyalty* (New York: Oxford University Press, 1993); Marcia Baron, *The Moral Status of Loyalty* (Dubuque, IA: Kendall/Hunt, 1984); and John H. Fielder, "Organizational Loyalty," *Business and Professional Ethics Journal* 11 (Spring 1992): 83.
7. The concept of dependent virtues is developed by Michael Slote in *Goods and Virtues* (Oxford: Clarendon Press, 1983).
8. Craig K. Ihara, "Collegiality as a Professional Virtue," in *Professional Ideals,* ed. Albert Flores (Belmont, CA: Wadsworth, 1988), p. 60.
9. William J. Frey, "Ethics of Team Work," *Ethics in Science and Engineering National Clearinghouse.* 334. Available at: https://scholarworks.umass.edu/esence/334
10. Joseph A. Pichler, "Power, Influence and Authority," in *Contemporary Management,* ed. Joseph W. McGuire (Englewood Cliffs, NJ: Prentice Hall, 1974), p. 428; Richard T. De George, *The Nature and Limits of Authority* (Lawrence, KS: University Press of Kansas, 1985).
11. Michael Davis, *Thinking Like an Engineer* (New York: Oxford University Press, 1998), pp. 130–31.
12. Douglas McGregor, *The Human Side of Enterprise* (New York: McGraw-Hill, 1960); R. J. Burke, "Methods of Resolving Superior-Subordinate Conflict: The Constructive Use of Subordinate Differences and Disagreements," *Organizational Behavior and Human Performance* 5 (1970): 393–411.
13. Hans J. Thamhain, *Engineering Management: Managing Effectively in Technology-Based Organizations* (New York: John Wiley & Sons, 1992), p. 454.
14. Marvin T. Brown, *Working Ethics: Strategies for Decision Making and Organizational Responsibility* (San Francisco: Jossey-Bass, 1990), pp. 55–69; Joseph P. Folger, Marshall Scott Poole, and Randall K. Stutman, *Working Through Conflict,* 2nd ed. (New York: HarperCollins, 1993), pp. 153–80.
15. Roger Fisher and William Ury, *Getting to Yes: Negotiating Agreement Without Giving In* (New York: Penguin, 1983), p. 11.

16. Danny Ertel, "How to Design a Conflict Management Procedure That Fits Your Dispute," *Sloan Management Review* 32 (Summer 1991): 29–42 (especially p. 31).

17. This case is a summary of Cindee Mock and Andrea Bruno, "The Expectant Executive and the Endangered Promotion," *Harvard Business Review* (January-February 1994): 16–18.

18. Michael S. Baram, "Trade Secrets: What Price Loyalty?" *Harvard Business Review,* November-December 1968. Reprinted in Deborah G. Johnson (ed.), *Ethical Issues in Engineering* (Englewood Cliffs, NJ: Prentice Hall, 1991), pp. 279–90.

19. Ibid., pp. 285–90.

20. Sissela Bok, *Secrets* (New York: Pantheon Books, 1982), pp. 116–35.

21. Michael Davis, "Conflict of Interest," *Business and Professional Ethics Journal* 1 (Summer 1982): 17–27; Paula Wells, Hardy Jones, and Michael Davis, *Conflicts of Interest in Engineering* (Dubuque, IA: Kendall/Hunt, 1986); Michael Davis and Andrew Stark (eds.), *Conflict of Interest in the Professions* (New York: Oxford University Press, 2001).

22. Joseph Margolis, "Conflict of Interest and Conflicting Interests," in *Ethical Theory and Business,* ed. T. Beauchamp and N. Bowie (Englewood Cliffs, NJ: Prentice Hall, 1979), p. 361.

23. National Academy of Engineering, *On Being a Scientist: A Guide to Responsible Conduct in Research* (3rd ed.) (Washington, DC: National Academies Press, 2009).

24. Cf. Michael S. Pritchard, "Bribery: The Concept," *Science and Engineering Ethics* 4, no. 3 (1998): 281–86.

25. Charles E. Harris, Michael S. Pritchard, Ray W. James, Elaine E. Englehardt, and Michael J. Rabins, *Engineering Ethics: Concepts and Cases* (6th ed.) (Boston, MA: Cengage, 2019).

26. George L. Reed, "Moonlighting and Professional Responsibility," *Journal of Professional Activities: Proceedings of the American Society of Civil Engineers* 96 (September 1970): 19–23.

27. Philip M. Kohn and Roy V. Hughson, "Perplexing Problems in Engineering Ethics," *Chemical Engineering* 87 (May 5, 1980): 102. Quotations in text used with permission of McGraw-Hill Book Co.

28. Philip L. Alger, N. A. Christensen, and Sterling P. Olmsted, *Ethical Problems in Engineering* (New York: John Wiley & Sons, 1965), p. 111. Quotations used with permission of the publisher.

29. Charles M. Carter, "Trade Secrets and the Technical Man," *IEEE Spectrum* 6 (February 1969): 54.

30. *NSPE Opinions of the Board of Ethical Review,* Case 75.10, National Society of Professional Engineers, Washington, DC, website: www.nspe.org.

31. Philip M. Kohn and Roy V. Hughson, "Perplexing Problems in Engineering Ethics," *Chemical Engineering* 87 (May 5, 1980): 104. Quotations in text used with permission of McGraw-Hill Book Co.

32. "The Condo," in *Teaching Engineering Ethics: A Case Study Approach,* ed. Michael S. Pritchard (Kalamazoo, MI: Center for the Study of Ethics in Society, Western Michigan University, 1993).

33. David W. Ewing, *Freedom Inside the Organization* (New York: McGraw-Hill, 1977), pp. 234–35.

34. Mordechai Mironi, "The Confidentiality of Personnel Records," *Labor Law Journal* 25 (May 1974): 289.

35. Catherine A. MacKinnon, *Sexual Harassment of Working Women* (New Haven, CT: Yale University Press, 1978), pp. 1, 57–82.

36. Ibid., p. 162.

37. Eva Kaplan-Leiserson, "Crossing the Line," *PE* Magazine (May/June, 2017). Available at: https://www.nspe.org/resources/pe-magazine/may-2017/crossing-the-line

38. Several articles and long excerpts from the rulings appeared in the *New York Times,* June 24, 2003, pp. A1 and A23–27.

39. Steven M. Cahn (ed.), *The Affirmative Action Debate* (New York: Routledge, 1995); George E. Curry (ed.), *The Affirmative Action Debate* (Reading, MA: Addison-Wesley, 1996).

40. Joseph DesJardins and Ronald Duska, "Drug Testing in Employment," *Business & Professional Ethics Journal* 6 (1987): 3–21.

41. This case is a summary of Cindee Mock and Andrea Bruno, "The Expectant Executive and the Endangered Promotion," *Harvard Business Review,* January-February 1994, pp. 16–18.

42. We adopt the fourth condition from Marcia P. Miceli and Janet P. Near, *Blowing the Whistle: The Organizational and Legal Implications for Companies and Employees* (New York: Lexington Books, 1992), p. 15.

43. Ernest Fitzgerald, *The High Priests of Waste* (New York: W. W. Norton, 1972); Berkeley Rice, *The C5-A Scandal* (Boston: Houghton-Mifflin, 1971).

44. John H. Fielder and Douglas Birsch (eds.), *The DC-10 Case* (Albany, NY: State University of New York Press, 1992); Paul Eddy, Elaine Potter, and Bruce Page, *Destination Disaster* (New York: Quadrangle, 1976); John Godson, *The Rise and Fall of the DC-10* (New York: David McKay, 1975); Vicki Golich, *The Political Economy of International Air Safety* (New York: St. Martin's Press, 1989), pp. 75, 115.

45. Adapted from Richard T. De George, "Ethical Responsibilities of Engineers in Large Organizations: The Pinto Case," *Business and Professional Ethics Journal* 1 (Fall 1981): 6.

46. Gene G. James, "Whistle Blowing: Its Moral Justification," in *Business Ethics,* 4th ed., ed. W. Michael Hoffman, Robert E. Frederick, and Mark S. Schwartz (Boston: McGraw-Hill, 2001), pp. 291–302.

47. Mike W. Martin, "Whistleblowing," chapter 9 of *Meaningful Work: Rethinking Professional Ethics* (New York: Oxford University Press, 2000).

48. Michael Davis, "Avoiding the Tragedy of Whistle-blowing," *Business and Professional Ethics Journal* 8 (1989): 3–19.

49. Peter Raven-Hansen, "Dos and Don'ts for Whistle-Blowers: Planning for Trouble," *Technology Review* 82 (May 1980): 44.

50. See the several essays and the cover story for "persons of the Year," in *Time Magazine* (December 30, 2002/January 6, 2003), pp. 8, 30–60.

51. Ibid., p. 42; Stephen H. Unger, *Controlling Technology: Ethics and the Responsible Engineer,* 2nd ed. (New York: Holt, Rinehart and Winston, 1992), pp. 179–81.

52. Kenneth Walters, "Your Employees' Right to Blow the Whistle," *Harvard Business Review* 53 (July 1975): 34; David W. Ewing, *Freedom Inside the Organization* (New York: McGraw-Hill, 1977), p. 113; Alan F. Westin (ed.), *Whistle-Blowing! Loyalty and Dissent in the Corporation* (New York: McGraw-Hill, 1981), pp. 163–64; James C. Petersen and Dan Farrell, *Whistle-Blowing* (Dubuque, IA: Kendall/Hunt, 1986), p. 20.

53. Stephen H. Unger, "How to Be Ethical and Survive," *IEEE Spectrum* 16 (December 1979): 56–57; Frederick Elliston, John Keenan, Paula Lockhart, and Jane van Schaick, *Whistle-Blowing Research: Methodological and Moral Issues* (New York: Praeger, 1985); Frederick Elliston, John Kennan, Paula Lockhart, and Jane van Schaick, *Whistle-Blowing: Managing Dissent at the Workplace* (New York: Praeger, 1985).

54. Carl Page, "Star Wars, Down but Not Out," *Newsletter of Computer Professionals for Social Responsibility* 14, no. 4 (Fall 1996). For additional cases see Robert Scheer, "The Man Who Blew the Whistle on 'Star Wars'," *Los Angeles Times Magazine,* July 17, 1988; and Karen Fitzgerald, "Whistleblowing: Not Always a Losing Game," *IEEE Spectrum,* December 1990, pp. 49–52.

55. Tim Healy, "Parallels between teaching ethics and teaching engineering," Markkula Center for Applied Ethics, Santa Clara University, Retrieved March 21, 2016, from https://www.scu.edu/ethics/focus-areas/more/technology-ethics/resources/teaching-ethics-andteaching-engineering/ (1997).

56. Kenneth Kipnis, "Engineers Who Kill: Professional Ethics and the Paramountcy of Public Safety," *Business and Professional Ethics Journal* 1 (1981): 82.

57. "Edgerton Case," Reports of the IEEE-CSIT Working Group on Ethics & Employment and the IEEE Member Conduct Committee in the matter of Virginia Edgerton's dismissal as information scientist of New York City, reprinted in *Technology and Society* 22 (June 1978): 3–10. See also *The Institute,* news supplement to *IEEE Spectrum,* June 1979, p. 6, for articles on her IEEE Award for Outstanding Public Service.

58. Peter Pae, "For Whistle-blowers, Virtue May Be the Only Reward," *Los Angeles Times,* June 16, 2003, pp. A-1 and A-14.

59. Frederick A. Elliston, "Anonymous Whistle-Blowing," *Business and Professional Ethics Journal* 1 (1982): 39–58.

60. Robert M. Anderson, Robert Perrucci, Dan E. Schendel, and Leon E. Trachtman, *Divided Loyalties* (West Lafayette, Indiana: Purdue University Press, 1980).

61. Gordon Friedlander, "Bigger Bugs in BART?" *IEEE Spectrum* 10 (March 1973): 32–37; Gordon Friedlander, "A Prescription for BART," *IEEE Spectrum* 10 (April 1973): 40–44.

62. Robert M. Anderson, Robert Perrucci, Dan E. Schendel, and Leon E. Trachtman, *Divided Loyalties* (West Lafayette, Indiana: Purdue University Press, 1980), p. 353.

63. Ibid., p. 351.

64. Susannah Grant, *Erin Brockovich: The Shooting Script* (New Market Shooting Script Series, 2001).

CHAPTER
7

HONESTY

In 1973 Spiro T. Agnew resigned as vice president of the United States amid charges of bribery and tax evasion related to his previous service as county executive of Baltimore County. A civil engineer and attorney, he had risen to influential positions in local government. As county executive from 1962 to 1966 he had the authority to award contracts for public works projects to engineering firms. In exercising that authority he functioned at the top of a lucrative kickback scheme.[1]

Lester Matz and John Childs were two of the many engineers who participated in the scheme. Their consulting firm was given special consideration in receiving contracts for public works projects so long as they made secret payments to Agnew of 5 percent of fees from clients. Even though their firm was doing reasonably well, they entered into the arrangement in order to expand their business, rationalizing that in the past they had been denied contracts from the county because of their lack of political connections.

Agnew, Matz, and Childs were dishonest in several ways. They were dishonest in communication, by engaging in deception of the public and other engineering firms. They were dishonest in matters of property, by accepting and extorting bribes that ultimately cost the public money and perhaps failed to provide quality services. Agnew was dishonest as a citizen, by cheating on his taxes. They may have even been intellectually dishonest—dishonest with themselves by engaging in self-deception and rationalizing their conduct as being excusable or even justifiable. And they were dishonest as engineers, betraying professional standards and participating unfairly in competitive practices.

We have discussed honesty in connection with topics such as avoiding conflicts of interest and maintaining confidentiality, as well as about corporate scandals such as Enron and WorldCom that eroded public trust in corporations, but the topic deserves fuller discussion. We have also discussed William LeMessurier, Roger Boisjoly, and other moral exemplars whose honesty was accompanied by great courage. We begin with the connection between the two main aspects of honesty, truthfulness and trustworthiness, and apply them to academic integrity—the starting point in becoming an ethical engineer. Then we discuss three contexts that raise special issues about honesty: conducting research, serving as an expert witness, and engaging in consulting engineering.

7.1 TRUTHFULNESS AND TRUSTWORTHINESS

7.1.1 Truthfulness

The standard of truthfulness in engineering is very high. It imposes what many consider an absolute prohibition on deception, and in addition it establishes a high ideal of seeking and speaking the truth.

Ethicists have devoted considerable attention to understanding the nuances of deception in everyday life. Most conclude that deception is sometimes a necessary evil, and, in moderation and prudence, can be a healthy part of living as a social being.[2] Sissela Bok, however, insists that our society has gone too far in creating a climate of dishonesty. She acknowledges the need for occasional lies, for example, to protect innocent lives, and for instances of withholding truths in order to protect privacy rights. Yet she urges us all to embrace what she calls the "principle of veracity": there is a strong presumption against lying and deception, although the presumption can occasionally be overridden by other pressing moral reasons in particular contexts.[3]

As we noted in chapter 1, professional life often requires that heightened importance be given to certain moral values, and that applies to truthfulness in engineering. Because so much is at stake in terms of human safety, health, and well-being, engineers are required and expected to seek and to speak the truth conscientiously and to avoid all acts of deception. To be sure, confidentiality requirements limit what can be divulged, but there is a much stronger presumption against lying than even Bok's principle of veracity.

Two of the six Fundamental Canons in the NSPE Code of Ethics concern honesty. Canon 3 requires engineers to "Issue public statements only in an objective and truthful manner," and Canon 5 requires them to "Avoid deceptive acts." We will refer to these requirements, taken together, as the *truthfulness responsibility:* Engineers must be objective and truthful and must not engage in deception. All other engineering codes set forth a statement of the truthfulness rule.

The truthfulness responsibility enters often into the cases discussed by the National Society of Professional Engineers (NSPE) in its *Opinions of the Board of Ethical Review*. Here are summaries of a few such cases, each of which the board viewed as violating the NSPE Code of Ethics.[4]

1. An engineer who is an expert in hydrology and a key associate with a medium-sized engineering consulting firm gives the firm their two-week notice, intending to change jobs. The senior engineer-manager at the consulting firm continues to distribute the firm's brochure, which lists them as an employee of the firm. (Case No. 90-4)

2. A city advertises a position for a city engineer/public works director, seeking to fill the position before the incumbent director retires in order to facilitate a smooth transition. The top candidate is selected after an extensive screening process, and on March 10 the engineer agrees to start April 10. By March 15 the engineer begins to express doubts about being able to start on April 10, and after negotiations the deadline is extended to April 24, based on the firm commitment by the engineer to start on that date. On April 23 the engineer says they have decided not to take the position. (Case No. 89-2)

3. An engineer working in an environmental engineering firm directs a field technician to sample the contents of storage drums on the premises of a client. The technician reports back that the drums most likely contain hazardous waste, and hence require removal according to state and federal regulations. Hoping to advance future business relationships with the client, the engineer merely tells the client the drums contain "questionable material" and recommends their removal, thereby giving the client greater leeway to dispose of the material inexpensively. (Case No. 92-6)

As these examples suggest, the truthfulness responsibility applies widely and rules out all types of deception. Certainly it forbids lying, that is, stating what one knows to be false with the intention of misleading others. It also forbids intentional distortion and exaggeration, withholding relevant information (except for confidential information), claiming undeserved credit, and other misrepresentations designed to deceive. And it includes culpable failures to be objective, such as negligence in failing to investigate relevant information and allowing one's judgment to be corrupted.

7.1.2 Trustworthiness

Exactly why is truthfulness so important, especially within engineering but also in general? One answer centers on respect for autonomy. To deceive other persons is to undermine their autonomy, their ability to guide their own conduct. Deceit is a form of manipulation that undermines their ability to carry out their legitimate pursuits, based on available truths relevant to those pursuits. In particular situations, this can cause additional harm. Deceivers use other people as "mere means" to their own purposes, rather than respecting them as rational beings with desires and needs. This amounts to a kind of assault on a person's autonomy. Bok makes the point in this way: "Deceit and violence—these are the two forms of deliberate assault on human beings. Both can coerce people into acting against their will. Most harm that can befall victims through violence can come to them also through deceit. But deceit controls more subtly, for it works on belief as well as action."[5]

Most moral theories defend truthfulness along these lines. Duty ethics, for example, provides a straightforward foundation for truthfulness as a form of respect for a person's autonomy. Rights ethics translates that idea into respect for a person's rights to exercise autonomy (or liberty). Rule-utilitarianism emphasizes the good consequences that flow from a rule requiring truthfulness. Virtue ethics affirms truthfulness as a fundamental virtue, and it underscores how honesty contributes to desirable forms of character for engineers, the internal good of the social practice of engineering, and the wider community in which that practice is embedded.

In addition, each of these ethical theories highlights additional wrongs in how deception harms others in specific ways. Dishonest engineering causes financial losses, injuries, and death. Especially important, violating the truthfulness responsibility undermines trust. As we have noted, honesty has two primary meanings: (1) truthfulness, which centers on meeting responsibilities about truth, and (2) trustworthiness, which centers on meeting responsibilities about trust. The meanings are interwoven because untruthfulness violates trust, and because violations of trust typically involve deception. In the Agnew case, the public's trust was violated and the public was deceived.

Engineering, like all professions, is based on exercising expertise within fiduciary (trust) relationships in order to provide safe and useful products. Untruthfulness and untrustworthiness undermine expertise by corrupting professional judgments and communications. They also undermine the trust of the public, employers, and others who must rely on engineers' expertise. Sound engineering is honest, and dishonesty is bad engineering.

7.1.3 Academic Integrity

Studies of colleges and universities reveal alarming statistics about academic integrity. According to one study, among schools lacking a strong honor code, three out of four students admitted to having engaged in academic dishonesty at least once during their college career.[6] Among the schools with an honor code, one in two students made the same admission.

Academic dishonesty includes dishonesty among students, faculty, and other members of academic institutions. Here we focus on students, and in the next section we discuss failures of integrity by researchers in academia and elsewhere. Academic dishonesty among students takes several forms.[7]

Cheating: intentionally violating the rules of fair play in any academic exercise, for example, by using crib notes or copying from another student during a test.

Fabrication: intentionally falsifying or inventing information, for example, by faking the results of an experiment.

Plagiarism: intentionally or negligently submitting others' work as one's own, for example, by quoting the words of others without using quotation marks and citing the source.

Facilitating academic dishonesty: intentionally helping other students to engage in academic dishonesty, for example, by loaning them your work.

Misrepresentation: intentionally giving false information to an instructor, for example, by lying about why one missed a test.

Failure to contribute to a collaborative project: failing to do one's fair share on a joint project.

Sabotage: intentionally preventing others from doing their work, for example, by disrupting their lab experiment.

Theft: stealing, for example, stealing library books or other students' property.

Why do engineering students engage in academic dishonesty such as cheating?[8] Studies in engineering education reveal a variety of motivations: inadequate job of the instructor teaching the course, unfairly written exams, unfair grading, too much material assigned, the perception that the instructor does not care about whether the student has learned the material. Faculty and administrators often think that faculty can serve as positive role models for students and students can benefit from observing faculty's ethical behaviors. However, students do not often find these role models helpful as they have observed faculty participating in unethical behavior or approving of unethical behavior.[9] These negative observation experiences of students may potentially become neutralizations that prevent them from making ethical judgments. We should also ask why students do not cheat, that is, why they meet standards of academic integrity. Here, too, there are many motives: the conviction that dishonesty is wrong and unfair; the conviction that cheating undermines the meaning of achievement; strict moral values; respect for the teacher; and fear of getting caught.

Explanation is one thing, justification is another. Is there any justification for cheating, or are proffered justifications simply rationalization—that is, biased and distorted reasoning? As authors, and like most educators, we take a firm stand. Academic dishonesty is a serious offense. It violates fair procedures. It harms other students who do not cheat by creating an undeserved advantage in earning grades. It is untruthful and deceives instructors and other members of an academic community. It violates trust—the trust of professors, other students, and the public who expect universities to maintain integrity. It undermines one's own integrity. And it renders dishonest and hollow any achievement or recognition based on the cheating.

Given the seriousness of academic dishonesty, and aware that we are all vulnerable to temptation, what can be done to foster academic integrity?[10] Researchers make several recommendations. Universities, as organizations, need to create and maintain a culture of honesty. Honor codes, which set forth firm standards and require students and faculty to report that cheating is going on, make a dramatic difference, even though they are not enough. Especially important, universities must support professors and students who follow university policies in reporting cheating, refusing to bow to the current market mentality in higher education that is more concerned about losing a paying "customer" than about ensuring academic integrity.

In addition, professors need to maintain a climate of respect, fairness, and concern for students. Course requirements and restrictions need to be explained clearly and then implemented. Tests and assignments need to be reasonable in terms of matching the material studied in class, and helpful feedback should be provided as the course progresses. Opportunities to cheat should be minimized. Firm and enforced disciplinary procedures are essential. Just as the Internet has made cheating easier, so has detecting plagiarism been made easier using new Web services.[11] And, we might add, teaching about academic integrity can be a valuable way to integrate an ethics component into courses.

DISCUSSION QUESTIONS

1. With regard to each of the three NSPE examples described earlier, discuss exactly what is at stake in whether the truthfulness responsibility is met. In doing so, identify the relevant rights, duties, and good and bad consequences involved.

2. Kermit Vandivier had worked at B. F. Goodrich for five years, first in instrumentation and later as a data analyst and technical writer. In 1968 he was assigned to write a report on the performance of the Goodrich wheels and brakes commissioned by the Air Force for its new A7-D light attack aircraft. According to his account, he became aware of the design's limitations and of serious irregularities in the qualification tests.[12] The brake failed to meet Air Force specifications. Upon pointing out these problems, however, he was given a direct order to stop complaining and write a report that would show the brake qualified. He was led to believe that several layers of management were behind this demand and would accept whatever distortions might be needed because their engineering judgment assured them the brake was acceptable.

   Vandivier then drafted a 200-page report with dozens of falsifications and misrepresentations. Yet, he refused to sign it. Later he gave as excuses for his complicity the facts that he was 42 years old with a wife and six children. He had recently bought a home and felt financially unable to change jobs. He felt certain that he would have been fired if he had refused to participate in writing the report.

   a. Assuming for the moment that Vandivier's account of the events is accurate, present and defend your view as to whether Vandivier was justified in writing the report or not. Which moral considerations would you cite in defending your view?

   b. Vandivier was a technical writer, not an engineer, to whom the truthfulness responsibility (as stated in an engineering code of ethics) applies. Does that matter morally? That is, would you answer section *a* of this question in the same manner, whether an engineer or technical writer were involved? Or does the applicability of a code of ethics alter the ethics of the situation?

   c. Vandivier's account of the events has been challenged. After consulting the record of congressional hearings about this case, John Fielder concluded that Vandivier's "claims that the brake was improperly tested and the report falsified are well-supported and convincing, but he overstates the magnitude of the brake's defects and, consequently, [exaggerates] the danger to the [test] pilot."[13] Comment on the difficulties in achieving complete honesty when Vandivier and other participants in such instances tell their side of the story. Also comment on the limitations and possible harm (such as to companies' reputations) in relying solely on the testimony of one participant.

3. A third-year engineering student has been placed on probation for a low grade point average, even though they are doing the best work they can. A concerned friend offers

to help by sitting next to them and "sharing" answers during the next exam. The engineering student has never cheated on an exam before, but this time they are desperate. What should they do?

4. A student receives a copy of a professor's previous midterm exam from a friend who took the class last year. (a) Is it ethical to accept the exam, without asking any questions? (b) The student decides to ask how the exam was obtained and learns that the professor required all copies of the exam sheet to be returned but had inadvertently missed this copy, which their friend circulated to selected other students. They decide to decline the exam, but do they have any additional responsibilities?

5. A university has an honor code but it is not taken seriously. Many students believe the administration cares more about avoiding bad publicity and not offending students (to keep enrollments high) than it does about deterring cheating. Does this mean that cheating is somewhat less blameworthy for students at the university?

6. A professor feels certain that a student cheated on an essay assignment by using an Internet site that sells essays, but the professor is either unable to prove it or unwilling to confront the student. Instead, the professor lowers the student's grade, both on the particular essay and at the end of the term. Is this procedure permissible? Why or why not?

7. What would you say to a professor who does not proctor their exams (nor have them proctored)? They believes that by staying in the room and taking basic steps to prevent cheating, it would signal a lack of trust in the students, and that trust is promoted by a willingness to show trust.

8. You are a professor who is asked to write a letter of recommendation for a student who is applying to graduate school. You know that letters of recommendation tend to be inflated, and you very much hope the student will be accepted in a good graduate program. Is it permissible to withhold negative information and accent positive information about the student?

7.2 RESEARCH INTEGRITY

Research in engineering takes place in many settings, including universities, government labs, corporations, and communities. The exact moral requirements vary somewhat, according to the applicable guidelines and regulations, but the truthfulness responsibility applies in all settings.

Research ethics has many facets, several of which we discuss: Defining research integrity and misconduct, conducting and reporting experiments, protecting research subjects, giving and claiming credit, and reporting misconduct.[14]

7.2.1 Excellence versus Misconduct

Truthfulness takes on heightened importance in research because research aims at discovering and promulgating truth. Research consists in trying to discover, express, and promulgate truth. Integrity in research is about promoting excellence (high quality) in these activities, and this positive emphasis on excellence should be kept paramount in thinking about honesty in research.

As Rosemary Chalk notes, an emphasis on quality and excellence in research broadens research ethics to include much more than avoiding fraud: "If shoddy research means sloppy science as well as dishonest exchanges, then a concern for quality in research may create incentives to address the type of wrongdoing that falls short of serious misconduct but still disturbs and creates a sense of unease and dissatisfaction with the practice of science"—and engineering.[15] This positive and broad understanding of research ethics invites dialogue about many neglected topics, such as leadership and enhancing organizational procedures, teamwork among researchers, fostering equal opportunity and diversity among researchers, environmental sensitivity (such as safety, conserving resources used in experimentation, and practicing recycling of materials used), and promote well-being of humans especially people living in underserved communities. An emphasis on excellence also invites greater attention to the personal commitments that promote creative endeavors.

The activity of reporting research is an important part of conducting research. Research results are useful when they are reported clearly, completely, in a timely manner, and honestly. Richard Feynman expressed the ideal of honesty in a famous commencement address at Caltech. In objecting to the superstition and pseudoscience rampant in our society, he highlighted

> a principle of scientific thought that corresponds to a kind of utter honesty—a kind of leaning over backwards. For example, if you're doing an experiment, you should report everything that you think might make it invalid—not only what you think is right about it: other causes that could possibly explain your results; and things you thought of that you've eliminated by some other experiment, and how they worked—to make sure the other fellow can tell they have been eliminated.[16]

These positive ideals and requirements of research ethics should be borne in mind as we attend to the concerns about research misconduct that rivet public attention. Indeed, even misconduct in research is given both wider and narrower definitions, developed in specific contexts and for different purposes. For example, if the purpose is to assure high-quality research, in all its dimensions, a wider definition might be adopted. Wide definitions typically emphasize honesty in conducting and reporting experiments, while also including theft, other misuses of research funds, and sexual harassment among researchers. If instead the purpose is to punish wrongdoers, a narrow and legalistic definition is likely to be favored.

It appears that narrow definitions tend to be favored by universities, corporations, and other groups whose members are liable to punishment for offenses, while government agencies have pushed for broader definitions. For example, the National Science Foundation (NSF) defines misconduct in science and engineering as:

> fabrication, falsification, plagiarism, or other serious deviation from accepted practices in proposing, carrying out, or reporting results from activities funded by NSF; or retaliation of any kind against a person who reported or provided information about suspected or alleged misconduct and who has not acted in bad faith.[17]

The clause "or other serious deviation from accepted practice" is controversial because it is vague and wide in scope, thereby causing understandable anxiety in the research community. It becomes less controversial when the word "intentional" is inserted before "deviation."

Historically, the clear-cut instances of scientific misconduct are intentional violations, as the nineteenth-century mathematician Charles Babbage emphasized. Babbage distinguished four types of deception and fraud in research.[18] *Forging* is deception intended to establish one's reputation as a researcher, whereas *hoaxing* is deception intended to last only for a while and then to be uncovered or disclosed, typically to ridicule those who were taken in by it. *Trimming* is selectively omitting bits of outlying data—results that depart furthest from the mean. His most famous category was *cooking,* a term still used today to refer to all kinds of selective reporting of results, falsifying of data, and massaging data in the direction that supports the result one prefers.

Although Babbage's emphasis on intentional misrepresentation is the most common way of defining research fraud, intent is sometimes difficult to prove. Moreover, what about *gross negligence,* in which a researcher unintentionally, but culpably, fails to meet the minimum standards for conducting and reporting research, and other forms of extreme incompetence? Most negligence results from lack of due care in setting up an experiment, for example, by failing to establish a reliable control group or failing to properly monitor an experiment as it unfolds. Negligence can also result from biases and even from self-deception, in which there is a purposeful evasion of the truth. These reflections lead to alternative definitions of research misconduct as violation of the basic standards for sound research.[19]

7.2.2 Bias and Self-Deception

At a hastily called news conference on March 23, 1989, the president of the University of Utah, Chase Peterson, made a stunning announcement.[20] A new and potentially limitless source of energy had been discovered by Stanley Pons, the chair of the university's chemistry department, and Martin Fleischmann, a Southhampton University professor who collaborated with Pons. Soon dubbed "cold fusion," the experiment outlined by Pons and Fleischmann was extraordinarily simple. It consisted of an electrochemical cell in which two electrodes, including one made of palladium, were immersed in a liquid containing the hydrogen isotope deuterium. According to Pons and Fleischmann, applying an electric current forced deuterium to concentrate in the palladium in a manner that caused hydrogen nuclei to fuse, thereby producing excess heat, radiation, and radiation byproducts such as tritium.

The Utah announcement generated frenetic research around the world, involving hundreds of researchers and tens of millions of dollars. Some researchers thought they confirmed the Pons-Fleischmann results, but there were also quick reversals, as hastily interpreted data were reexamined. The most careful experiments failed completely to replicate the results of the Pons-Fleischmann experiment. Today a small number of researchers continue to study cold fusion, but the consensus in the scientific community is that cold fusion does not occur.

There is also a consensus that the cold fusion episode is a cautionary tale about how bias and self-deception, bolstered by external pressures, can undermine sound research.

Pons and Fleischmann were well-respected electrochemists. They had made mistakes, and Fleischmann especially was known for his daring hypotheses that sometimes failed and other times succeeded dramatically. But mistakes and creative daring are integral to research. What was objectionable was their highly unorthodox step of announcing the results of research that had not yet been published in peer-reviewed journals. If they had simply published their results and allowed other researchers to confirm or refute their conclusions, we would be dealing with science as usual. Instead, they became so caught up in the excitement of extraordinary achievement, including the prospect of Nobel prizes, that they allowed their judgment to be distorted and their work to become careless.

Their failing lies somewhere between deliberate deception (fraud) and unintentional error (simple sloppiness). The term *self-deception* is often applied to them.[21] At the same time, there is disagreement about exactly what self-deception is.[22] According to one view, self-deception is motivated irrationality—that is, unreasonable belief that is motivated by biases and for which one is responsible. More fully, self-deception is allowing one's judgment to be biased by what one wants to believe and by one's emotions—especially wishes, hopes, self-esteem, and fears. The word *allowing* implies negligence, that is, the failure to take sufficient care to prevent biases from distorting one's thinking and observations.

According to another view, self-deception is sometimes motivated irrationality but other times it constitutes a more purposeful evasion. For example, researchers suspect an unpleasant reality, perhaps sensing that the data are going against what they want to believe. Then, instead of confronting the data honestly, they purposefully disregard the evidence or downplay its implications. The purpose and intention involved is typically unconscious or less than fully conscious.

We believe that both forms of self-deception—motivated irrationality and purposeful evasion—occur and probably occurred in the cold fusion case. The important point is that the truthfulness responsibility requires trying to overcome both forms of self-deception. That is what Feynman meant in speaking of "a kind of utter honesty—a kind of leaning over backwards" to adjust for possible bias and distortion. At the very least, Pons and Fleischmann fell short of rigor and caution when they announced their results to the press before submitting it to scrutiny by peers. As for Chase Peterson, who was himself a physician with an understanding of scientific protocols, he apparently allowed his judgment to be biased by his ambitions for his university, including fears that a rival Utah institution, Brigham Young University, was about to claim undeserved credit for similar research before his university made its announcement. Additional errors occurred elsewhere as competing universities made hasty attempts to confirm cold fusion.[23]

Individuals are influenced by their institutions, and institutions of higher education are under increasing economic pressures. Financially strapped universities now eagerly seek commercial ties, and corporations seek the expertise and prestige of university researchers. This combination can and often does lead to creative partnerships, but it also poses risks. Derek Bok, the former president of Harvard, warns that universities must find ways to place reasonable limits on commercial pressures in athletics, the marketing of products on campus, and especially in technology transfers from universities to industry.[24] The "commercialization of research" in universities threatens objective judgment by engineers and scientists in several areas: secrecy, conflicts of interest, attempts to manipulate research results, and other ethically questionable practices such as ghost authorship (e.g., a petroleum engineer who drafts a technical report or an article for an oil company is not credited for the work).

Corporations that fund research have a legitimate interest in keeping experimental results confidential until they have time to file for patents. But there are limits. Corporations continue to push for longer periods of secrecy and wider control over the amounts of information obtained. Universities need to exert counterpressure, to ensure the timely dissemination of knowledge, but in fact the promise of economic rewards to universities often leads them to submit to commercial pressures. Bok recommends that secrecy agreements should be limited to a few months following the end of experiments.

Conflicts of interest arise when individual researchers and universities themselves become heavily invested in lucrative research projects. One example is owning large shares of stock in companies for which one does research. An even more common example is doing research for companies that promise large additional funding if research results favor their products. Thus, it is no coincidence that 94 percent of researchers with ties to the tobacco industry found no harmful effects when they studied second-hand smoke, while only 13 percent of researchers without such ties reported similar results.[25] In this connection, Bok calls for stringent requirements for researchers to openly acknowledge even potential conflicts of interest, as well as limits on the extent to which they can be invested in companies for which they do research. Nevertheless, it is worth distinguishing *conflicts of interest* and *conflicts of commitment*. In contrast to conflicts of interest that refer to situations in which personal interests of engineers interfere with their professional judgment, conflicts of commitment are moments when engineers have to decide how to balance their time between their role as engineers and other social roles such as members of professional societies, employees, and family members. Strategies for tackling conflicts of commitment are different from those for dealing with conflicts of interest.

Finally, sometimes corporations exert direct pressures to influence research results. For example, a pharmaceutical company funded research aimed at showing its drug is superior to a cheaper generic drug.[26] Initially the researchers believed the experimental results would support the company's hopes, but in fact the results showed there was no difference in the effectiveness of the drug. When the researchers attempted to publish the result, the company accuses them of sloppy research and had their attorneys threaten legal action if they continue to seek publication of the results. Bok calls for greater university backing of responsible researchers who seek to safeguard the rigor of experimental procedures and interpretations.

7.2.3 Protecting Research Subjects

Research in fields like medicine and psychology extensively involve human beings as research subjects as well as animals. Research in engineering can also involve experimental subjects especially when it overlaps with biomedical research and other fields on human-technology interaction such as robotics. The standards for experimenters are extensive and detailed. The most helpful document specifying them is the book *On Being a Scientist,* which the National Academy of Engineering (NAE) codeveloped with other branches of the National Research Council (NRC).[27] In what follows we limit discussion to human subjects.

Experiments on humans are permissible only after obtaining the voluntary consent of human subjects. That means giving to experimental subjects (or their surrogate decision makers) all information about the risks, possible benefits, alternatives, exact procedures involved, and all other information a reasonable person would want to know before participating in an experiment. In addition, there must be no coercion, threats, or undue pressure. And the individual must have the capacity to make a reasonable decision about whether to participate. However, consent alone is insufficient to make a wrong act ethically acceptable. For instance, people may consent to be murdered or allow all their organs to be harvested for money for their families. Nevertheless, receiving consent from these people does not change the wrongness of murdering or organ trafficking.[28]

Special safeguards are taken when experimental subjects other than competent adults are the research subjects. When children participate in experiments, an appropriate surrogate decision maker, usually the parents, must give voluntary informed consent, and usually it is required that the child can reasonably be expected to benefit from the procedure. Despite that children may be too young to consent (e.g., making independent, rational, and autonomous decisions based on information provided by the researcher), they may be able to *assent,* which means that children are able to *express the willingness* to participate in studies.[29] If the child has expressed unwillingness to participate, the researcher should not include the child as a research participant, despite that the parent or guardian may express a strong interest in having the child involve in the study. Also, the absence of the child's objection to participating does not mean that the child assents to the study. Experimentation on institutionalized persons, for example in prisons or mental health facilities, is either forbidden or requires especially high standards. That is because of the inherently coercive nature of institutions that control all aspects of a person's life. Conducting research in other cultures can encounter extra ethical challenges especially in indigenous cultures or countries where no ethical review boards exist. Traditional research ethics frameworks are challenged by increasingly prevalent online research (driven by artificial intelligence and data science) conducted by Internet companies such as Facebook.[30]

The Nuremburg Code, written immediately after World War II, is the most important historical document requiring informed consent in research. It was developed in light of the Nazi horrors, and it has been flagrantly violated under authoritarian regimes. In addition, the United States has occasionally violated informed consent. During World War II, the U.S. government conducted biological, chemical, and nuclear experiments on unsuspecting individuals. Some instances were more egregious than others, but the following example illustrates the abuses that sometimes occurred.

> On March 24, 1945, a Black, 53-year-old cement worker named Ebb Cade had a car accident near Oak Ridge, Tennessee. Suffering from broken bones in his right arm and both legs, Cade was taken to the nearby Manhattan Project Army Hospital. Because Cade's injuries required several operations to properly set the bones, he was kept in the hospital for a few weeks. It was long enough, also, for Cade—code-named "HP [human product] 1"—to become the first of eighteen patients to be injected with plutonium, 4.7 micrograms.[31]

Even after the United States officially embraced the Nuremberg Code guidelines, there were occasional abuses. One of the worst was the Tuskegee Syphilis Study that was ended only in 1972, after four decades of exploitation of African-American men who were uneducated and earned low incomes. The study began with the aim of learning more about the progression of syphilis, thereby contributing to finding a cure for it. Informed consent was not obtained, and the experiments were continued for many years after penicillin was discovered to be an effective treatment. Members of the experimental control group were not informed of the discovery of penicillin nor given the option of taking it. At least 40 men died.

7.2.4 Giving and Claiming Credit

Often there are pressures on researchers to varnish the truth when competing for professional recognition, not only because it brings ego gratification but also because it might involve winning jobs, promotions, and income. Outright fraud of the following types also occurs.

Plagiarism, as defined earlier, is intentionally or negligently submitting others' work as one's own. In research, it is claiming credit for someone else's ideas or work without acknowledging it, in contexts where one is morally required to acknowledge it. The latter clause, about what is morally required, is important. In a novel, where footnoting is not customary, a brief quotation without quotation marks and a reference might be an acceptable "literary allusion," but in an essay by a student or professor it would be considered theft.

Failure to give credit occurs in many different settings within engineering, and the NSPE Board of Ethical Review frequently comments on them. In Case No. 92-1, for example, a city hires an engineer to design a bridge, and the engineer in turn subcontracts some key design work to a second engineer.[32] Months after the bridge is completed, the first engineer submits the design to a national design competition where it wins an award, but he fails to credit the work of the second engineer. As we might expect, the board ruled that such conduct violates the truthfulness rule. More recently, scholars in research ethics have been debating about a unique form of plagiarism "self-plagiarism," which refers to reusing or recycling one's own previously published text and assume it to be completely new.

Misrepresenting credentials is a second type of deception. Occasionally researchers forge credentials, creating widely publicized scandals. Fabrications about articles and credentials are relatively easy to uncover. Misrepresentations of credentials can take more subtle forms, however. One of the earliest cases discussed by the NSPE Board of Ethical Review (Case 79-5) was about an engineer who received a Ph.D. from a "diploma mill" organization that required no attendance or study at its facilities. The engineer then listed the degree on all professional correspondence and brochures. The NSPE board reasoned that listing a Ph.D., especially without listing where it is from, is widely understood to convey that it constitutes an earned doctorate, and that hence the engineer was indeed using unprofessional deception.

Misleading listing of authorship, whether of articles or other documents, is another area where subtle deception occurs. Obviously it is unethical to omit a coauthor who makes a significant contribution to the research. But the order of authors in many disciplines, including engineering, is also usually understood to convey information about the relative contributions of the authors, with the earlier listing indicating greater contributions. To be sure, customs vary—and on this topic respect for customs is important in order to ensure truthfulness. In some disciplines, the listing order is not considered important, for example, in mathematics, economics, and high energy physics where alphabetical listing is common, or in some sciences such as high energy physics where there can be dozens of coauthors. But in an engineering paper written by several coauthors, the order of names is significant.

7.2.5 Reporting Misconduct

There is a growing consensus that researchers have a responsibility to report misconduct by other researchers when the misconduct is serious and when they are in a position to document it. Yet typically there are strong pressures—from supervisors, colleagues, and others—not to report misconduct, and hence most instances fall into the category of whistleblowing. Measures to protect individuals who responsibly report research misconduct are being implemented at research facilities, and, as noted earlier, the concept of research misconduct now applies to punitive measures taken against these individuals. More needs to be done, however, and there is still a stigma against "turning in" a colleague.

"The Baltimore case" illustrates the conflicts that can arise.[33] The case is named after Nobel laureate David Baltimore, who coauthored a paper published in 1986 in the journal *Cell* reporting experiments concerning antibody production when foreign genes were inserted into mice. The lead investigator, MIT scientist Thereza Imanishi-Kari, was charged with manufacturing data and omitting other data from her report. Baltimore did not participate in the fraud. The case is famous, however, because of Baltimore's astonishing attempt to block investigation of the issue, as well as two universities' botched investigations.

Margot O'Toole, a Tufts-educated postdoctoral student working at MIT under Imanishi-Kari, identified the problem. First she notified several faculty members at Tufts, where Imanishi-Kari was applying for a job. Tufts conducted a cursory investigation based on experimental notes she provided (which turned out to be fabricated). O'Toole then reported the matter to the dean at MIT, who conducted an equally cursory investigation and dropped the matter. O'Toole persisted. She next confronted both Baltimore and Imanishi-Kari in a meeting attended by others, and where Imanishi-Kari admitted she had not accurately reported the experiment.

Baltimore insisted the problem was minor. When O'Toole mentioned the possibility of informing *Cell* about the problem, Baltimore said he would write an opposing letter. The meeting was so discouraging that O'Toole was ready to let the matter drop, but by then the National Institutes of Health (NIH) had informally heard of the matter and began investigating. Representative John Dingell, who chaired the House Subcommittee on Oversight and Investigations, was also notified and began an investigation. Baltimore dug in. He began a national campaign, calling on support from hundreds of other scientists, protesting government involvement in science, and directly attacking O'Toole. An incident of scientific misconduct escalated to a congressional inquiry, two investigations by the National Institutes of Health, and damage to the reputations of two distinguished universities. It caused great harm to a responsible scientist, Margot O'Toole, whose whistleblowing made it difficult for some time to find a job, but whose courageous action held a mirror up to the research community. As if all this were not complicated enough, a federal panel ultimately found no evidence of fraud by Imanishi-Kari. In the future, that community would have to engage in more honest investigation of charges of misconduct.

DISCUSSION QUESTIONS

1. In Case 95-7, the NSPE Board of Ethical Review discusses a case involving an engineer who served as project manager for a bridge built by a structural engineering firm, UVW Consultants. After completing the bridge, the engineer moves to a new company. He publishes an article about the bridge design in an international engineering journal, listing his affiliation with the new company. UVW Consultants is mentioned only at the end of the article, as "Engineer of Record." (a) Was the engineer untruthful and fail to give proper credit to UVW? (b) After you work out your initial answer, consult the website, www.onlineethics.org, to see how the NSPE board ruled on the case. Do you agree or disagree with their ruling?

2. You are a senior professor who mentored a younger colleague who is coming up for tenure and promotion next year. The colleague had assisted you with some suggestions on a research project, but was not sufficiently involved in the project to warrant coauthorship. In your view, the colleague clearly deserves tenure and promotion, but you know that last year a key committee turned down an equally deserving candidate. With

this in mind, you offer to add the colleague's name at the end of two articles about to be submitted for publication. Is this deception permissible? Compare and contrast how utilitarianism, duty ethics, rights ethics, and virtue ethics apply to the case.

3. You are part of a research team designing a new artificial heart. In an informal discussion, another team member mentions that they think the lead physician has been something less than fully truthful in explaining the risks of the new technology to potential experimental subjects. In particular, the lead physician conveyed much greater optimism than is warranted, by comparison with the alternatives available, probably inadvertently as part of their eagerness to have the device tested. What should you do?

4. The use of Agent Orange defoliants in Vietnam has only recently been officially recognized by the United States as a health hazard as former U.S. soldiers began to show symptoms of ill effects, long after scientists warned of its effects on farmers and their animals in the war zones of Vietnam. More recently, the use of depleted uranium to improve the penetration capacity of shells used in the Balkans and in the Middle East has raised the level of background radiation in these war zones. When, if ever, does a war justify exposing soldiers and noncombatants to substances that can affect humans in ways that can have long-term effects?

7.3 CONSULTING ENGINEERS

Consulting engineers work in private practice. They are compensated by fees for the services they render, not by salaries received from employers. Because of this, they tend to have greater freedom to make decisions about the projects they undertake. Yet their freedom is not absolute: They share with salaried engineers the need to earn a living.

We will raise questions in three areas where honesty plays a key role—advertising, competitive bidding, and contingency fees. We will also note how in safety matters consulting engineers may have greater responsibility than salaried engineers, corresponding to their greater freedom.

7.3.1 Advertising

Some corporate engineers are involved in advertising because they work in product sales divisions. But within corporations, the advertising of services, job openings, and the corporate image are left primarily to advertising executives and the personnel department. By contrast, consulting engineers are directly responsible for advertising their services, even when they hire consultants to help them.

Prior to a 1976 Supreme Court decision, competitive advertising in engineering, beyond the simple notification of the availability of one's services, was considered a moral issue and was banned by professional codes of ethics. It was deemed unfair to colleagues to win work through one's skill as an advertiser rather than through one's earned reputation as an engineer. It was also felt that competitive advertising caused friction among those in the field, lessened their mutual respect, and damaged the profession's public image by placing engineering on a par with purely money-centered businesses. However, the Supreme Court disagreed and ruled that general bans on professional advertising are improper restraints of competition. They keep prices for services higher than they might otherwise be, and they reduce public awareness of the range of professional services available, particularly from new firms.

The ruling shifted attention away from whether professional advertising as such is acceptable toward whether an advertisement is honest. Deceptive advertising normally occurs when products or services are made to look better than they actually are. This can be done in many ways, including: (1) by outright lies, (2) by half-truths, (3) through exaggeration, (4) by making false innuendos, suggestions, or implications, (5) through obfuscation created by ambiguity, vagueness, or incoherence, (6) through subliminal manipulation of the unconscious.[34] Another way is to impress with performance data that is meaningless because it has no reference standards.

There are notorious difficulties in determining whether specific ads are deceptive or not. Clearly it is deceptive for a consulting firm to claim in a brochure that it played a major role in a well-known project when it actually played a very minor role. But suppose the firm makes no such claim and merely shows a picture of a major construction project in which it played only a minor role? Or, more interestingly, suppose it shows the picture along with a footnote that states in fine print the true details about its minor role in the project? What if the statement is printed in larger type and not buried in a footnote?

As another example, think of a photograph of an electronic device used in an ad to convey the impression that the item is routinely produced and available for purchase, perhaps even "off the shelf," when in fact the picture shows only a preliminary prototype or mockup and the item is just being developed. To what extent, then, should the buyer—as the subject or participant of an "experiment" conducted by the manufacturer—be protected from misleading information about a product?

Advertisers of consumer products are generally allowed to suppress negative aspects of the items they are promoting and even to engage in some degree of exaggeration or "puffery" of the positive aspects. Notable exceptions are ads for cigarettes and saccharin products, which by law must carry health warnings. By contrast, norms concerning the advertising of professional services tend to be stricter in forbidding deception.

7.3.2 Competitive Bidding

For many years, codes prohibited consulting engineers from engaging in competitive bidding, that is, from competing for jobs on the basis of submitting priced proposals (as contrasted with a fee structure to be applied to the contract). However, in 1978 the Supreme Court ruled that professional societies were unfairly restraining free trade by banning competitive bidding. The ruling still left several loopholes, though. In particular, it allowed state registration boards to retain their bans on competitive bidding by registered engineers. It also allowed individual consulting firms to refuse to engage in competitive bidding. Thus fee competition where creative design is involved has remained a lively ethical issue. Is it in the best interests of clients and the public to encourage the practice?

If the use of competitive bidding is widely rejected by engineering firms, clients will have to rely almost exclusively on reputation and proven qualifications in choosing between them. This raises the problem of how the qualifications are to be determined in an equitable way. Is the younger, but still competent, consulting engineer placed at an unfair disadvantage? Or is it reasonable to view this disadvantage as justifiable, given the general importance of experience in consulting work?

7.3.3 Contingency Fees

Consulting engineers essentially make their own arrangements about payment for their work. Naturally this calls for exercising a sense of honesty and fairness. But what is involved specifically?

As one illustration of the kinds of problems that may arise, consider the following entry in the code of the National Society of Professional Engineers:

> Engineers shall not request, propose, or accept a commission on a contingent basis under circumstances in which their judgment may be compromised. (NSPE Code of Ethics, III, Sec. 6a.)

A contingency fee or commission is dependent on some special condition beyond the normal performance of satisfactory work. Typically, under a contingency fee arrangement, the consultant is paid only if she or he succeeds in saving the client money. Thus a client may hire a consultant to uncover cost-saving methods that will save 10 percent on an already contracted project. If the consultant does not succeed in doing so, no fee is paid. The fee may be either an agreed-upon amount or a fixed percentage of the savings to be realized.

In many contingency fee situations, the consultant's judgment may easily become biased. For example, the prospects of winning the fee may tempt the consultant to specify inferior materials or design concepts in order to cut construction costs. Hence the point of the NSPE code entry. But even allowing for this problem, is the thoroughgoing ban on such fees in the NSPE code warranted? There is, after all, a point to their use. They are intended to help stimulate imaginative and responsible ways of saving costs to clients or the public, and presumably this consideration deserves some weight.

Resolving the issue calls for balancing the potential gains against the potential losses that result from allowing or banning the practice. In this respect it is like many other issues in engineering ethics that call for reasonable judgments based on both experience and foresight. And ethical theories can be useful in making those judgments by providing a general framework for assessing the morally relevant features of the problems under consideration.

7.3.4 Safety and Client Needs

The greater amount of job freedom enjoyed by consulting engineers as opposed to salaried engineers leads to wider areas of responsible decision making concerning safety. It also generates special difficulties concerning truthfulness.

Very often, for example, consulting engineers have the option of accepting or not accepting "design-only" projects. In a design-only project, the consultant contracts to design something, but not to have any role, even a supervisory one, in its construction. Design-only projects are sometimes problematic because of difficulties encountered in implementing the designing engineer's specifications and because that engineer is often the only person well qualified to identify the areas of difficulty. For example, clients or contractors may lack adequately trained inspectors of their own. In fact, when novel projects are being undertaken, clients may not even know that their own inspectors are unsatisfactory. Again, a contractor may be unable or unwilling to spot areas where the original design needs to be modified to best serve the client. Thus, while the designer is often the person best able to ensure that the client's needs (and safety needs) are met, he or she may not be around to do so.

The importance of having the designer involved in on-site inspection is illustrated by the following example:

> An engineering firm designed a flood control project for the temporary retention of storm water in a nearby city. Included in the project were some high reinforced concrete retaining walls to support the earth at the sides of the retention basin. Although the consulting engineer had no responsibility for site visits or inspection, one of the designers decided to visit the site, using part of his lunch hour to see how it was progressing. He found that the retaining wall footings had been poured and the wall forms were placed. He was shocked, however, to find that the reinforcing steel extended from the footings into the walls only a small fraction of the specified distance. He immediately returned to his office and the client was notified of the situation. The inspectors responsible were disciplined and corrective measures taken with respect to the steel reinforcement. There is no question that in the first heavy rain the walls would have collapsed had the designer not discovered this fault. The result would have been heavy property loss, waste of resources, environmental damage and possible injury, even loss of life.[35]

It is thus important to determine when consulting engineers should accept design-only projects. And when they do accept them, are they not obligated to make at least occasional on-site inspections later, in order "to monitor the experiment" they have set in motion? That is, are there at least some minimal moral responsibilities in this context that reach beyond the legal responsibilities specified in the contract?

In the course of making on-site inspections, consulting engineers may notice unsafe practices that endanger workers. For example, they may notice an insufficient number of secondary support struts for a building or bridge. They may know from their experience that this could cause a partial collapse of the structure while construction workers are on the job. Of course, job safety is the primary responsibility of the contractor who has direct control over the construction. Yet for the consultant to do nothing would be negligent, if not callous. But how far do the consultant's responsibilities extend? Is a letter to the construction supervisor sufficient? Or is the consultant morally required to follow through by checking to see that the problem is corrected? It should be noted that an engineer who does point out construction deficiencies on one occasion, even if not contractually required to, but refrains from doing so on other occasions can be held liable for complicity in any damages that result from unreported deficiencies.

DISCUSSION QUESTIONS

1. Find three advertisements: one for a technological product, one for the services of a consulting engineer, and one for positions in an engineering firm. Critique each ad in terms of whether the information or pictures included are misleading or deceptive in any way (be specific).
2. State why you agree or disagree with the following positions regarding advertising in local telephone directories. The first three examples comprise Case No. 72-1 of the *NSPE Opinions of the Board of Ethical Review* (which you may wish to consult for NSPE's rulings).[36] The fourth is our adaptation of a case involving a dentist and the dentists' association in San Francisco.

    a. "Are boldface listings in the classified advertising section of local telephone directories consistent with the Code of Ethics?" The NSPE says no.

    b. "Are boldface listings in the regular section of local telephone directories consistent with the Code of Ethics?" The NSPE says yes.

    c. "Are professional card-type listings (set off by lines or blank space) in the classified section of local telephone directories consistent with the Code of Ethics?" The NSPE says no.

    d. Mr. Zebra is a consulting engineer whose firm, Zebra Associates, appears last in the telephone directory's classified listing of engineers. In order to gain a more advantageous position in the yellow pages and in other directories, he changes the name of his firm to Aardvark and Zebra. Aardvark is a purely fictitious partner. Is this ethical?

3. Is there anything unethical about the conduct of the engineer employee described in the following?

    A firm in private practice handles many small projects for an industrial client, averaging 20 to 30 projects a year. The firm has a signed agreement with the industrial client which does not obligate the client to give the firm any work, but does establish the respective responsibilities, terms of payment and other contractual details when the client does use the firm's services. The actual assignments are made by means of purchase orders referring to the agreement. An engineer employee of the firm resigns their employment and establishes their own firm and then actively solicits the industrial client of their former employer without any prior indication of interest by the client.[37]

4. Is the decision of the consulting firm in the following example the morally obligatory one?

    A large civil engineering consulting firm completed a comprehensive arterial highway plan for a large, midwestern metropolitan area. The area appropriated funds for the next phase of the program: the preparation of a design report covering the recommended highways. The officials concerned, not knowing what a reasonable fee would be for the design report work, felt that it was necessary to invite proposals from various consultants. The consulting firm explained that it could not participate if the selection were to be made on the basis of engineering fees. The officials replied that, although price would not be the only consideration, it would be a very important one. They could not explain to their constituents why the work was awarded to Engineer X if Engineer Y offered to do it for 2 percent less. The civil engineering company declined to participate.[38]

5. In the following case, are Doe's presentation and offer entirely ethical?

John Doe, P.E., a principal in a consulting engineering firm, attended a public meeting of a township board of supervisors which had under consideration a water pollution control project with an estimated construction cost of $7 million. Doe presented a so-called "cost-saving plan" to the supervisors under which his firm would work with the engineering firm retained for the project to find "cost-saving" methods to enable the township to proceed with the project and thereby not lose the federal funding share because of the township's difficulty in financing its share of the project.

Doe further advised the supervisors that his company contemplated providing his "cost-saving" services on the basis of being paid 10 percent of the savings; his firm would not be paid any amount if it did not achieve a reduction in the construction cost. Doe added that his firm's value engineering approach would be based on an analysis of the plans and specifications prepared by the design firm and that his operation would not require that the design firm be displaced.[39]

7.4 EXPERT WITNESSES AND ADVISERS

Engineers increasingly are asked to serve as consultants who provide expert testimony in adversarial or potentially adversarial contexts. Indeed, in one study of a federal court, it was estimated that 20 to 30 percent of cases involved significant scientific or engineering issues.[40] The focus of the case might be on the past, as in explaining the causes of accidents, malfunctions, and other events involving technology. Or the focus might be on the future, as in public planning, policy-making that involves technology, and the potential value of patents. Usually engineers are hired by one adversary in the dispute, and that raises special ethical concerns about their proper roles. Should they function as impartial seekers and communicators of truth, or do they essentially become "hired guns," paid to tell one side of the story? Without becoming hired guns, may engineers function as advocates for attorneys (and their clients), for public officials, or for private organizations who hire them, much like corporate engineers are advocates for their corporation's interests?

In between the hired gun and the disinterested observer lie a range of forms of advocacy that, while morally ambiguous and open to abuse, may be valuable to judges and juries, to planners and policymakers, and even to the general public. Moreover, the line between the role of the impartial analyst, who states and assesses facts, and the advocate, who makes recommendations about responsibility and preferable options, is not altogether precise, any more than the line between facts and values is crystal clear.

7.4.1 Expert Witnesses in the Courts

Let us begin with the court system, where engineers may be hired by either the plaintiff or the defense, usually in civil lawsuits but also in criminal proceedings. Some engineers serve only occasionally as expert witnesses, while others do so routinely and become specialists in forensic engineering: the application of engineering skills within the justice system.[41] Testimony may concern a wide variety of cases: defective products, personal injury, property damage, traffic accidents, or airplane crashes. At stake are opposing views of legal liability and competing economic interests, not to mention the reputations of corporations and professionals. Typically the main issue is who will be required to pay compensatory damages for injuries, loss of property, or violation of rights. In addition, there may be the issue of exemplary damages when rights were violated under circumstances of fraud, malice, or other wrongdoing.

At first glance, it might seem permissible for engineers to adopt an unqualified adversarial role within the legal system. After all, that is the role of attorneys, and attorneys hire engineers to serve the interests of their clients. Why not regard the engineer as an extension of the attorney, who is permitted and required to tell the side of the story favorable to the client? The opposing side can then hire its own expert to emphasize its side of the case.

Certainly engineers do have special responsibilities to those who hire them, in adversarial contexts as elsewhere. They have obligations to represent their qualifications accurately, to perform thorough investigations, and to present a professional demeanor when called to testify in court. Equally important, they have a responsibility of confidentiality, just as they do in other consulting and employee roles. They cannot divulge the contents of their investigations to the opposing side of a controversy until required to do so by the courts or by the attorney who hired them. Perhaps most important, when called as witnesses they are not required to volunteer evidence favorable to the other side. They must answer questions truthfully, but it remains the responsibility of the attorney for the opposing side to ask pertinent questions. In these respects, they are not entirely disinterested participants in adversarial proceedings.

It does not follow that they will function as mouthpieces paid to slant the truth according to who pays them. Their primary responsibility is to be objective in discovering the truth and communicating it honestly, as honesty is understood within the court system. The appropriate role of expert witnesses is not determined in the abstract, but instead depends on the shared understanding created within society. In particular, the role must be understood in terms of the aims of a (morally justified) legal system, consistent with professional standards (as promulgated in codes of ethics).

What is the aim of our legal system? We like to think of the aim as discovering the truth about events that disputing parties perceive differently. Guided by attorneys trained in court procedures, each conflicting party presents its viewpoint, and then a judge or a jury arrives at a verdict about the truth. In fact, however, the primary purpose of the court system is to administer a complex system of legal rights that define legal justice. Legal justice is achieved through adversarial relationships, with sophisticated rules about admissible forms of evidence and permissible forms of testimony that may not in each situation ideally serve the truth.

What role, then, do the courts give to expert witnesses? We can easily imagine alternative judicial systems that would give expert witnesses wide latitude to emphasize one side of a case in adversarial contexts, regarding them as playing roles much like attorneys. The courts have chosen not to do so, however. Given the complexity of modern technology and science, the court system must rely on experts who earnestly try to be impartial in identifying and interpreting complicated data. Ideally, expert witnesses would be paid by the courts, rather than opposing attorneys, in order to counter potential biases. In practice, the high costs require that the parties to disputes pay for consultants. The possible biases resulting are then balanced by allowing both sides to hire consultants and then making it the responsibility of opposing attorneys to cross-examine expert witnesses.

The legal system distinguishes between eyewitnesses and expert witnesses. Eyewitnesses testify in matters of perceived facts, whereas expert witnesses are permitted wider latitude in testifying on facts in their areas of expertise, on interpreting facts (especially in terms of cause-effect relationships), in commenting on the views of the opposing side's expert witnesses, and in reporting on the professional standards—especially the standard of care applicable at the time of making a product or providing a service.[42] The role of expert witnesses is to identify the truth about the causes of accidents, not to directly serve attorneys' clients. Although they work within an adversarial context, expert witnesses are not themselves adversaries, at least not to the degree attorneys are.

Attorneys hire and pay engineers for their services in impartially investigating the truth; they do not pay them to testify in a way favorable to their clients. Indeed, engineers who slant the truth may do great harm to attorneys who hire them. Attorneys need an objective appraisal of the facts in order to prepare the best defense for their clients. At the same time, in adversarial contexts there are both subtle and not-so-subtle pressures to distort the truth.

Codes of ethics have only recently begun to clarify the roles of engineers in adversarial contexts, and as a result there has been little shared understanding about the appropriate role.[43] There is wide consensus that engineers must not become "hired guns" who engage in outright lies and distortions according to who pays their consulting fee. There are moral nuances, however, in what the responsibility of impartiality requires within adversarial contexts. One must not lie, but how far may one go in shading the truth by interpreting facts in a manner favorable to one's client, or even withholding information from the opposing side? Not surprisingly, there are many abuses.

7.4.2   Abuses

HIRED GUNS. The most flagrant abuse is the unscrupulous engineer who makes a living by not even trying to be objective, but instead in helping attorneys to portray the facts in a way favorable to their clients. A small minority of engineers do become hired guns who violate the standards of honesty and due care in conducting investigations. Unfortunately, this minority has tainted the entire practice of serving as an expert witness.

Consider a simple case. A roofer falls while climbing down a wooden ladder and is seriously injured.[44] The roofer sues the manufacturer of the ladder for medical costs and lost wages. Witnesses of the accident offer conflicting testimony about whether the accident was caused by a crack in the ladder, raising the question of a product defect, or by the carelessness of the roofer who was descending the ladder too quickly and perhaps caused the ladder to crack when falling on it. A structural engineer hired by the manufacturer writes a report favoring the manufacturer, selecting and emphasizing facts in the opposite way the engineer would have done if hired by the plaintiff's attorney.

The engineer acted improperly. A truthful report would express the best personal judgment of the engineer, and presumably that judgment should yield the same report whether it was paid for by the plaintiff or defense attorney. Suppose, however, that the engineer states at the outset of the report their intentions to accent one side of the case. Would that render the report acceptable? Whatever our answer, such a report would be useless in court, since a judge or jury would immediately dismiss it as a distortion. Hence the enormous pressure for abuse: The role of the expert witness is to be impartial, but there is the temptation to tell one side of the story in order to earn a living as a hired gun.

The most common abuses involve more subtle biases resulting from money, ego, and sympathy.[45]

FINANCIAL BIASES. Merely being paid by one side can exert some bias, however slight. This bias might influence one's investigations, testimony, and even the presentation of one's qualifications. Obviously the bias would increase substantially if engineers were hired on the basis of contingency fees paid only if the case is won. Attorneys are permitted to accept contingency fees because the fees are believed to strengthen their determination to serve their clients. But contingency fees in adversarial contexts would tend to bias the judgment of expert witnesses. That is why they are unethical, even if they are sometimes permitted by law.

Money biases in more subtle ways. Forensic engineers who make all or much of their living from serving as expert witnesses will naturally tend to want the attorneys who hire them to win their cases. If the engineer gains a reputation for helping to win cases, then future income (for the same or other interested attorneys) will be more likely.

EGO BIASES. Most of us know from experience that adversarial situations evoke competitive attitudes that can influence judgment. Engineers can easily be influenced by identifying with their "own" side of the dispute. The other side comes to be seen as the guilty party, and one's own side as the innocent victim. There is also a combination of desires to serve the interests of one's client and to be well regarded by the client.

SYMPATHY BIASES. The courts are filled with human drama in which people's suffering is all too poignant. It is easy to identify with the plight of victims. Indeed, one may feel great sympathy for the opposing attorney's client. Such biases are capable of upsetting a purely disinterested investigation of the facts.

To overcome these biases, engineers must make a special effort to maintain their integrity when serving as expert witnesses. Given the risks of unintentional biases and purposeful self-deception, the courts must also rely on the balance provided by having expert witnesses on both sides of the case, combined with the responsibility of opposing attorneys to examine expert witnesses for any possible biases.

7.4.3 Advisers in Planning and Policy-Making

We turn now to the role of expert advisers in public policy-making and planning, a role played by engineers as well as economists, sociologists, urban planners, and other professionals. Technology is always involved in decisions about public policy-making (forming general strategies for society) and public planning (forming projects that affect communities). In policy-making, public officials and the general public need objective studies about the costs and benefits of alternative systems of transportation, housing, energy use, land use, and national defense. In planning, they need expert advice about the feasibility, risks, and benefits of particular technological projects that affect local communities. For that reason, many laws and government policies have been adopted that require objective studies before public funds are committed to projects.

Is it realistic to expect impartiality from engineers who work as consultants in conducting these studies, especially within the highly charged atmosphere of public policy-making? Consider the debate over whether nuclear energy should be expanded or whether the government should invest taxpayers' money in supporting traditional energy sources such as fossil fuels and alternative forms such as wind and solar energy. Engineers hired by pronuclear corporations or antinuclear groups will invariably feel pressure to accent one side of the case. Sometimes there is direct and overt pressure applied to write a report pleasing to their clients. More often, they simply wish to please their clients, if only because of the hope of additional work in the future and the wish to be respected by one's clients. The pressures of money and ego, and a desire to serve the client's needs, play a role here as they do in serving as an expert witness in court.

Even if engineers manage to set aside their own political and self-serving economic biases, as in most situations they must try to do, there are additional factors that make it easy for them to bend the truth in one direction.[46]

TECHNICAL COMPLEXITY AND THE NEED FOR ASSUMPTIONS. The scale of public policy decisions can be immense, with considerable resources, potential benefits, and uncertainties involved. A variety of assumptions must be made, including highly controversial ones. In looking to the future, there is usually a higher degree of uncertainty than in forensic investigations of past failures. This invites each adversary in political controversies to accent assumptions and estimates favorable to its case, all the while appearing to be in good faith.

For example, with regard to energy decisions, some assumptions concern the extent of demands for future energy, and hence assumptions about population increases and lifestyle. Other assumptions involve economic estimates about the projected costs of developing alternative forms of energy. Still others involve political assumptions about the risks acceptable to the public. As a result, clients will always pressure engineers to limit studies to assumptions favorable to them. In addition, some bias arises merely by stipulating the range of options that the client is willing to pay to have studied. Entirely objective studies would take pains to make forecasts under a range of alternative assumptions, but often clients are not willing to pay for such exhaustive studies and instead pressure engineers to use a set of assumptions favorable to their side of the case. If engineers are completely forthright in their reports about the restrictions imposed by these assumptions, their reports might be impartial only within limits.

DIFFUSED RESPONSIBILITY. The usual sharing of responsibility within corporations is multiplied in public policy-making. Policy forecasts are usually made by consulting corporations that work for government or other corporations and ultimately must make their case in the arena of public opinion. As a result it is easy to rationalize and to pass the buck in thinking about personal responsibility for complete impartiality. Corporate managers and engineers might tell themselves that it is the responsibility of the public, through its officials and public referendums, to make adjustments for corporate partiality in studies. Politicians can rationalize how they exert pressure for studies favorable to their positions on the grounds that the studies will help them make their case for funding projects they believe are in the public interest. Then, if things go wrong, and overly optimistic estimates result in huge cost overruns that the public must pay, politicians easily blame the consultants for failing to be sufficiently impartial.

What, then, is the role of engineering consultants in policy matters? Is it to chart all realistic options, carefully assessing each, and doing so under a range of alternative assumptions about future contingencies? Or is it permissible for them to base their studies on particular assumptions about future contingencies that are favorable to their client's case? As in the context of court testimony, the answer turns largely on the shared understanding about the role of policy analysts. In particular, it turns on a shared understanding about how to balance potentially conflicting responsibilities both to clients and to the general public. We can distinguish three normative (value-laden) models for how to balance these responsibilities.

HIRED GUNS. This model makes the obligation to clients paramount, if not exclusive. Studies conform to clients' wishes, whatever they may be. Facts favorable to the client are dramatically highlighted and unfavorable facts downplayed. Assumptions about uncertainties are slanted in a direction favorable to the client's case. The responsibilities to the public are regarded as the minimal ones of avoiding outright lies, fraud, and direct harm.

VALUE-NEUTRAL ANALYSTS. This model insists that engineers should be completely impartial. Not only should they conscientiously avoid any taint of bias and favoritism, but they should avoid any form of advocacy. Their role is to identify all options and analyze the factual implications of each option. If they engage in weighing options, in particular by making cost-benefit analyses, they do so according to value criteria that are stipulated by someone else and made explicit and overt.

VALUE-GUIDED ADVOCATES. According to this model, engineering consultants may adopt partisan views in controversial issues, but they remain honest and independent in their professional judgment. Unlike value-neutral analysts, they understand that values are interwoven with facts, and they also affirm the help provided by value-oriented technological studies. Unlike hired guns, value-guided advocates make their responsibility to the public paramount and maintain honesty about both technical facts and the values that guide their studies. Even though reports are value-laden, they are written in essentially the identical way they would be if one's client came from the opposing side of a controversy.

Rosemarie Tong defends this last model as often the most appropriate way of thinking about policy analysts.[47] Engineers and other policy analysts stand in a fiduciary relationship (a relationship of trust) with their clients, yet they also have responsibilities to the public. Both clients and the public often stand in need of more than recitations of facts. They need help and guidance from technical advisers who make recommendations in the spirit of meeting these needs. Such recommendations, however, must express the best judgment of independent experts who have moral integrity, not the distortions of hired guns.

As Tong notes, honesty is essential, both in the negative sense of avoiding deception and in the positive sense of being candid in stating all relevant facts and in being truthful in how the facts are interpreted. In addition to honesty about the technical data, it is also essential to be honest about one's role and about the values guiding one's study. One tries to recognize and reveal any political, economic, and social values that influence one's recommendations.

DISCUSSION QUESTIONS

1. Expert witnesses are morally and legally obligated not to lie, and hence not to withhold information requested under cross-examination from opposing attorneys. Yet expert witnesses are not legally required in all situations to volunteer information and opinions that damage the case of the attorney who hired them. For example, an engineer might know of a test that could shed further light in a liability case, but which was not conducted because of its prohibitive cost. The engineer is required to state this fact if directly questioned but is not required to volunteer the information otherwise. This opens the door to expert witnesses being trained in subtle ways to avoid being led by the opposing side into being asked directly to reveal information damaging to their attorney's case. Should the legal system be modified to require expert witnesses hired by one party in the trial to produce all relevant information about a case with complete impartiality?

2. A forensic engineer hired by an attorney conducts an impartial study that turns out not to support the attorney's case. The attorney pays the engineer and then terminates the engineer's services. Later the engineer is approached by the opposing counsel and asked to serve as an expert witness on the other side of the case. The engineer agrees. Did the engineer do anything wrong?[48]

3. Martin Wachs once asked the president of a large consulting firm why the firm repeatedly underestimated the costs of large public works projects. Wachs suspected that the firm was acting to please government officials who were eager to gain public approval for their projects from funding agencies. Comment on the following reply made by the president.

> He said that he was against the wasting of public funds, and consequently he saw it as his moral duty to estimate that projects would cost less than their critics thought they would. By underestimating project costs he insisted that he was providing public officials with an incentive to meet those low cost estimates and thereby to save the public's money. "Higher cost estimates," he said, "would merely be an incentive for wasteful contractors to spend more of the taxpayers' money."[49]

4. A transportation engineer has been hired as a consultant by a large development firm to make a study of the feasibility of a proposed toll road. The engineer quickly learns that the toll road would have a very negative impact, especially in terms of pollution and economics, on the lives of a low-income rural population and that the developers had no intention of divulging that information during public hearings. The engineer believes the rural population has a right to informed consent and also that the road places an unfair burden on them.[50] Is it all right for the engineer to say nothing and continue with the study? Does confidentiality require saying nothing?

5. In 1969 Daniel D. McCracken and some 800 other computer specialists sent a petition to Congress that argued against the technical feasibility of the antiballistic missile (ABM) system, whose funding was then being debated. Most participants identified their affiliation in professional societies, and it was clear they were trying to influence Congress more strongly than they could have in their role as ordinary citizens. Similar events have occurred with "Star Wars"—the Strategic Defense Initiative—and other major military initiatives. Is there anything morally suspect about such petitions, especially when the intent is to use technical arguments against systems that the participants view as unethical on other grounds (such as that ABM systems would escalate the nuclear arms race)? Specifically, do you agree or disagree with the following two criticisms that were raised against McCracken: (a) "[W]hen presumed or actual expertise in one area is employed toward coercion of judgment in another area, I must object. The subtle shift in your [McCracken's] discussion from 'we don't think it can be done' to 'we don't think it ought to be done' is irresponsible in the extreme." (b) "I think you and a lot of other respected people are on soft ground both philosophically and politically. . . . I am dubious about technical judgments made by people who don't have access to the determinative information" that only the military supposedly had fully available then.[51]

KEY CONCEPTS

—**Principle of veracity** (for everyday life): there should be a strong presumption against lying and deception, although the presumption can occasionally be overridden by other pressing moral reasons in particular contexts.

—**Truthfulness responsibility** (for engineers): Engineers must be objective and truthful, and they must not engage in deception.

—**Academic integrity:** Maintaining standards of honesty (truthfulness and trustworthiness) and avoiding cheating, fabrication, plagiarism (intentionally or negligently submitting others' work as one's own), facilitating dishonesty in others, misrepresentation, not doing one's fair share on collaborative projects, sabotage, and theft.

—**Research misconduct:** violation of the basic standards for sound research, for example, by intentional fabrication, falsification, plagiarism, or (in wider definitions) by gross negligence.

—**Self-deception:** fooling oneself, either (a) motivated irrationality in which one allows biases to distort judgment, or (b) purposeful (though not fully conscious) evasion of unpleasant realities such as data that goes against what one wants to believe.

—**Informed consent** (in experimentation): (a) the requirement of researchers to give to experimental subjects (or their surrogate decision makers) all information about the risks, possible benefits, alternatives, exact procedures involved, and all other information a reasonable person would want to know about the experiment, (b) the absence of coercion, threats, or undue pressure, and (c) the capacity of the experimental subject (or their surrogate) to make a reasonable decision.

—**Giving credit in research:** being truthful in reporting research, especially by avoiding plagiarism, misrepresentations of credentials, and improper listing of authors.

—**Honesty in consulting engineering:** requires maintaining truthfulness in such areas as advertising, competitive bidding, fee arrangements, and reporting safety infractions, in addition to expected truthfulness for technical opinion.

—**Honesty as expert witnesses and advisors:** requires maintaining objectivity while working in adversary contexts; avoiding distortions in judgment from financial arrangements, ego, sympathy, and political interests.

REFERENCES

1. Richard M. Cohen and Jules Witcover, *A Heartbeat Away: The Investigation and Resignation of Vice President Spiro T. Agnew* (New York: Viking, 1974).
2. For example, see David Nyberg, *The Varnished Truth* (Chicago: University of Chicago Press, 1992); Loyal Rue, *By the Grace of Guile* (New York: Oxford University Press, 1994); and Jeremy Campbell, *The Liar's Tale: A History of Falsehood* (New York: W. W. Norton, 2001).
3. Sissela Bok, *Lying: Moral Choice in Public and Private Life* (New York: Vintage Books, 1979), p. 32.
4. National Society of Professional Engineers, *Opinions of the Board of Ethical Review*, vol. 7 (Alexandria, VA: NSPE, 1994); also see the website at www.nspe.org/eh-home.asp.
5. Sissela Bok, *Lying: Moral Choice in Public and Private Life,* p. 19.
6. D. L. McCabe and L. K. Trevino, "Academic Dishonesty: Honor Codes and Other Contextual Influences," *Journal of Higher Education* 64 (1993): 522–38.
7. This is a slightly modified version of the categories assembled by Bernard E. Whitley, Jr. and Patricia Keith-Spiegel, *Academic Dishonesty: An Educator's Guide* (Mahwah, NJ: Lawrence Erlbaum Associates, 2002), p. 17.
8. Donald D. Carpenter, Trevor S. Harding, Cynthia J. Finelli, Susan M. Montgomery, and Honor J. Passow, "Engineering Students' Perceptions of and Attitudes Towards Cheating," *Journal of Engineering Education* 95, no. 3 (2006): 181–94.
9. Matthew A. Holsapple, Donald D. Carpenter, Janel A. Sutkus, Cynthia J. Finelli, and Trevor S. Harding, "Framing Faculty and Student Discrepancies in Engineering Ethics Education Delivery," *Journal of Engineering Education* 101, no. 2 (2012): 169–86.

10. Bernard E. Whitley, Jr. and Patricia Keith-Spiegel, *Academic Dishonesty: An Educator's Guide* (Mahwah, NJ: Lawrence Erlbaum Associates, 2002), pp. 43–155.

11. Wilfried Decoo, *Crisis on Campus: Confronting Academic Misconduct* (Cambridge, MA: The MIT Press, 2002), pp. 44–45, 49–50.

12. Kermit Vandivier, "Engineers, Ethics and Economics," in *Conference on Engineering Ethics* (New York: American Society of Civil Engineers, 1975), pp. 20–34.

13. John Fielder, "Give Goodrich a Break," *Business and Professional Ethics Journal* 7 (1988): 3–25; "Air Force A7-D Brake Problem," Hearing before the Subcommittee on Economy in Government of the Joint Economic Committee, Congress of the United States, Ninety-First Congress, First Session, August 13, 1969. LC card 72-606996.

14. In what follows we are influenced by Caroline Whitbeck's excellent discussion of the ethics of research in engineering in *Ethics in Engineering Practice and Research* (New York: Cambridge University Press, 1998), especially chapters 6, 7, and 9.

15. Rosemary Chalk, "Integrity in Science: Moving into the New Millennium," *Science and Engineering Ethics* 5, no. 2 (1999): 181.

16. Richard P. Feynman, *Surely You're Joking, Mr. Feynman!* (New York: Bantam, 1986), p. 311. Also see Richard P. Feynman, *The Meaning of It All* (Reading, MA: Perseus Books, 1998), pp. 106, 109–10.

17. Caroline Whitbeck insightfully discusses the controversy about this definition in *Ethics in Engineering Practice and Research* (New York: Cambridge University Press, 1998), p. 201.

18. Charles Babbage, "Reflections on the Decline of Science in England and on Some of Its Causes," in M. C. Kelly (ed.), *The Works of Charles Babbage,* vol. 7 (London: Pickering, 1989), pp. 90–93.

19. David H. Guston, "Changing Explanatory Frameworks in the U.S. Government's Attempt to Define Research Misconduct," *Science and Engineering Ethics* 5, no. 2 (1999): 137–54.

20. Gregory N. Derry, *What Science Is and How It Works* (Princeton: Princeton University Press, 1999), pp. 174–80; John R. Huizenga, *Cold Fusion: The Scientific Fiasco of the Century* (New York: Oxford University Press, 1993); Gary Taubes, *Bad Science: The Short Life and Weird Times of Cold Fusion* (New York: Random House, 1993); Bart Simon, *Undead Science: Science Studies and the Afterlife of Cold Fusion* (New Brunswick, NJ: Rutgers University Press, 2002).

21. See, for example, John R. Huizenga, *Cold Fusion,* pp. 201–14, who also dubs cold fusion an instance of "pathological science," a term that is even less pellucid than "self-deception."

22. The substantial literature on self-deception includes Mike W. Martin, *Self-Deception and Morality* (Lawrence, KS: University Press of Kansas, 1986); Brian P. McLaughlin and Amelie Oksenberg Rorty (eds.), *Perspectives on Self-Deception* (Berkeley, CA: University of California Press, 1988); Herbert Fingarette, *Self-Deception* (Berkeley: University of California Press, 2000); and Alfred R. Mele, *Self-Deception Unmasked* (Princeton: Princeton University Press, 2001).

23. For example, at Texas A&M an inquiry concluded that sloppy lab procedure rather than fraud was most likely the explanation for erroneous findings of tritium during cold fusion experiments. See Robert Pool, "Cold Fusion at Texas A&M: Problems, But No Fraud," *Science* 250 (December 14, 1999): 1507–8.

24. Derek Bok, *Universities in the Marketplace: The Commercialization of Higher Education* (Princeton: Princeton University Press, 2003), especially pages 57–78, 139–56. (Incidentally, Derek Bok is married to Sissela Bok.)

25. Ibid., p. 76.

26. Ibid., p. 72.

27. The National Research Council's Committee on Science, Engineering and Public Policy, *On Being a Scientist*, 2nd ed. (Washington, DC: National Academy Press, 1992).

28. Heather Widdows, *Global Ethics: An Introduction* (Abington: Routledge, 2011), pp. 210–11.

29. Caroline Whitbeck, *Ethics in Engineering Practice and Research*, 2nd ed. (New York, NY: Cambridge University Press, 2011), p. 317.

30. Anna Lauren Hoffmann and Anne Jonas, "Recasting Justice for Internet and Online Industry Research Ethics," in Michael Zimmer and Katharina Kinder-Kurlanda (eds.), *Internet Research Ethics for the Soical Age,* (New York: Peter Lang, 2017), pp. 3–19.

31. Jonathan D. Moreno, *Undue Risk: Secret State Experiments on Humans* (New York: Routledge, 2001), p. 120.

32. National Society of Professional Engineers, *Opinions of the Board of Ethical Review*, vol. VII (Alexandria, VA: National Society of Professional Engineers, 1994), p. 61.

33. Philip J. Hilts, "The Science Mob: The David Baltimore Case—And Its Lessons," *The New Republic,* May 18, 1992, pp. 25, 28–31. Reprinted in Deni Elliott and Judy E. Stern (eds.), *Research Ethics: A Reader* (Hanover, NH: University Press of New England, 1997), pp. 43–51; Tim Beardsley, "Profile: Thereza Imanishi-Kari," *Scientific American* 275 (November, 1996): 50–52.

34. Burton M. Leiser, "Truth in the Marketplace: Advertisers, Salesmen, and Swindlers," in Burton M. Leiser (ed.), *Liberty, Justice and Morals,* 2nd ed. (New York: Macmillan, 1979), pp. 262–97.

35. Frank E. Alderman and Robert A. Schultz, "Ethical Problems in Consulting Engineering," p. 11 (unpublished manuscript). Quotations in text used with permission of the authors.

36. *NSPE Opinions of the Board of Ethical Review,* vol. IV (Washington, DC: National Society of Professional Engineers, 1994). Quotations in the text are used with permission of NSPE.

37. *NSPE Opinions of the Board of Ethical Review,* vol. IV, Case 73–7, 1976.

38. Philip L. Alger, N. A. Christensen, and Sterling P. Olmsted, *Ethical Problems in Engineering* (New York: John Wiley and Sons, 1965), p. 49. Quotation used with permission of the publisher.

39. *NSPE Opinions of the Board of Ethical Review,* Case 77–10. *Professional Engineer,* vol. 48 (June 1978), p. 52.

40. R. A. Meserve, "Science in the Courtroom," in *Trying Times: Science and Responsibilities after Daubert,* ed. Vivian Weil (Chicago, IL: Illinois Institute of Technology, 2001), p. 23.

41. Joseph S. Ward, "What Is a Forensic Engineer?" in Kenneth L. Carper (ed.), *Forensic Engineering: Learning from Failures* (New York: American Society of Civil Engineers, 1986), pp. 1–6.

42. Richard C. Vaughn, *Legal Aspects of Engineering,* 3rd ed. (Dubuque, IA: Kendall/Hunt, 1977), pp. 21–22.

43. For example, see "EXPERT: A Guide to Forensic Engineering and Service as an Expert Witness, The Association of Engineering Firms Practicing in the Geo-sciences," in Jack V. Matson (ed.), *Effective Expert Witnessing: A Handbook for Technical Professionals* (Chelsea, MI: Lewis Publishers, 1990), pp. 133–38.

44. Max Schwartz and Neil Forrest Schwartz, *Engineering Evidence,* vol. 2 (Colorado Springs, CO: Shepard's/McGraw-Hill, 1987), pp. 268–69, 809.

45. Kenneth L. Carper, "Ethical Considerations for the Forensic Engineer Serving as an Expert Witness," *Business & Professional Ethics Journal* 9, nos. 1 and 2 (1990): 21–34.

46. Martin Wachs, "Ethics and Advocacy in Forecasting for Public Policy," *Business & Professional Ethics Journal* 9, nos. 1 and 2 (1990): 141–57. Also see Martin Wachs (ed.), *Ethics in Planning* (New Brunswick, NJ: Center for Urban Policy Research, Rutgers University, 1985).

47. Rosemarie Tong, *Ethics in Policy Analysis* (Englewood Cliffs, NJ: Prentice Hall, 1986), especially pp. 99–112.

48. See *NSPE Opinions of the Board of Ethical Review,* Case 85-4, vol. VI (Alexandria, VA: National Society of Professional Engineers, 1989).

49. Martin Wachs, "Ethics and Advocacy in Forecasting for Public Policy," *Business & Professional Ethics Journal* 9, nos. 1 and 2 (1990): 152.

50. See A. John Simmons, "Consent and Fairness in Planning Land Use," *Business & Professional Ethics Journal* 6 (1987): 5–20.

51. Daniel D. McCracken, *Public Policy and the Expert: Ethical Problems of the Witness* (New York: The Council on Religion and International Affairs, 1971), p. 17.

CHAPTER

8

ENGINEERING AND
ENVIRONMENTAL ETHICS
IN THE ANTHROPOCENE

For decades a 55-mile waterway through the Napa Valley in California had been a battleground between humans and nature. A series of earthen levees were used to redirect and constrain the natural flow of water, and low concrete bridges spanned the river, but these makeshift controls failed to prevent periodic floods. A 1986 flood alone caused $100 million in property damage, killed three people, and forced the evacuation of 5,000 others. In July 2000, groundbreaking ceremonies took place on a project to restore the waterway into a "living river," with natural floodplains, wetlands, and other natural habitat.[1] The restoration project was innovative in the way it combined ecological and economic goals. Instead of imposing further constraints on the river, the project was to restore the river to something closer to its natural state. Furthermore, although initial costs would be much higher, even after the savings from conforming to regulations about preserving wetlands and endangered species, there were counterbalancing long-term economic benefits. They included increased tourism because of more scenic countryside, elevated property values, and lower home insurance costs due to lessened flood risks.

The solution was also innovative in the role played by the U.S. Army Corps of Engineers. Initially, the Corps proposed a deeper concrete channel with higher concrete walls and concrete steps. This proposal reflected traditional Corps thinking.[2] The Corps is comprised of talented and conscientious engineers, with some 30,000 civilian employees directed by about 200 Army officers who have enormous power, reporting directly to Congress through the Office of Management and Budget. Yet, the Corps had acquired a controversial tradition of reengineering nature with a preference for straight-line, concrete structures that placed environmental concerns, not to mention the desires of local communities, a distant second.

The local community rejected the Corps' initial plan. Members of the community were asked to pay higher taxes to help fund a concrete eyesore slicing through miles of beautiful farmland and also the middle of Napa City. Activists rallied support from citizens and various state agencies for an alternative plan involving multiple compromises. Eventually the plan also gained the support of the Corps, which was essential because its federal money funded most of the project. Indeed, the new plan helped boost the Corps' reputation: "The approved plan made national news, and Corps officials embraced it with fanfare. 'What we will be doing in Napa is radically different from anything we have ever done before,' Corps spokesman Jason Fanselau told reporters. 'It's going to totally change the way we do business.'"[3]

According to the compromise plan, the Corps will destroy current levees and nine bridges, rebuilding five bridges at higher levels after rerouting the water closer to its original state. The local community will accept a one-half cent increase in sales taxes, and major sacrifices will be required of some homes and businesses that require relocation. But the long-term economic benefits promise to outweigh the sacrifice. Of course, no one can foresee the exact costs and benefits, and critics argue that global climate changes might lead to greater flood problems than anyone can predict. The project is truly a social—and environmental—experiment. But there is a basis for hope in this "collaboration with nature," especially because of how it integrates enlightened self-interest, ecological awareness, and engineering expertise.

The Napa Valley case is but one of many instances where engineers are increasingly responsive to concerns about the environment. Indeed, many engineers and their corporations are now showing leadership in advancing ecological awareness. In this chapter, we discuss some ways in which this responsibility for the environment can be shared by engineers, industry, government, and the public. We also introduce some perspectives developed in the new field of environmental ethics that enter into engineers' personal commitments and ideals. In taking account of what many view as an environmental crisis, we highlight the many positive steps that are now being taken.

Since the early 2000s, scientists and environmentalists have argued for the use of the term "Anthropocene" to describe the current environment we live in. Anthropocene refers to the current geological age in which human activity has been the dominant influence on climate and the global environment. An environmental ethic for engineers in the Anthropocene therefore requires them to be sensitive to the extremely complex, global environmental issues and be critical about the role of their expertise in shaping human futures.

8.1 ENGINEERING, ECOLOGY, AND ECONOMICS

Like the word *ethics,* the expression *environmental ethics* can have several meanings. We use the expression to refer to (1) the study of moral issues concerning the environment, and (2) moral perspectives, beliefs, and attitudes concerning those issues.

8.1.1 The Invisible Hand and the Commons

Two powerful metaphors have dominated thinking about the environment: the invisible hand and the tragedy of the commons. Both metaphors are used to highlight unintentional impacts of the marketplace on the environment, but one is optimistic and the other is cautionary about those impacts. Each contains a large part of the truth, and they need to be reconciled and balanced.

The first metaphor was set forth by Adam Smith in 1776 in *The Wealth of Nations,* the founding text of modern economics. Smith conceived of an invisible (and divine) hand governing the marketplace in a seemingly paradoxical manner. According to Smith, businesspersons think only of their own self-interest: "It is not from the benevolence of the butcher, the brewer, or the baker, that we expect our dinner, but from their regard to their own interest."[4] Yet, although "he intends only his own gain," he is "led by an invisible hand to promote an end which was no part of his intention. . . . By pursuing his own interest he frequently promotes that of the society more effectually than when he really intends to promote it. I have never known much good done by those who affected to trade for the publick [sic] good."[5]

In fact, professionals and many businesspersons do profess to "trade for the public good," claiming a commitment to hold paramount the safety, health, and welfare of the public. Although they are predominantly motivated by self-interest, they also have genuine moral concern for others.[6] Nevertheless, Smith's metaphor of the invisible hand contains a large element of truth. By pursuing self-interest, the businessperson, as entrepreneur, creates new companies that provide goods and services for consumers. Moreover, competition pressures corporations to continually improve the quality of their products and to lower prices, again benefiting consumers. In addition, new jobs are created for employees and suppliers, and the wealth generated benefits the wider community through consumerism, taxes, and philanthropy.

Despite its large element of truth, the invisible hand metaphor does not adequately take account of damage to the environment. Writing in the eighteenth century, with its seemingly infinite natural resources, Adam Smith could not have foreseen the cumulative impact of expanding populations, unregulated capitalism, and market "externalities"—that is, economic impacts not included in the cost of products. Regarding the environment, most of these are negative externalities—pollution, destruction of natural habitats, depletion of shared resources, and other unintended and often unappreciated damage to "common" resources.

This damage is the topic of the second metaphor, which is rooted in Aristotle's observation that we tend to be thoughtless about things we do not own individually and which seem to be in unlimited supply. William Foster Lloyd was also an astute observer of this phenomenon. In 1833 he described what the ecologist Garrett Hardin would later call "the tragedy of the commons."[7] Lloyd observed that cattle in the common pasture of a village were more stunted than those kept on private land. The common fields were themselves more worn than private pastures. His explanation began with the premise that individual farmers are understandably motivated by self-interest to enlarge their common-pasture herd by one or two cows, especially given that each act taken by itself does negligible damage. Yet, when all the farmers behave this way, in the absence of laws constraining them, the result is the "tragedy" of overgrazing that harms everyone.

The same kind of competitive, unmalicious but unthinking, exploitation arises with all natural resources held in common: air, land, forests, lakes, oceans, endangered species, and indeed the entire biosphere. Hence, the tragedy of the commons remains a powerful image in thinking about environmental challenges in today's era of increasing population and decreasing natural resources. Its very simplicity, however, belies the complexity of many issues concerning ecosystems and the biosphere. Ecosystems are systems of living organisms interacting with their environment—for example, within deserts, oceans, rivers, and forests. The biosphere is the entirety of the land, water, and atmosphere in which organisms live. Ecosystems and the biosphere are themselves interconnected and do not respect national boundaries.

As an illustration of that complexity, consider acid rain. Over several decades, hundreds of lakes in large parts of the northeastern United States and eastern Canada were "dying."[8] In the higher elevations of the Adirondack Mountains, more than half the lakes that were once pristine could no longer support fish. Forests were also dying, and larger animals were suffering dramatic decreases in population, while some farmlands and drinking water sources were being damaged. The proximate cause was quickly identified. "Acid shock" from snow melt containing elevated levels of acid caused annual mass killings of fish. Longer-term effects of the acidic rain harmed fish eggs and food sources. Deadly quantities of aluminum, zinc, and many other metals leached from the soil by the acid rain also take a toll as they wash into streams and lakes.

It took longer to identify the deeper cause: the burning of fossil fuels that release large amounts of sulfur dioxide (SO_2)—the primary culprit—and nitrogen oxides (NO_x). Major sources of the locally observed pollutants are often located hundreds and even thousands of miles away, in the Ohio Valley, for example, with winds supplying a deadly transportation system to the damaged ecosystems. The costs of this damage to the ecosystem was external to (not included in) the cost of products supplied to the market. Hence, even after the damage was identified, it took some time for the U.S. and Canadian governments to agree on how best to handle the costs and to embed them as much as possible into market exchanges. Much remains to be learned about the mechanisms involved in the processes pictured in figure 8-1. It is still impossible to link specific sources with specific damage, and more research into shifting wind patterns and the air transport of acids is needed.

Today there is a wide consensus that we need concerted responses to ecological concerns that combine economic realism (the invisible hand) with ecological awareness (sustainable use and development of the commons). Engineers

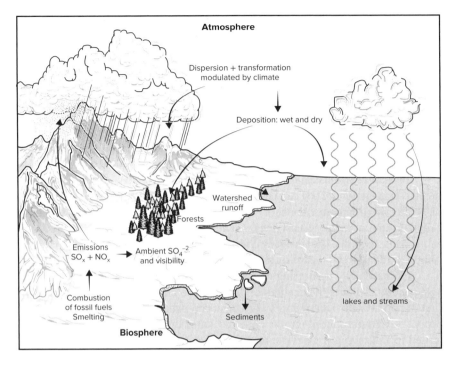

FIGURE 8-1
Acid deposition: sources and affected ecosystems.

play a key role in that consensus, as the Napa Valley compromise illustrates. So do corporations, government, local communities, market mechanisms, and social activists. Let us discuss the positive steps underway among each of these groups.

8.1.2 Engineers: From Sustainable Development to Geoengineering

Ali Ansari, a scholar in India, suggests that there is a "standard engineering worldview—that of a mechanical universe," which is at odds with mainstream "organic" environmental thought.[9] According to Ansari, central to the engineering view is "technothink," which "implicitly assumes that things can be understood by analyzing them and, if something goes wrong, can be fixed." In contrast, "green philosophy" "demands humility, respect and sensitivity towards the natural world."

We believe there is at most a tension, not a dichotomy, between technothink and green philosophy. It is true that historically engineers were not as responsible concerning the environment as they should have been, but in that respect they simply reflected attitudes predominant in society. The U.S. environmental movement that emerged from the 1960s began a social transformation that has influenced engineers as much as other populations, and more than most professions. Individual engineers, like individuals in all professions, differ considerably in their views, including their broader holistic views about the environment. What is important is that all engineers should reflect seriously on environmental values and how they can best integrate them into understanding and solving problems. In doing so, as Sarah Kuhn points out in replying to Ansari, engineers should also be able to "work in an organizational context in which an eco-friendly approach is valued and supported with the tools, information, and incentives necessary to succeed. Beyond that, they must work in a market that rewards sustainable products and processes, and in a policy context that encourages, or at least does not discourage, environmental protection."[10]

In many respects, engineers are singularly well-placed to make environmental contributions. They can encourage and nudge corporations in the direction of greater environmental concern, finding ways to make that concern economically feasible. At the very least, they can help ensure that corporations obey applicable laws. In all these endeavors, they benefit from a supportive code of ethics stating the shared responsibilities of the profession.

Increasingly, engineering codes of ethics explicitly refer to environmental responsibilities under the heading of "sustainable development."[11] In the U.S., a first important step occurred in 1977 when The American Society of Civil Engineers (ASCE) introduced into its code the statement "Engineers should be committed to improving the environment to enhance the quality of life." "Should" indicates the desirability of doing so, although (in contrast to "shall") it does not indicate something mandatory or enforceable. Still, the mere mention of the environment was a breakthrough. Two decades later, in 1997, ASCE's fundamental canon has changed from recommendations ("should") to requirements ("shall"): "Engineers shall hold paramount the safety, health and welfare of the public and shall strive to comply with the principles of sustainable development in the performance of their professional duties." Additional requirements are added that require notifying "proper authorities" when the principles of sustainable development are violated by employers, clients, and other firms. In 2006, the NSPE approved a change to its code of ethics and added Section III.2.d. to the code of ethics. The new section explicitly emphasized that engineers should take the responsibility to practice the principles of sustainable development and protect the environment for future generations.

What is "sustainable development" (sometimes shortened to "sustainability")? The term was introduced in the 1970s, but it became popular during the 1980s and 1990s, especially since the publication in 1987 of *Our Common Future,* produced by the United Nations in its World Commission on Environment and Development (also called The Brundtland Report).[12] Put negatively, the term was invented to underscore how current patterns of economic activity and growth cannot be sustained as populations grow, technologies are extended to developing countries, and the environment is increasingly harmed. Put positively, the term implies the crucial need for new economic patterns and products that are sustainable, that is, compatible with both ongoing technological development and protection of the environment. As such, the term suggests a compromise stance between advocates of traditional economic development that neglected the environment, and critics who warned of an environmental crisis: economic development is essential, but it must be sustainable into the future. The compromise is somewhat uneasy, however, for different groups understand its meaning in different ways.

In *Our Common Future,* sustainable development is defined as "development that meets the needs of the present without compromising the ability of future generations to meet their own needs."[13] This statement emphasizes *inter*generational justice—balancing the needs of living populations against those of future generations. The document also calls for greater *intra*generational justice—justice in overcoming poverty among living populations, for conserving natural resources, and for keeping populations at sustainable levels. In tune with these themes, ASCE defines *sustainable development* this way:

> Sustainable development is a process of change in which the direction of investment, the orientation of technology, the allocation of resources, and the development and functioning of institutions [is directed] to meet present needs and aspirations without endangering the capacity of natural systems to absorb the effects of human activities, and without compromising the ability of future generations to meet their own needs and aspirations.[14]

Similar to ASCE, in 2006 NSPE added a footnote to its code of ethics to further clarify the term "sustainable development" from the perspective of engineering practice. Sustainable development is "the challenge of meeting human needs for natural resources, industrial products, energy, food, transportation, shelter, and effective waste management while conserving and protecting environmental quality and the natural resources base essential for future development."

In their everyday practice, engineers employ various methods (which often are quantitative) to assess and mitigate environmental impacts of technological innovation. One of these methods used by engineers to assess the impact of technology on the environment from a more *holistic* perspective is Life Cycle Assessment (LCA). As a quantitative modeling framework, LCA estimates emissions that are generated throughout the life cycle (from the extraction of raw materials to the end-of-life treatment and disposal) of a product.[15] LCA can provide comprehensive information about environmental impacts of technological development that can be useful for decision-making in corporate and public policy contexts.

More recently, due to the increasingly globalized nature of environmental problems (e.g., climate change), engineers have been developing technologies that can mitigate these environmental problems at a much larger scale. As one of these technologies, geoengineering aims to deliberately modify the climate to achieve some specific effects such as cooling. Most geoengineering schemes use natural or mechanical ways to remove carbon dioxide from the atmosphere. Some engineers have developed the technique of ocean fertilization that dumps iron dust into the open ocean to trigger algal blooms. Others studied how to genetically modify crops to increase biotic carbon uptake. Other geoengineering schemes include blocking or reflecting incoming solar radiation (e.g., stratospheric aerosol injection, space-based sun shields).[16] Nevertheless, these geoengineering technologies are often considered as large-scale, social experiments and thus have generated considerable ethical controversies such as those involve cross-cultural, geopolitical, or intergenerational conflicts.

8.1.3   Corporations: Environmental Leadership

In the present climate, it is good business for a corporation to be perceived by the public as environmentally responsible, indeed as a leader in finding creative solutions. Compaq Computer Corporation, now merged with Hewlett-Packard, is only one of a great many encouraging examples.[17] After being founded in 1982, it grew with astonishing success, making the *Fortune 500* after only four years. As it did so, it made environmental commitments central to its mission, as recognized with a series of awards, including the 1997 World Environment Center Gold Medal for International Corporate Environmental Achievement.

Three features of Compaq's commitments are especially noteworthy as aspects of its "global" perspective on how its products affected the environment. First, Compaq developed a life-cycle strategy for its products that it dubbed "Design for Environment." Priorities were set for the efficient use of resources, the design of energy-efficient products, easy disassembly for recycling, and waste minimization. For example, it set a timetable for eliminating CFC emissions in its manufacturing process that was ahead of government requirements, and then met its goal two years ahead of schedule.

Second, Compaq developed unified standards that would apply throughout its operations. This was no minor feat, given that Compaq not only markets its products in over 100 countries, but also has subsidiaries in dozens of countries in North America, Latin America, Europe, the Middle East, Africa, and Asia. Rather than exploiting lower standards in other countries as an excuse to engage in cost savings, Compaq established consistent policies that serve as an exemplar for other companies and industries.

Third, in choosing its suppliers, Compaq places a high priority on companies with a record of environmental concern. Doing so tends to serve its business interests, since some of its costs are shifted to suppliers who already factor in part of the life cycle concerns. But it also expresses Compaq's genuine and systematic commitments to make environmental responsibility a priority.

As a second example, we cite the Solar Electric Light Fund (SELF), which provides solar energy in developing countries, beginning with rural areas in South Africa and China.[18] Neville Williams had several core convictions when he founded SELF. He was convinced that replicating the traditional fossil fuel and grid-distribution system of energy distribution would do enormous environmental damage. He also knew that marketing environmentally friendly sources of power was crucial now, not only because of enormous demand but also because the first system in place could shape the future trend. And he knew that although financing was a huge obstacle for families, it was nevertheless important for them to pay for the energy in order to accept responsibility for taking care of the technology once it was implemented. A simple solar technology, combined with reasonable financing plans, proved realistic. Photovoltaic units for homes that provided 20 years of energy could be marketed at $500. An innovative funding system was devised whereby grant money provided loans for initial sales and then payments on the loans were used to finance additional loans.

8.1.4   Government: Technology Assessment

Government laws and regulations are understandably the lightning rod in environmental controversies. Few would question the need for the force of law in setting firm guidelines regarding the degradation of the "commons," especially in limiting the excesses of self-seeking while establishing fair "rules to play by." Yet, how much law, of what sort, and to what ends are matters of continual disagreement. In the United States, landmark environmental legislation at the national level in the United States began in 1969 with passage of the National Environmental Policy Act, which requires environmental impact statements for federally funded projects affecting the environment. Other key legislation quickly followed, including the Occupational Safety and Health Act (1970), the Clear Air Act (1970), the Clean Water Act (1972), and the Toxic Substances Control Act (1976). These and subsequent legislative acts involved heated controversy at several states: in passing laws, in developing and managing enforcement procedures, and in modifying laws to take account of unforeseen problems.

Until 1995, the U.S. Congress had an Office of Technology Assessment. It prepared studies on the social and environmental effects of technology in areas such as cashless trading (via bank card), nuclear war, health care, or pollution. At the federal and state levels, many large projects must be examined in terms of their environmental impact before they are approved. The purpose of all this activity is praiseworthy, but it needs to be complemented by good-faith commitments of engineers and their corporations.

Engineers, it is sometimes said, are apt to find the right answers to the wrong questions. The economist Robert Theobald observed that many of "the questions we should be answering are not yet known. Unfortunately the process required for discovering the right questions is totally different from the process of discovering the right answers."[19] With regard to the increasing complexities of the global economy affecting the biosphere, it is often difficult to know what questions to ask. And technology assessment and other forecasting methods suffer because of this.

When scientists conduct experiments, they try to distill some key concepts out of their myriad observations. As shown in figure 8-2, a funnel can be used to portray this activity. At the narrow end of the funnel we have the current wisdom, the state of the art. Engineers use it to design and build their projects. These develop in many possible directions, as shown by the shape of the lower, inverted funnel. The difficult task of technology assessment and environmental impact analyses is to explore the extent of this spread and to separate the more significant among the possibly adverse effects.

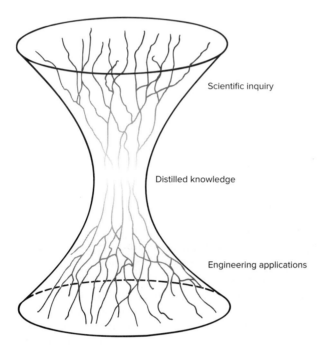

Scientific inquiry

Distilled knowledge

Engineering applications

FIGURE 8-2
Distilling and applying knowledge

The danger in any assessment of technology is that some serious risks can easily be overlooked while the studies and subsequent reports, properly authenticated by the aura of scientific methodology, lull the decision maker into believing that nothing is amiss—or perhaps that perceived risks are more serious than they really are. Such efforts are surely worthwhile, if only because of those questions they raise—and answers they uncover—that otherwise might not have surfaced. But there is a danger in believing that no further action is required once the reports have been approved and filed. There is also a danger that legalistic detail and finesse can shade the issues in directions that favor the narrow interests of corporations. Our contention remains that engineering must be understood as social experimentation and that the experiment continues, indeed enters a new phase, when the engineering project is implemented. Only by careful monitoring will it be possible to gather a more complete picture of the tangled web of effects encompassed in figure 8-2 within the inverted, lower funnel.

8.1.5 Communities: Preventing Natural Disasters

Communities at the local and even state level have special responsibility to conserve natural resources and beauty for future generations. They have special responsibility, as well, for preventing natural events—such as hurricanes, floods, fires, and earthquakes—from becoming disasters. There are four sets of measures communities can take to avert or mitigate disasters.

One set of defensive measures consists of restrictions or requirements imposed on human habitat. For instance, homes should not be built in floodplains, homes in prairie country should have tornado shelters, hillsides should be stabilized to prevent landslides, structures should be able to withstand earthquakes and heavy weather, roof coverings should be made from nonflammable materials, and roof overhangs should be fashioned so flying embers will not be trapped. These are not exorbitant regulations, but merely reminders to developers and builders to do what their profession expects them to do anyway. Sometimes, as in the Napa Valley case, holistic thinking about entire ecosystems leads to more dramatic restructuring of communities.

A second set of measures consists of strengthening the lifelines for essential utilities such as water (especially for fire fighting) and electricity. A third category encompasses special-purpose defensive structures that would include dams, dikes, breakwaters, avalanche barriers, and means to keep floodwaters from damaging low-lying sewage plants placed where gravity will take a community's effluents. A fourth set of measures should assure safe exits in the form of roads and passages designed as escape routes, structures designated as emergency shelters, adequate clinical facilities, and agreements with neighboring communities for sharing resources in emergencies.

When disasters do occur, lessons can be learned, rather than shrugged aside by a disbelief that the event could occur again—"Lightning never strikes twice in the same place," and "Another 100-year flood is about that far away"—or by a belief that government would once more hand out disaster relief payments. For example, the weaknesses in steel structures found after the 1994 Northridge earthquake have in many cases still not been reported to building departments because the owners claim that any inspection reports they had commissioned should be the owners' private property and concern. The ethical issues of how to enforce the needed retrofitting of weakened steel structures are being examined closely by the engineering profession after Northridge. A far greater tragedy occurred on August 16th, 1999, when an earthquake hit Izmit and neighboring cities in northwestern Turkey, including Istanbul. The damage to structures and the resulting death toll in the tens of thousands were unusually heavy. Why? Because during a building boom, multistory apartment houses had been built and inspected without serious attention to seismic dangers even though this region was known to be part of an active earthquake belt.

Turning to a different example, communities show leadership when they develop programs that encourage recycling, often in conjunction with state governments. In May 2003, the California Department of Conservation launched an awareness campaign to encourage the recycling of plastic bottles. In one year, four billion plastic bottles were sold in California, but only a third of those were recycled. The rest ended up in landfills where they are not biodegradable. Worse, they were often incinerated, which released toxic fumes. A human-made disaster is in the making: "If the problem continues, enough water bottles will be thrown in the state's trash dumps over the next five years to create a two-lane, 6-inch-deep highway of plastic along the entire California coast."[20] Cities can alleviate the problem by making recycle bins readily available. In addition, the legislature is seeking to raise the cash refund on bottles, a solution that brings us to the next component in responding to environmental challenges, market mechanisms.

8.1.6 Market Mechanisms: Internalizing Costs

Democratic controls take many forms beyond passing laws. One such option is internalizing costs of harm to the environment. When we are told how efficient and cheap many of our products and processes are—from agriculture to the manufacture of plastics—the figures usually include only the direct costs of labor, raw materials, and the use of facilities. If we are quoted a dollar figure, it is at best an approximation of the price. The true cost would have to include many indirect factors such as the effects of pollution, the depletion of energy and raw materials, disposal, and social costs. If these, or an approximation of them, were internalized (added to the price) then those for whose benefit the environmental degradation had occurred could be charged directly for corrective actions.

Taxpayers are revolting against higher levies, so the method of having the user of a particular service or product pay for all its costs is gaining more favor. The engineer must join with the economist, the scientist, the lawyer, and the politician in an effort to find acceptable mechanisms for pricing and releasing products so that the environment is protected through truly self-correcting procedures rather than adequate-appearing yet often circumventable laws.

A working example is the tax imposed by governments in Europe on products and packaging that impose a burden on public garbage disposal or recycling facilities. The manufacturer prepays the tax and certifies so on the product or wrapper.

Fortunately, good design practices may in themselves provide the answers for environmental protection without added real cost. For example, consider the case of a lathe that was redesigned to be vibration-free and manufactured to close tolerances. It not only met occupational safety and health standards for noise, which its predecessor had not, but it also was more reliable, more efficient, and had a longer useful life, thus offsetting the additional costs of manufacturing it.[21]

8.1.7 Social Activists

We would be remiss not to underscore the importance of social activism by concerned citizens in raising public awareness. As examples, we cite Rachel Carson and Sherwood Rowland.

In the United States the environmental movement had many roots, but its catalyst was Rachel Carson's 1962 book *Silent Spring*. Carson made a compelling case that pesticides, in particular DDT, were killing creatures beyond their intended target, insects. DDT is a broad-spectrum and highly toxic insecticide that kills a variety of insects. It also persists in the environment by being soluble in fat, and hence storable in animal tissue, but not soluble in water, so that it is not flushed out of organisms. As a result, DDT enters into the food chain at all levels, with increasing concentrations in animals at the higher end of the chain.

To the public, Carson had scientific credibility, having earned a bachelor's degree in biology and a master's degree in zoology and then spending a career working for the Fish and Wildlife Service. But many other scientists with stronger credentials had been warning of the dangers of DDT for nearly two decades. Carson was unique because her prose combined scientific precision, poetic expression, and a trenchant argument understandable by the general public. Critics, especially chemical companies, were less sympathetic. She was patronized, mocked, and reviled as a sinister force that threatened American industry.[22] If today she is an American icon, it is in large measure because of the courage of her convictions in confronting a hostile establishment. At the same time, we have since gained new knowledge that balances Carson's insights, in particular an appreciation that DDT remains a valuable way to fight malaria by killing the mosquitoes that spread it. DDT has been banned in the United States and other western nations since the 1970s, but when its use in Madagascar was suspended in 1986, 100,000 deaths from malaria occurred, leading to its immediate reuse.[23]

Our second example is Professor Sherwood Rowland at the University of California, Irvine, who also confronted the wrath of an entire industry following the publication of a 1974 essay in *Nature,* coauthored with Mario Molina, identifying the depletion of the ozone layer by chlorofluorocarbons (CFCs).[24] Rowland and Molina, building on the work of Paul Crutzen, argued that CFCs were rising 15 miles to the stratosphere and beyond, where they broke down ozone (O_3): $CFC + O_3 = CFC + O_2 + O$. The ozone layer protects the entire planet from deadly ultraviolet (UV) radiation; although it is a relatively thin and diffuse layer, it is critical for protecting nearly all life forms. CFC gases, such as freon, are synthetic chemicals that since the 1930s had been widely used in refrigerators and air conditioners, and also as propellants in aerosol spray cans. Hence, in setting forth a scientific argument, the authors set off a firestorm of protest from industry, and Rowland in particular spent much of the next decade countering criticism. NASA tests in 1987 confirmed what Rowland and Molina had argued by identifying huge areas of thinning of the ozone layer. In the same year, with unprecedented speed, the Montreal Protocol, signed by the main producers and users of CFCs, mandated the phase-out of CFCs by 2000. The danger persists, however, because the CFCs already produced will go on interacting with ozone for decades, requiring addition UV protection by sunbathers in order to prevent deadly skin cancers. Rowland, Molina, and Paul Crutzen (a Dutch scientist who showed that NO and NO_2 react catalytically with ozone) were awarded the Nobel Prize in Chemistry in 1995, the first time the prize was given for applied environmental science.

To conclude, the environment is no longer the concern of an isolated minority. Engineers, corporations, federal and state laws, local community regulations, market mechanisms, and social activists are among the many influences at work. Given the complexity of the issues, we can expect controversy among viewpoints, and nowhere is there a greater need for ongoing dialogue and mutual respect. There is no longer any doubt, however, about the urgency and importance of the issues confronting all of us.

DISCUSSION QUESTIONS

1. Identify and comment on the importance of each of the environmental impacts described in the following passage.

   The Swedish company IKEA, the world's largest furniture and home furnishings retailer, has adopted a global corporate policy that prohibits the use of old-growth forest wood or tropical wood in its furniture. All timber must come from sustainably managed forests. IKEA has eliminated the use of chlorine in its catalog paper, uses 100 percent recycled paper fibers, and is committed to eliminating waste in its retail stores. The "Trash is Cash" program has transformed the thinking of retail store workers to see trash as a revenue-generating resource.[25]

2. As regards the U.S. Army Corps of Engineers, compare and contrast the Napa Valley case (at the start of this chapter) with the following case. As contrast points, include shifting attitudes, understandings, and approaches.

   The great marshes of southern Florida have attracted farmers and real estate developers since the beginning of the century. When drained, they present valuable ground. From 1909 to 1912 a fraudulent land development scheme was attempted in collusion with the U.S. Secretary of Agriculture. Arthur Morgan blew the whistle on that situation, jeopardizing not only his own position as a supervising drainage engineer with the U.S. Department of Agriculture, but also that of the head of the Office of Drainage Investigation. An attempt to drain the Everglades was made again by a Florida governor from 1926 to 1929. Once more Arthur Morgan, this time in private practice, stepped in to reveal the inadequacy of the plans and thus discourage bond sales.

   But schemes affecting the Everglades did not end then. Beginning in 1949, the U.S. Army Corps of Engineers started diverting excess water from the giant Lake Okeechobee to the Gulf of Mexico to reduce the danger of flooding to nearby sugar plantations. As a result, the Everglades, lacking water during the dry season, were drying up. A priceless wildlife refuge was falling prey to humanity's appetite. In addition, the diversion of waters to the Gulf and the ocean also affected human habitations in southern Florida. Cities that once thought they had unlimited supplies of fresh groundwater found they were pumping salt water instead as ocean waters seeped in.[26]

3. Consider the following example of environmental side effects cited by Garrett Hardin:

   The Zambesi River . . . was dammed . . . to create the 1700-square-mile Lake Kariba. The effect desired: electricity. The "side-effects" produced: (1) destructive flooding of rich alluvial agricultural land above the dam; (2) uprooting of long-settled farmers from this land to be resettled on poorer hilly land that required farming practices with which they were not familiar; (3) impoverishment of these farmers . . . [and various other social disorders]; (6) creation of a new biotic zone along the lake shore that favored the multiplication of tsetse flies.[27]

   Similar problems have occurred when dams were built in the United States and when the Aswan Dam was erected on the Nile. One might ask if the original purpose may not itself begin to look like merely a side effect. If so, Hardin asks, can we never do anything? Describe under what conditions you think a dam such as the one on the

Zambesi River should be built and operated. To whom is the engineer in charge of its construction ultimately responsible?

4. Human life is possible because of "the greenhouse effect," in which atmospheric gases such as water vapor and carbon dioxide block solar energy from escaping, after being reflected from the earth's surface. Evidence is mounting that since the nineteenth century, owing to human burning of fossil fuels that increase levels of greenhouse gases, the earth's climate is warming. The change is small, but even a few degrees of "global warming" could melt enough of the polar ice caps to raise the oceans a few feet and thereby cause severe flood damage. Other effects include major disruptions in weather patterns, such as increased drought and increased storms. The complexity of the issue has divided scientific opinion, but an emerging consensus led to the 1997 Kyoto agreement, signed by 150 governments, to reduce carbon emissions to 5.2 percent below 1990 levels by 2012. In 2001, President Bush announced the United States would abandon the agreement because it was antithetical to American business and also unfair because it did not place stronger requirements on developing countries. Research the current state of the scientific and political debate. Present and defend your view as to whether or not the United States should do more to show leadership on the issue.

5. Most companies want to have a reputation for environmental responsibility, but there are different "shades of green" in their commitments.[28] They include (1) "light green"—compliance with the law; (2) "market green"—seeking competitive advantage by attending to customer preferences; (3) "stakeholder green"—responding to and fostering environmental concern in the stakeholders of the corporation, including suppliers, employees, and stockholders; and (4) "dark green"—creating products and using procedures that include respect for nature as having inherent worth. Which of these shades of green would you ascribe to Compaq and SELF?

6. Discuss one of the following topics with an eye to how individual choices in everyday life affect the environment: (a) drinking from disposable cups for coffee or soda pop, (b) driving a sports utility vehicle that gets low gas mileage, (c) eating beef, (d) going the extra mile to dispose of your spent dry cell at a collection point.

7. Using California as an example, we discussed calls for stronger programs for recycling plastic bottles. William A. McDonough, who is an architect, designer, and cofounder of McDonough Braungart Design Chemistry, proposes an alternative solution: "Plastic bottles could easily be redesigned so that they don't contain questionable substances and could safely replenish the soil. Right now they may contain antimony, catalytic residues, UV stabilizers, plasticizers, and antioxidants. . . . Why not design a bottle so that when you finish with it you toss it into the compost or it biodegrades by the roadside, or it can be used as fuel for needy people to cook with?" What types of obstacles do you see in pursuing McDonough's proposal, and how might they be confronted?

8.2   ETHICAL FRAMEWORKS

Individual engineers can make a difference. Although their actions are limited—within corporations, they share responsibility with many others—they are uniquely placed to act as agents of change, as responsible experimenters. Doing so requires personal commitments that are often rooted in wider moral or religious frameworks. Here we provide an overview of some of the environmental ethics that are currently being explored, in order to stimulate further reflection on wider moral frameworks concerning the environment.[29]

8.2.1   Human-Centered Ethics

Human-centered, or anthropocentric, environmental ethics focuses exclusively on the benefits of the natural environment to humans and the threats to human beings presented by the destruction of nature. Each of the ethical theories we examined in chapter 3—utilitarianism, rights ethics, duty ethics, and virtue ethics—provides a framework for exploring the moral issues concerning the environment. In their classic formulations, all of them assume that, among the creatures on earth, only human beings have inherent moral worth and hence deserve to be taken into account in making moral decisions concerning the environment (or anything else). Other creatures and ecosystems have at most "instrumental value"—as means to promoting human interests.

Utilitarianism says maximize good consequences for human beings. In developing an environmental ethic, the relevant goods consist of human pleasures and interests linked to nature. Many of those pleasures and interests concern engineered products made from natural resources. In addition, we have aesthetic interests, as in the beauty of plants, waterfalls, and mountain ranges, and recreational interests, as in hiking and backpacking in wilderness areas. We have scientific interests, especially in the study of "natural labs" of ecological preserves, such as the rain forests. And most basic, we have survival interests, which are linked directly to conserving resources and preserving the natural environment.

Rights ethics typically argues that the basic rights to life and to liberty entail a right to a livable environment. The right to a livable environment did not generally enter into people's thinking until the end of the twentieth century, at the time when pollution and resource depletion reached alarming proportions. Nevertheless, it is directly implied by the rights to life and liberty, given that these basic rights cannot be exercised without a supportive natural environment. As such, according to William T. Blackstone, it is itself a basic human right.

> Each person has this right [to a livable environment] *qua* being human and because a livable environment is essential for one to fulfill his human capacities. And given the danger to our environment today and hence the danger to the very possibility of human existence, access to a livable environment must be conceived as a right which imposes upon everyone a correlative moral obligation to respect.[30]

Duty ethics, which makes duties rather than rights fundamental, urges that respect for human life implies far greater concern for nature than has been traditionally recognized. Kant believed that we owe duties only to rational beings, which in his view excluded all nonhuman animals, although of course he did not have access to recent scientific studies showing striking parallels between humans and other primates. Nevertheless, he condemned callousness and cruelty toward conscious animals because he saw the danger that such attitudes would foster inhumane treatment of persons. In any case, a duty-centered ethics would emphasize the need for conserving the environment because doing so is implied by respect for human beings who depend on it for their very existence.

Finally, virtue ethics draws attention to such virtues as prudence, humility, appreciation of beauty, and gratitude toward the natural world that makes life possible, and also the virtue of stewardship over resources that are needed for further generations. Thomas E. Hill, Jr., offers an anecdote:

> A wealthy eccentric bought a house in a neighborhood I know. The house was surrounded by a beautiful display of grass, plants, and flowers, and it was shaded by a huge old avocado tree. But the grass required cutting, the flowers needed tending, and the man wanted more sun. So he cut the whole lot down and covered the yard with asphalt.[31]

The man's attitudes, suggests Hill, are comparable to the callousness shown in strip mining, the cutting of redwood forests, and other destruction of ecosystems with blinkered visions of usefulness.

All these human-centered ethics permit and indeed require a long-term view of conserving the environment, especially because the human beings who have inherent worth will include future generations. Not everything of importance within a human-centered ethics fits neatly into cost-benefit analyses with limited time horizons; much must be accounted for by means of constraints or limits that cannot necessarily be assigned dollar signs. Yet, some have argued that all versions of human-centered ethics are flawed and that we should widen the circle of things that have inherent worth, that is, value in themselves, independent of human desires and appraisals. Especially since 1979, when the journal *Environmental Ethics* was founded, philosophers have explored a wide range of *nature-centered ethics* that, for example, affirm the inherent worth of all conscious animals, of all living organisms, or of ecosystems. Let us consider each of these approaches.

8.2.2   Sentient-Centered Ethics

One version of nature-centered ethics recognizes all sentient animals as having inherent worth. Sentient animals are those that feel pain and pleasure and have desires. Thus, some utilitarians extend their theory (that right action maximizes goodness for all affected) to sentient animals as well as humans. Most notably, Peter Singer developed a revised act-utilitarian perspective in his influential book, *Animal Liberation*. Singer insists that moral judgments must take into account the effects of our actions on sentient animals. Failure to do so is a form of discrimination akin to racism and sexism. He labels it "speciesism": "a prejudice or attitude of bias toward the interests of members of one's own species and against those of members of other species."[32] In Singer's view, animals deserve equal consideration, in that their interests should be weighed fairly, but that does not mean equal treatment with humans (since their interests are different from human interests). Thus, in building a dam that will cause flooding to grasslands, engineers should take into account the impact on animals living there. Singer allows that sometimes animals' interests have to give way to human interests, but their interests should always be considered and weighed.

Singer does not ascribe rights to animals, and hence it is somewhat ironic that *Animal Liberation* has been called the bible of the animal rights movement. Other philosophers, however, do ascribe rights to animals. Most notably, Tom Regan contends that conscious creatures have inherent worth not only because they can feel pleasure and pain, but because more generally they are subjects of experiences who form beliefs, memories, intentions, and preferences, and they can act purposefully.[33] In his view, their status as subjects of experiments makes them sufficiently like humans to give them rights.

Singer and Regan tend to think of inherent worth as all-or-nothing. Hence they think of conscious animals as deserving equal consideration. That does not mean they must be treated in the identical way we treat humans, but only that their interests should be weighed equally with human interests in making decisions. Other sentient-ethicists disagree. They regard conscious animals as having inherent worth, though not equal to that of humans.[34]

8.2.3 Biocentric Ethics

A life-centered ethics regards all living organisms as having inherent worth. Albert Schweitzer (1875–1965) set forth a pioneering version of this perspective under the name of "reverence for life." He argued that our most fundamental feature is not our intellect but instead our will to live, by which he meant both a will to survive and a will to develop according to our innate tendencies. All organisms share these instinctive tendencies to survive and develop, and hence consistency requires that we affirm the inherent worth of all life. More than an appeal to logical consistency, however, Schweitzer appealed to what has been called "bioempathy"—our capacity to experience a kinship with other life, to experience other life in its struggle to survive and grow. Empathy, if we allow it to emerge, grows into sympathy and compassion, gradually leading us to accept "as good preserving life, promoting life, developing all life that is capable of development to its highest possible value."[35]

Schweitzer often spoke of reverence for life as the fundamental excellence of character, and hence his view is a version of nature-centered virtue ethics. He refused to rank forms of life according to degrees of inherent worth, but he believed that a sincere effort to live by the ideal and virtue of reverence for life would enable us to make inevitable decisions about when life must be maintained or has to be sacrificed. More recent defenders of biocentric ethics, however, have developed complex sets of rules for guiding decisions.

Paul Taylor, for example, provides extensive discussion of four duties: (1) nonmaleficence, which is the duty not to kill other living things; (2) noninterference, which is the duty not to interfere with the freedom of living organisms; (3) fidelity, which is the duty not to violate the trust of wild animals (as in trapping); and (4) restitution, which is the duty to make amends for violating the previous three duties.[36] These are prima facie duties, which have exceptions when they conflict with overriding moral duties and rights, such as self-defense.

8.2.4　Ecocentric Ethics

A frequent criticism of sentient-centered and biocentered ethics is that they are too individualistic, in that they locate inherent worth in individual organisms. Can we seriously believe that each microbe and weed has inherent worth? By contrast, ecocentered ethics locates inherent value in ecological systems. This more holistic approach was voiced by the naturalist Aldo Leopold (1887–1948), who urged that we have an obligation to promote the health of ecosystems. In one of the most famous statements in the environmental literature, he wrote: "A thing is right when it tends to preserve the integrity, stability, and beauty of the biotic community. It is wrong when it tends otherwise."[37] This "land ethic," as he called it, implied a direct moral imperative to preserve (leave unchanged), not just conserve (use prudently), the environment, and to live with a sense that we are part of nature, rather than that nature is a mere resource for satisfying our desires.

More recent defenders of ecocentric ethics have included within this holistic perspective an appreciation of human relationships. Thus, J. Baird Callicott writes that an ecocentric ethic does not "replace or cancel previous socially generated human-oriented duties—to family and family members, to neighbors and neighborhood, to all human beings and humanity."[38] That is, locating inherent worth in wider ecological systems does not cancel out or make less important what we owe to human beings.

8.2.5　Religious Perspectives

Each world religion reflects the diversity of outlooks of its members, and the same is true concerning environmental attitudes. Moreover, these religions have endured through millennia in which shifting attitudes have led a mixed legacy of concern and callousness, with large gaps between ideals and practice. Nevertheless, the potential for world religions to advance ecological understanding is enormous, and we briefly take note of several examples.

Judeo-Christian traditions begin with two contrasting images in *Genesis*. The first chapter portrays God as commanding human dominion over the earth: "'Be fruitful and multiply, and fill the earth and subdue it; and have dominion over the fish of the sea and over the birds of the air, and over every living thing that moves upon the earth.'" The second chapter commands "stewardship over all the earth," suggesting the role of caretaker. In principle the two roles are compatible and mutually limiting, especially if "dominion" is interpreted to mean stewardship rather than dominance. In practice, the message of dominance has predominated throughout most of human history in sanctioning unbridled exploitation.[39] Islam also contains a mixed heritage on the environment, with the Koran containing passages that alternate between themes of exploitation of nature for human pleasure and themes of responsible stewardship over what ultimately remains the property of God, not humans.[40]

Today, many concerned Christians, Jews, and Muslims are rethinking their traditions in light of what we have learned. For example, Ian Barbour, a physicist and ecumenical religious thinker, urges that we keep before us the lunar astronauts' pictures of the earth as "a spinning globe of incredible richness and beauty, a blue and white gem among the barren planets," while at the same time exploring "its natural environments and its social order" as we seek together "a more just, participatory, and sustainable society on planet earth."[41]

Asian religions emphasize images of unity with nature, which is distinct from both stewardship and domination. Zen Buddhism, flourishing in Japan, stresses unity of self with nature through immediate, meditative experience. It calls for a life of simplicity and compassion toward the suffering of humans and other creatures. Daoism, rooted in Chinese thought, also accents themes of unity with nature and the universe. The Dao (The Way) is the way of harmony attained by experiencing ourselves as being at one with nature. And Hinduism, the predominant religion in India, promulgates an ideal of oneness with nature and the doctrine of *ahimsa*—non-violence and non-killing. It also portrays the sacred and the natural as fused, symbolized in the idea of divinities being reincarnated in living creatures.

Themes of unity are familiar in nineteenth-century English Romanticism and American Transcendentalism. The most deeply rooted American themes of unity, however, are found in American Indian thinking and rituals. Nonhuman animals have spirits. They are to be killed only out of necessity, and then atoned for and apologies made to the animal's spirit. In addition, the identity of tribes was linked to features of the landscape. Unity was understood in terms of interdependence and kinship among types of creatures and natural systems.[42]

Many additional approaches could be cited, including forms of spirituality not tied exclusively to particular world religions. For example, feminist outlooks—"ecofeminism"—might or might not be tied to specific religions. They draw parallels between traditional attitudes of dominance and exploitation of men over women and humans over nature. Many, although not all, build on an "ethics of care" that emphasizes themes of personal responsibility, relationships, and contextual reasoning.[43]

We have set forth these environmental ethics in connection with the reflections of individuals, not engineering corporations. Engineering would shut down if it had to grapple with theoretical disputes about human- and nature-centered ethics. Fortunately, at the level of practical issues the ethical theories often converge in the general direction for action, if not in all specifics. Just as humanity is part of nature, human-centered and nature-centered ethics overlap extensively in many of their practical implications.[44] Thus, nature-centered ethics will share with human-centered ethics the justification of human beings' rights to survive, defend themselves, and pursue their self-fulfillment in reasonable ways. Just as it is important for individuals to explore their personal beliefs on this topic, it is equally important for them to seek out and build upon areas of overlap, so as to participate in developing responsible social policies and projects.

8.2.6 Environmental Ethics and the Anthropocene

Albeit with some controversies, some philosophers and scientists argue that we are now situated in a geological epoch named the Anthropocene. The term "Anthropocene" consists of two parts "anthropo- (meaning human)" and "-cene (meaning new or recent)." Despite that, so far neither the International Commission on Stratigraphy (ICS) nor the International Union of Geological Sciences (IUGS) has officially affirmed the existence of the Anthropocene as a subdivision of geological time, the term Anthropocene has been informally used in scientific contexts (e.g., a key concept at academic conferences and in technical reports and peer-reviewed publications). Scholars have been debating about when the Anthropocene actually started and various possible start dates have been proposed, such as the beginning of Agricultural Revolution during the Neolithic period and the 1960s. The Anthropocenic perspective invites us to critically examine the role of human influence on global environmental challenges such as climate change. Compared with traditional ethical frameworks, environmental ethical theories in the Anthropocene examine environmental issues from much larger spatial and temporal scales.

Philosopher J. Baird Callicott argues that an environmental ethic in the Anthropocene needs to be both *holistic* and *affective*.[45] An Anthropocenic environmental ethic is concerned mainly with holistic moral agents and patients such as governments and organizations. It examines whether governments *per se* have preferences, wills, welfare, dignity, intrinsic values, and rights. Cooperation and collaboration between governments exemplify their moral agency and are important for the efficacy in combating global environmental issues. Acknowledging that non-individual entities such as governments can be moral agents is critical for achieving international justice. Taking a holistic approach to environment ethics is also to recognize that the well-being and environmental responsibility of oneself are closely associated with the socio-environmental whole. Such holistic approach to the Anthropocenic environmental ethics is aligned with some core ideas of Asian religions.

The affective aspect of environmental ethics in the Anthropocene values the unique role of moral sentiments such as love in motivating the current generation of humans to take on collective and cooperative political actions to address global environmental challenges. More specifically, the idea of love in an Anthropocenic environmental ethic has two meanings: (1) a universal love toward oneself, kith, and kin; and (2) care for the artifacts of human civilization and the processes that cultivated them.

DISCUSSION QUESTIONS

1. What ethical theory would you apply to our relation to the environment? Explain why you favor it, and also identify how extensively its practical implications differ from at least two alternative perspectives, selected from those discussed in this section.
2. Do you agree or disagree, and why, with Peter Singer's claim that it is a form of bigotry—"speciesism"—to give preference to human interests over the interests of

other sentient creatures? Also, should we follow Albert Schweitzer in refusing to rank life forms in terms of their importance?

3. Exxon's 987-foot tanker *Valdez* was passing through Prince William Sound on March 24th, 1989, carrying 50 million gallons of oil when it fetched up on Bligh Reef, tore its bottom, and spilled 11 million gallons of oil at the rate of a thousand gallons a second.[46] The immediate cause of the disaster was negligence by the ship's captain, Joseph J. Hazelwood, who was too drunk to perform his duties. Additional procedural violations, lack of emergency preparedness, and a single- rather than double-hull on the ship all contributed in making matters worse. This was one of the worst spills ever, not in quantity, but in its effect on a very fragile ecosystem. No human life was lost, but many thousands of birds, fish, sea otters, and other creatures died.

Discuss how each of the human-centered and nature-centered ethical theories would interpret the moral issues involved in this case, and apply your own environmental ethic to the case.

4. Discuss the "last person scenario": You are the last person left on earth and can press a button (connected to nuclear bombs) destroying all life on the planet.[47] Is there a moral obligation not to press the button, and why? How would each of the environmental ethics answer this question?

5. Evaluate the following argument from W. Michael Hoffman. In most cases, what is in the best interests of human beings may also be in the best interests of the rest of nature. . . . But if the environmental movement relies only on arguments based on human interests, then it perpetuates the danger of making environmental policy and law on the basis of our strong inclination to fulfill our immediate self-interests. . . . Without some grounding in a deeper environmental ethic with obligations to nonhuman natural things, then the temptation to view our own interests in disastrously short-term ways is that much more encouraged.[48]

6. Buckminster Fuller (in the epigraph for chapter 9) compared the earth to a spaceship. Compare and contrast the moral implications of that analogy with the Gaia Hypothesis set forth by James Lovelock in the passage that follows. What are the strengths and weaknesses of each analogy?

"We have . . . defined Gaia as a complex entity involving the Earth's biosphere, atmosphere, oceans, and soil; the totality constituting a feedback or cybernetic system which seeks an optimal physical and chemical environment for life on this planet. The maintenance of relatively constant conditions by active control may be conveniently described by the term 'homoeostasis.' "[49]

7. Write an essay on one of the following topics: "Why Save Endangered Species?" "Why Save the Everglades?" "What are corporations' responsibilities concerning the environment?" In your essay, explain and apply your environmental ethics.

8. "Malthus was wrong, so were William Vogt and Paul Ehrlich" writes Ron Gray of the Christian Heritage Party in its CHP Communiques of 22 October 2001 (www.chp.ca). Robert Malthus proposed in 1798 that the arithmetically increasing food production (in the ratio 1, 2, 3, 4, . . .) could in the long run not keep up with population growth following a geometric progression (1, 2, 4, 8, . . .). Two centuries later we still find the world's population doubling every 70-year life span (the 1980 population of 4.5 Billion is expected to grow to 9 Billion by 2050). Considerations of this sort, along with the greater per capita demand for natural resources accompanied by growing competition for them, gave rise to cautionary writings by William Vogt (*Road to Survival,* 1948) and Paul Ehrlich (*The Population Bomb,* 1960), and Garrett Hardin. Ron Gray rebuts the "doom-sayers" in the pages of the pro-life CHP by pointing to the wonders of modern

technological developments accompanied by a gradual drop in birth rates which should be able to sustain mankind in relative comfort on earth forever. Discuss these contrasting views in a paper or arrange a discussion on this topic with fellow students or colleagues.

KEY CONCEPTS

—**Environmental ethics:** (1) the study of moral issues concerning the environment, and (2) a moral perspective, belief, or attitude concerning those issues.

—**Anthropocene:** the current geological age in which human activity has been the dominant influence on climate and the global environment.

—**Invisible hand:** the ways in which pursuing self-interest in the competitive marketplace promotes the public good, for example, by providing quality products at lower cost, jobs, and wealth and philanthropy.

—**Tragedy of the commons:** the ways in which the marketplace harms public goods (such as clean air and water) by creating unintended "externalities," that is, harmful effects such as pollution that are not factored into the cost of products.

—**Internalizing costs:** the cost of products and services is made to include indirect costs such as the effects of pollution.

—**Sustainable development:** economic and technological patterns that are compatible with preserving environmental capacities to sustain future generations.

—**Human-centered ethics:** the view that only humans have inherent worth and that other creatures and ecosystems have at most "instrumental value" as means to promoting human interests.

—**Sentient-centered ethics:** the view that all conscious animals have inherent worth.

—**Biocentric ethics:** the view that all living organisms have inherent worth.

—**Ecocentric ethics:** the view that ecosystems have inherent worth.

REFERENCES

1. Gretchen C. Daily and Katherine Ellison, *The New Economy of Nature: The Quest to Make Conservation Profitable* (Washington, DC: Island Press, 2002), pp. 87–108.
2. P. Aarne Vesilind, "Decision Making in the Corps of Engineers: The B. Everett Jordan Lake and Dam," in P. Aarne Vesilind and Alastair S. Gunn, *Engineering, Ethics, and the Environment* (Cambridge: Cambridge University Press, 1998), pp. 171–77; Arthur E. Morgan, *Dams and Other Disasters* (Boston: Porter Sargent, 1971), pp. 370–89.
3. Gretchen C. Daily and Katherine Ellison, *The New Economy of Nature: The Quest to Make Conservation Profitable,* p. 103.
4. Adam Smith, *An Inquiry Into the Nature and Causes of the Wealth of Nations,* vol. 1 (New York: Oxford University Press, 1976), pp. 26–27.
5. Ibid., p. 456.
6. In *The Theory of Moral Sentiments,* Adam Smith also expressed this view. Scholars continue to struggle with how to reconcile the seeming contradiction in Smith's reflections on personal ethics versus the marketplace. See Patricia H. Werhane, *Adam Smith and His Legacy for Modern Capitalism* (New York: Oxford University Press, 1991).
7. Garrett Hardin, *Exploring New Ethics for Survival* (New York: Viking, 1968), p. 254.
8. Donald D. Adams and Walter P. Page (eds.), *Acid Deposition* (New York: Plenum, 1985); and Jurgen Schmandt and Hilliard Roderick (eds.), *Acid Rain and Friendly Neighbors: The Policy Dispute Between Canada and the United States* (Durham, North Carolina: Duke University Press, 1985).

9. Ali Ansari, "The Greening of Engineers: A Cross-Cultural Experience," *Science and Engineering Ethics* 7, no. 1 (2002): 105, 115.

10. Sarah Kuhn, "Commentary On: The Greening of Engineers: A Cross-Cultural Experience," *Science and Engineering Ethics* 7, no. 1 (2001): 124.

11. The following discussion is based on P. Aarne Vesilind and Alastair S. Gunn, *Engineering, Ethics, and the Environment* (New York: Cambridge University Press, 1998), pp. 48–65.

12. Alan Holland, "Sustainability," in Dale Jamieson (ed.), *A Companion to Environmental Philosophy* (Malden, MA: Blackwell Publishing, 2003), pp. 390–401.

13. World Commission on Environment and Development, *Our Common Future* (Oxford: Oxford University Press, 1987).

14. American Society of Civil Engineers, "The Role of the Engineer in Sustainable Development," www.asce.org. Discussed by Charles E. Harris, Jr., Michael S. Pritchard, and Michael J. Rabins, *Engineering Ethics: Concepts and Cases,* 2nd ed. (Belmont, CA: Wadsworth, 2000), pp. 209–10.

15. Matthew J. Eckelman, John Basl, Christopher Bosso, Jacqueline A. Isaacs, and Kathleen Eggleson, "Case Studies of Product Life Cycle Environmental Impacts for Teaching Engineering Ethics," in *Next-Generation Ethics: Engineering a Better Society*, ed. Ali E. Abbas (New York, NY: Cambridge University Press, 2020).

16. Alan Robock, "20 Reasons Why Geoengineering May Be a Bad Idea," *Bulletin of the Atomic Scientists* 64, no. 2 (2008): 14–18, 59.

17. Noel M. Tichy, Andrew R. McGill, and Lynda St. Clair (eds.), *Corporate Global Citizenship* (San Francisco: New Lexington Press, 1997), pp. 230–44.

18. Michael E. Gorman, Matthew M. Mehalik, Schott Sonenshein, and Wendy Warren, "Toward a Sustainable Tomorrow," in Laura Westra and Patricia H. Werhane, *Business Consumption: Environmental Ethics and the Global Economy* (Lanham, MD: Rowman & Littlefield, 1998), pp. 333–38.

19. Quoted in Charles A. Thrall and Jerold M. Starr (eds.), *Technology, Power, and Social Change* (Lexington, MA: Lexington, 1972), p. 17.

20. Miguel Bustillo, "Water Bottles Are Creating a Flood of Waste," *Los Angeles Times,* May 28, 2003, pp. B-1 and B-7.

21. Seymour Melman, "A Note on: Safety Improvements as a Zero Defect Problem," in *Designing for Safety: Engineering Ethics in Organizational Contexts,* ed. Albert Flores (Troy, NY: Rensselaer Polytechnic Institute, 1982), p. 176.

22. Frank Graham, Jr., *Since Silent Spring* (Boston: Houghton Mifflin, 1970); Lisa H. Newton and Catherine K. Dillingham, *Watersheds 3: Ten Cases in Environmental Ethics* (Belmont, CA: Wadsworth, 2002), pp. 100–114.

23. D. R. Roberts, S. Manguin, and J. Mouchet, "DDT House Spraying and Re-emerging Malaria, *Lancet* 356 (2000): 330–32. Cited by Greg Pearson and A. Thomas Young (eds.), *Technically Speaking* (Washington, DC: National Academy Press, 2002), p. 19.

24. Mary H. Cooper, "Ozone Depletion," *CQ Researcher* (April 3, 1992).

25. Andrea Larson, "Consuming Oneself: The Dynamics of Consumption," in Laura Westra and Patricia H. Werhane (eds.), *The Business of Consumption: Environmental Ethics and the Global Economy,* pp. 320–21.

26. Arthur E. Morgan, *Dams and Other Disasters* (Boston: Porter Sargent, 1971), pp. 370–89.

27. Garrett Hardin, *Exploring New Ethics for Survival* (New York: Viking, 1968), p. 68.

28. R. Edward Freeman, Jessica Pierce, and Richard Dodd, "Shades of Green: Business, Ethics, and the Environment," in Laura Westra and Patricia H. Werhane (eds.), *The Business of Consumption: Environmental Ethics and the Global Economy* (Lanham, MD: Rowman & Littlefield Publishers, 1998), pp. 339–53.

29. An especially valuable resource, one that includes illuminating discussions as well as anthologized readings, is P. Aarne Vesilind and Alastair S. Gunn, *Engineering, Ethics, and the Environment* (New York: Cambridge University Press, 1998). A wide range of views are represented in Joseph R. Des Jardins (ed.), *Environmental Ethics: An Introduction to Environmental Philosophy,* 3rd ed. (Belmont, CA: Wadsworth, 2001).

30. William T. Blackstone, "Ethics and Ecology," in W. M. Hoffman and J. M. Moore (eds.), *Business Ethics,* 2nd ed. (New York: McGraw-Hill, 1990), p. 473.
31. Thomas E. Hill, Jr., "Ideals of Human Excellence and Preserving Natural Environments," in *Autonomy and Self-Respect* (Cambridge, MA: Cambridge University Press, 1991), p. 104.
32. Peter Singer, *Animal Liberation*, rev. ed. (New York: Avon Books, 1990), p. 6.
33. Tom Regan, *The Case for Animal Rights* (Berkeley, CA: University of California Press, 1983).
34. Mary Midgley, *Animals and Why They Matter* (Athens, GA: University of Georgia Press, 1984).
35. Albert Schweitzer, *Out of My Life and Thought*, trans. A. B. Lemke (New York: Henry Holt and Company, 1990), p. 157. Relevant readings are anthologized in Marvin Meyer and Kurt Bergel (eds.), *Reverence for Life* (Syracuse, NY: Syracuse University Press, 2002).
36. Paul W. Taylor, *Respect for Nature* (Princeton, NJ: Princeton University Press, 1986).
37. Aldo Leopold, *A Sand County Almanac* (New York: Ballantine, 1970), p. 262.
38. J. Baird Callicott, "Environmental Ethics," in *Encyclopedia of Ethics*, vol. 1, ed. L. C. Becker (New York: Garland, 1992), pp. 313–14.
39. Lynn White, "The Historical Roots of Our Ecological Crisis," *Science* 155 (March 10, 1967): 1203–7.
40. Mawil Y. Izzi Deen (Samarrai), "Islamic Environmental Ethics, Law and Society," in J. R. Engel and J. G. Engel (eds.), *Ethics of Environment and Development* (Bellhaven Press, London, 1990).
41. Ian G. Barbour, *Technology, Environment, and Human Values* (New York: Praeger, 1980), p. 316. On Islam, see Mawil Y. Izzi Deen (Samarrai), "Islamic Environmental Ethics, Law, and Society," in *Ethics of Environment and Development,* ed. J. Ronald Engel and Joan Gibb Engel (London: Bellhaven Press, 1990).
42. John Neihardt, *Black Elk Speaks* (Lincoln: University of Nebraska Press, 1961).
43. Karen J. Warren (ed.), *Ecological Feminist Philosophies* (Bloomington, IN: Indiana University Press, 1996).
44. James P. Sterba, "Reconciling Anthropocentric and Nonanthropocentric Environmental Ethics," in *Ethics in Practice,* ed. Hugh LaFollette (New York: Blackwell, 1997), pp. 644–56.
45. J. Baird Callicott. "Environmental Ethics in the Anthropocene," *Transtext(e)s Transcultures* 13 (2018). Available at: http://journals.openedition.org/transtexts/1064; DOI: 10.4000/transtexts.1064
46. Art Davidson, *In the Wake of the Exxon Valdez* (San Francisco: Sierra Club Books, 1990).
47. Richard Routley and Val Routley, "Human Chauvinism and Environmental Ethics," *Environmental Philosophy,* Monograph Series, no. 2, ed. Don Mannison, Michael McRobbie, and Richard Routley (Australian National University, 1980), p. 121.
48. W. Michael Hoffman, "Business and Environmental Ethics," in W. Michael Hoffman, Robert E. Frederick, and Mark S. Schwartz (eds.), *Business Ethics: Readings and Cases in Corporate Morality,* 4th ed. (Boston: McGraw-Hill, 2001), p. 441.
49. James Lovelock, *Gaia: A New Look at Life on Earth* (New York: Oxford University Press, 2000), p. 10.

CHAPTER
9

ENGINEERING ETHICS IN
THE GLOBAL CONTEXT

On September 11, 2001, at 8:46 A.M., Al-Qaeda terrorists flew a hijacked American Airlines Boeing 767 into floors 94 to 98 of the 110-story North Tower of the World Trade Center. Seventeen minutes later, the World Trade Center was hit again as more terrorists flew a United Airlines Boeing 767 into floors 78 to 84 of its 110-story South Tower. The impact of the airplanes did not collapse the twin towers, but the firestorm set off by the full loads of jet fuel, together with the tons of combustible office material, created intense heat that weakened the steel supports. First the steel floor trusses weakened and began to tear away from the exterior and interior steel columns, and then the compromised columns gave way. Once the top floors collapsed, it took only 12 seconds for the pancake-like cascade of the South Tower to occur, followed a short time later by the North Tower. A third hijacked airliner was flown into the Pentagon, and a fourth crashed outside Shankesville, Pennsylvania, as its passengers fought the hijackers. All passengers and crews on the jets were killed, and the overall death toll was more than 3000 people, including hundreds of firefighters and police officers.

Engineers were prescient in designing the twin towers to withstand impacts from jumbo jets, but they only envisioned jets that were moving slowly and, with depleted fuel, making emergency landings; they had not imagined the possibility of a terrorist attack like the one on 9-11, nor had anyone else. Since airplanes had crashed into tall buildings before, engineer James Sutherland had warned in 1974 of the vulnerability of hundreds of skyscrapers to further crashes, but it was only in 1994 that the warning was taken seriously after a terrorist plot to hijack an Algerian airliner to attack Paris was foiled.[1] In addition, a bold decision (costing $300,000) was made during construction of the towers to replace asbestos insulation, whose health dangers were only then becoming clear and which had already been used on the first 34 floors, with new fireproofing material coming on the market.[2] The impact of the crash, however, stripped the insulation from the steel beams, leaving them unprotected from temperatures over 1100 degrees Fahrenheit. Nor was there a safe exit for people above the impact area, as sprinkler systems, emergency elevators, and stairways were damaged by the crash. Fortunately, the buildings stood long enough for some 25,000 people to escape. Preliminary studies of the exact failure mode that led to the collapse left questions that ongoing studies will attempt to answer.[3] These findings will shape future engineering design, adding additional complexities to future social experiments in engineering tall buildings and other vulnerable structures.

As the tallest buildings in New York City and as centers of international commerce, the Twin Towers symbolized the global economy and America's dominance within that economy. The terrorists were fanatics who opposed Western capitalism, democracy, and moral pluralism. Politicians portrayed the attack as an assault on civilization, but perhaps a more accurate statement is that the violence expressed "tensions built into a single global civilization as it emerges against a backdrop of traditional ethnic and religious divisions."[4]

Globalization refers to the increasing integration of nations through trade, investment, transfer of technology, and exchange of ideas and culture. Daniel Yergin and Joseph Stanislaw distinguish a narrow and broader sense of "globalization."

> In a more narrow sense, it represents an accelerating integration and interweaving of national economies through the growing flows of trade, investment, and capital across historical borders. More broadly, those flows include technology, skills and culture, ideas, news, information, and entertainment—and, of course, people. Globalization has also come to involve the increasing coordination of trade, fiscal, and monetary policies among countries.[5]

Today's interdependence among societies—economic, political, and cultural—is unprecedented in its range and depth. So are the possibilities for increased unity and increased fractures during the process of globalization. Global interdependency affects engineering and engineers in many ways, including the environmental issues discussed in chapter 8. As engineering work becomes ever more global, growing numbers of educational institutions, programs, and initiatives are grappling with how to better prepare their engineering graduates to work more effectively across geographic boundaries. Nevertheless, students typically receive very little guidance on how to act ethically and professionally when working with people from cultures different from their own. Philosophers and engineering educators have recently explored different efforts to situate engineering ethics in the global context. The discrepancies in these different efforts often derive from different understandings of what constitutes the *global*. This chapter discusses four different approaches to addressing engineering ethics issues in the global context: (1) global ethical codes, or developing a code of ethics that is expected to be applied across cultures; (2) functionalist theory, which posits some fundamental, shared characteristics internal to the engineering profession that apply globally and might prove foundational for creating ethical codes; (3) cultural studies, which emphasizes the importance of cultural differences in formulating effective ethical decisions in the global context; and (4) global ethics and justice, which engages students and professionals in ideas and practices aiming to promote global justice.

9.1 GLOBAL ETHICAL CODES

Codes of ethics assume a most critical role in the education and professional practice of engineers especially in the United States. Therefore, in developing tools for tackling ethical issues arising from engineering practice in the global context, it is quite natural that some engineers and engineering educators have been striving to build up a code of ethics that is expected to be applicable across cultures. To a large extent, their major goal is to create a *globalized* engineering profession. It is worth noting that proponents embracing this approach often emphasize the importance of *coordination* among professional societies from different countries in creating a global code of ethics. In this sense, creating a global code of ethics involves achieving agreement among organizations, cultures, and countries.

Since the beginning of the 21st century, countries in the same region or with similar cultural traditions have been exploring regional codes of ethics for their engineering societies. These codes of ethics can be seen as "globalizing" efforts to seek common ground among cultures. For instance, in November 2004, the Chinese Academy of Engineering, together with two other academies of engineering from Japan and South Korea, issued a "Declaration on Engineering Ethics" that included the "Asian Engineers' Guideline of Ethics." This guideline emphasized "cherishing the Asian cultural heritage of harmonious living with neighboring people and nature."[6] This code was intended to be shared by practicing engineers in three countries deeply influenced by the Confucian culture. This declaration stands as a recent effort to build up a code of ethics for engineers in countries that embrace some shared cultural values.

Saif alZahir and Laura Kombo additionally compare the IEEE code of ethics with 32 international codes of ethics of professional engineering societies in Africa, Asia, Australia, Europe, and Latin America.[7] They found that only four countries completely adopted the IEEE code of ethics, while the other 28 countries have embraced variations of the IEEE code. These variations were caused mainly by sociopolitical and cultural differences in these countries. Nevertheless, they argue that a global code of ethics is conceivable since at least some articles of the IEEE code of ethics are shared among the countries under investigation. If professional societies can work together to accommodate the sociopolitical and cultural differences in these countries (although it is unclear whether this "accommodation" process is easy or difficult to undertake), a global code of ethics is possible. It is also worth noting that the 32 international codes compared to the IEEE code of ethics are in different branches of engineering, i.e., not all in electrical, computer, and other allied fields of engineering that have historically been the primary purview of the IEEE. Disciplinary differences may therefore represent another source of variation in the codes.

As this overview suggests, creating a global code of ethics often requires inter-organizational or international governance and coordination. A typical example in this regard is the code of ethics developed by the World Federation of Engineering Organizations (WFEO). Under the auspices of the United Nations Educational, Scientific and Cultural Organization (UNESCO), the WFEO was founded in 1968 by regional engineering organizations from more than 90 nations. According to the WFEO's official website:

> The WFEO is the sole Body representing the engineering profession of all kind and disciplines at World Level . . . [It] encourages all of its national and international members to contribute to global efforts to establish a sustainable, equitable and peaceful world by providing an international perspective and enabling mechanisms . . . [It] is the internationally recognized and chosen leader of the engineering profession and cooperates with national and other international professional institutions in being the lead profession in developing and applying engineering to constructively resolve international and national issues for the benefit of humanity.[8]

WFEO's code of ethics is viewed as a model code for its member organizations to formulate their own codes of ethics. Therefore, "the values and principles in the WFEO Model Code of Ethics are those which are deemed to be applicable universally to the practice of engineering" and "Member organisations of WFEO are encouraged to develop a Code of Ethics for their organisation based on the values and principles set down in the Model Code and to impart the values and principles that individuals need to assist their decision making processes through ethics support programs."[9]

WFEO's model code of ethics is more specifically organized around four different themes: "demonstrate integrity," "practise competently," "exercise leadership," and "protect the natural and built environment." Taking a closer look at the specific articles and their interpretations, one finds that many of these articles (e.g., "practise in a careful and diligent manner in accordance with their areas of competence") are highly similar (if not identical) to articles in existing codes of ethics of many Western, and especially American, professional societies such as the U.S.-based National Society of Professional Engineers (NSPE). Nevertheless, it is difficult to conclude the specific reasons why the WFEO's model code of ethics is similar to these other codes. For instance, it is possible that the American codes of ethics were seen as the "best practices" which WFEO intentionally "learned" from and expected other countries to "learn" as well. Additionally, it might be that American organizations had overwhelming influence when WFEO was drafting its model code of ethics. Finally, it is possible that engineering societies outside of America and other Western countries lacked a comparable professional tradition, and therefore had little choice but to follow the American societies, which were globally known for a tradition of professionalism in which ethical codes have historically had *the* central role.

However, philosophers such as Michael Davis counter that it is unnecessary to construct these global codes of ethics. To some extent, he argues that building up such codes is somewhat similar to "reinventing the wheel" as there are already "global" standards of engineering practice in the profession.[10] Many of the major professional societies such as IEEE, ASME, and NSPE encourage their members to apply their codes of ethics globally.[11] Some of these societies, such as IEEE and ASME, are essentially international organizations, and the members of these societies are bound by the codes of ethics of these societies wherever they live and work. Additionally, legal acts in the United States such as the Foreign Corrupt Practices Act (1977) require U.S. engineering firms to follow all federal anti-bribery and related laws when performing engineering and related services abroad.[12] Nevertheless, as pointed out by Arthur E. Schwartz, deputy executive director and general counsel of NSPE, "there is a limited exception for 'facilitation payment' (e.g., connecting to the electric grid, water, sewage hookup, installation of internet services) to acknowledge local compensation customs and practices."[13]

In light of this overview, the global ethical codes approach still leaves two questions unresolved. First, it is unclear whether it is realistic to expect that the global codes of ethics be used to guide engineering practice in different countries, beginning with their introduction to students in formal coursework and later as guidelines for conduct in the workplace. Second, given that American societies encourage their members to apply their codes of ethics universally, their foreign colleagues might know very little (if anything) about these codes. It might therefore be ineffective or unfair for two sides of a collaboration to have an unbalanced understanding of the codes of ethics that are supposed to guide their collaborative engineering practice. And third, there is the question of how to avoid Western "paternalism" in formulating and implementing these global codes of ethics in ways that override diversity in local values and practices, especially given that many countries do not fully embrace Western ideals of professionalism, expectations of social contracts, closely following laws and regulatory requirements, etc.

9.2 FUNCTIONALIST THEORY

The functionalist theory posits some fundamental, shared characteristics internal to the engineering profession that apply globally and might prove foundational for creating ethical codes. These shared characteristics make engineers and engineering in different countries "function" in similar ways. The functionalist theory assumes that engineering itself serves as a *culture* that is more prominent than national cultures in global engineering practice. Such a theory is represented mainly by two philosopher engineering ethicists, namely Michael Davis and Heinz C. Luegenbiehl.

Davis argues that engineers, no matter where they are from, are identified by "a common curriculum imparting a common discipline (a culture, that is, a shared way of doing certain things, the distinctive way of doing things we call 'engineering')."[14] Thus, this shared culture between Chinese engineers and engineers elsewhere allows Chinese engineers to work well with engineers elsewhere. In this sense, engineering as a culture is more powerful than national culture—it is easier for a Chinese engineer to move from China to the United States than to change their profession from engineering to medicine. The shared culture among engineers from all over the world allows them to understand a common set of ethical values that is linked to the very nature of the engineering profession.

Scholars taking a "cultural studies" approach (which we return to below) may agree that it is easier for an engineer to move from one country to another than for them to move from engineering to medicine. However, they may counter that we cannot conclude there are no differences in teaching and practicing engineering between countries. Without such differences, it is difficult to understand why international accreditation agreements such as the Washington Accord struggle to establish and enforce common accreditation standards for signatory nations. If engineering is a global discipline with a common culture, it would seem that mechanisms like the Washington Accord would not be necessary. Yet indeed, we find non-Western countries like China and India working hard to reform their engineering curricula to meet the requirements of joining the Washington Accord. Cultural studies scholars may further argue that engineers from two different countries might eventually be able to work well with each other, but they need time (sometimes extensive) to figure out how to coordinate such collaboration. Similar to Davis, Luegenbiehl points out,

> In thinking about engineering ethics independently of a particular cultural background, it is instead helpful to consider engineers to be a community with a shared set of values. A bond is created among engineers based on these values, whatever their nationality or cultural background, just as there typically exists a set of common core values in other types of societies.[15]

Furthermore, Luegenbiehl argues that a global ethic should be built upon the nature of engineering. Specifically, he proposes six foundational principles of engineering ethics that are independent of any particular cultural context:

- The principle of public safety
- The principle of human rights
- The principle of environmental and animal preservation
- The principle of engineering competence
- The principle of scientifically founded judgment
- The principle of openness and honesty[16]

To say some of these principles are independent of any particular cultural context is to say these principles or the values embedded in these principles are *shared* among engineers from most (if not all) cultures. However, some of these values are not equally important in the moral life of different countries. For instance, the idea of honesty in Confucian culture has more contextual and pragmatic connotations. For instance, engineers influenced by the Confucian culture may be more inclined to ask questions such as honest *for what* and *for whom* and *under what context*.

To a large extent, Luegenbiehl's approach is quite similar to the "principlist" approach in biomedical ethics. However, a limitation with this approach is that a given set of principles might generate conflicts in certain situations.[17] There are no general guidelines available for how to deal with such conflicts. For instance, if engineers strictly follow the principle that engineers should only engage in "engineering activities which they are competent to carry out,"[18] they may be facing situations in which people in underdeveloped countries are losing some of their human rights, such as the right to drink safe and clean water, as these engineers working in underdeveloped countries are not "competent" in civil and environmental engineering. Philosophers argue that these principles often lack systematic relationships to one another, and there is no unified moral theory from which the principles are derived.[19] As the reader can tell, these principles are not novel and can be found in the existing codes of ethics of most professional societies. For instance, very few "local" professional engineering societies would omit "the principle of public safety" in their codes of ethics. In this sense, Luegenbiehl is close enough to Davis in implying that completely creating a new global code of ethics through coordination among professional engineering societies is not necessary. The reason is simple: engineering itself is a *globalized* profession and our current existing codes of ethics are already global, as engineering societies in different countries share a lot of common values which are central to the engineering profession.

9.3 CULTURAL STUDIES

Scholars trained in fields emphasizing the importance of culture to professional practice (e.g., anthropology, business) and those who have had extensive experience teaching and researching in other countries often pay closer attention to cultural differences in engineering practice. Thus, they often hold an *anthropological* or a *cultural studies* view of engineering ethics in global context. In particular, cultural studies scholars claim that cultural differences exist in at least three different senses or contexts: *professional*, *practical*, and *sociocultural*.

First, the cultural studies approach argues that concepts central to the Western (mainly American) engineering profession and engineering codes of ethics, such as professional autonomy, are often less valued or even peripheral in other cultural contexts. Luegenbiehl points out that not all societies value moral autonomy to the degree that the United States does, and in some cultures such as Japan, moral autonomy is discouraged in society and at work.[20] In fact, Luegenbiehl's observation is mostly valid in other countries with Confucian cultural heritage, including China, South Korea, and Singapore. In Confucian culture, it is often impossible to view a person as truly autonomous. Everyone has a variety of roles and has different relationships with others in the society. Ethical decision-making is always influenced by the specific relationships one has with others. In other words, for a professional engineer, their moral judgment can hardly be "autonomous" or "independent" as it needs to incorporate considerations about the relationships they have with others and expectations of how others might respond to their moral actions.

Another interesting case is related to the idea of nepotism, which in most codes of ethics is completely prohibited. The principle of avoiding nepotism also applies when engineers are working in the global context.[21] Some scholars might call the involvement of nepotism in the hiring processes as a sort of conflict of interest. Certainly, the hiring manager needs to be specifically careful about whether a candidate is qualified. However, we are arguing that some kind of special relationship between the hiring manager and the candidate is not necessarily the reason that prevents the manager from being involved in the process and the candidate from being considered. Arguably, from the Confucian perspective, it is the manager that knows much more about the candidate than anybody else. A good professional should be able to make sound judgments about the credentials of a person with whom this professional shares a special relationship by considering but not being "distracted" by such a relationship. As Confucianists have argued, "*juxian bu biqin* (selecting virtuous people does not avoid relatives)."

Second, scholars in engineering management tend to emphasize cultural differences in implementing and managing specific engineering projects. A vast majority of their theories and methods are often drawn from the literature in international business and management. For instance, Wang and Thompson compare cultural differences in business ethics in Europe, the United States, and Asia.[22] They have found that business organizations (e.g., companies) have varied understandings of (1) moral agents in business responsibility; (2) key actors in business ethics; (3) key guidelines for ethical behavior; (4) key issues in business ethics; and (5) the dominant stakeholder management approach.[23] As engineers are often employed in such firms, these differences in organizational cultures may affect engineers' judgment in making professional decisions in the workplace. Thus, engineering students are suggested to acquire global experience through *interactions with people in or from other cultures*. Pedagogies for building up global competency in ethical decision-making include field trips to foreign engineering sites, meetings with professionals at work in other countries, and interactions with exchange students and scholars.

Brent Jesiek and his research group at Purdue University have developed scenario-based tools to assess and develop the practical competency of engineers in navigating ethical issues arising from cross-cultural engineering practice. Table 9-1 demonstrates a sample cross-cultural engineering ethics scenario.[24]

More than 70 total scenarios like the example scenario were created covering six national/culture contexts (China, Japan, India, France, Germany, and Mexico), and with about a third of the scenarios falling in the "ethics, standards and regulations" category. When using this type of scenario for instructional or assessment purposes, respondents are typically asked to evaluate each item (*a–g*) according to their relative effectiveness in addressing the ethical problem (e.g., on a scale from 1 = not at all effective to 10 = very effective). In the sample scenario, the knowledge about engineering practice in the local Chinese context (e.g., the line manager often has moral influence over workers as the line manager

TABLE 9-1
Sample cross-cultural ethics scenario

Your work as an industrial engineer for a major North American OEM automotive parts supplier has landed you at a plant your firm recently acquired outside of Shanghai, China. As a member of an acquisition transition team, you are assigned to work on safety and compliance issues. For several weeks, you have been encouraging workers at the plant to wear eye protection when using certain machines. Yet even after posting signs, making safety glasses widely available, and talking to individual workers, you find that most employees continue to ignore the requirement. What would you do in this situation?

a. Ask the Chinese plant manager to work with the line managers to enforce compliance
b. Ask the Chinese line manager(s) to announce the requirement and enforce compliance
c. Continue to encourage the workers to wear eye protection, as you have been doing
d. When workers are found not wearing eye protection, scold them in front of their peers
e. Propose a new system that acknowledges and rewards individuals and groups who comply with the requirement
f. Report the issue back to management at your company's U.S. headquarters
g. Perform a study to find out why the employees are not wearing eye protection

will often take the major responsibility for work safety and other compliance requirements) is critical for making a good judgment.

In the classroom or other professional development settings, individuals could also be asked to "rehearse" each of the response options and anticipate the possible consequences of each. The moral deliberation process may additionally require that participating individuals acquire and utilize contextual and cultural knowledge about a "typical" Chinese manufacturing firm, including: (1) the power dynamic at the workplace among different employees, including workers, engineers, line managers, and the plant manager; (2) who makes decisions; (3) how decisions are made and implemented; and (4) knowledge about the typical character of individual interactions and social norms in Chinese society (e.g., the culture of saving face, tensions between natives and foreigners, etc.). Jesiek's approach to assessing and developing ethical competency in the context of global engineering places a strong emphasis on the value of knowledge about engineering practices in a given cultural context. More scenarios and associated teaching guides developed by Jesiek's group can be found at their website (https://geec. info/gec-about).

A third group of scholars in the cultural studies approach are interested in the "particularities" or "localities" of the broader sociocultural contexts in which engineers are educated, establish their professional identities, do their work, and organize themselves. Representative work in this vein has been carried out by Downey, Lucena, and Mitcham. By comparing engineering ethics in France, Germany, and Japan, they argue that the relationship between the identity of the engineer and the responsibility of engineering work has varied significantly over time and from place to place in the global context.[25] To a large extent, the idea of responsibility is extensively contextualized, and it is a cultural and historical concept. Responsible to whom, to what degree, and in what sense always matters to engineers and their everyday practices. For instance, American codes of ethics place significant emphasis on the idea of "contractarianism," which often makes less sense in national/cultural contexts such as China and India. This is not to say that China and India do not have contracts, but these countries view contracts very differently than their American colleagues do. For instance, Chinese engineers tend to view contracts only as the starting point of building up business relationship rather than legally binding documents. The Chinese engineers are often willing to conduct services that are not included in the contract if they think these services are conducive to their partners.

9.4   GLOBAL ETHICS AND JUSTICE

Scholars who embrace the global ethics and justice approach to engineering ethics in the global context have often been inspired by the idea of *minimal moral realism*. German philosopher Hans Küng's concept "global ethic" is a good example of minimal realism. What Küng means by the concept of global ethic is the "necessary minimum of common values, standards, and basic attitudes." In other words, it is "a minimal basic consensus relating to binding values, irrevocable standards, and moral attitudes, which can be affirmed by all religions despite their undeniable dogmatic or theological differences and should also be supported by non-believers."[26]

The United Nation's list of human rights has been widely considered as a resource for making ethical decisions in engineering practice across cultural boundaries.[27] Similar to Küng, these human rights are seen as minimal standards of living for people living globally. Another relevant concept embracing the philosophy of minimal moral realism is the "human capabilities" framework as originally advocated by philosophers Martha Nussbaum and Amartya Sen. Human capabilities are basic capabilities that "a person needs to be able to satisfy in order to live in a reasonable quality of life."[28] Engineering ethicists have pointed out that engineers working in the developing context have both a *negative* duty not to interfere with these human rights as well as a *positive* duty to help others achieve these rights.[29]

Philosophers of technology have additionally been exploring ways to assess whether engineering designs in developing countries have extended or hindered the human capabilities of local people. For instance, Oosterlaken, Grimshaw, and Janssen studied the introduction of information and computer technologies (ICTs) such as podcasting devices in local villages in Zimbabwe.[30] They argue that a successful development project is not merely about giving local community members access to resources such as podcasting devices and MP3 players but also involves asking to what extent these ICTs contribute to the expansion of human capabilities, i.e., "the freedom to do some basic things that are necessary for survival and to escape poverty."[31] From the perspective of human capabilities, the most important approach to evaluating a specific engineering design is to evaluate to what extent and in what ways a given design or solution enhances the human capabilities of the users.

An underlying assumption of this global ethics and justice approach for engineering ethics is that engineering design is critical and essential for promoting (basic) human well-being. This assumption in reality has been articulated in many engineering codes of ethics. For example, the NSPE code of ethics states that "engineering has a direct and vital impact on the quality of life for all people." Philosophically, this idea of portraying "the formation of engineers as contributing directly to human progress" and equating "the technical contents of engineering practices" with "material advancements" throughout the world for human benefit is what Gary Downey calls "normative holism."[32]

Nevertheless, it is worth noting that technology cannot automatically promote human well-being. In many cases, "conversion factors" exist between technological design and human well-being.[33] Conversion factors are "personal, environmental, or social factors" that may "hinder the conversion of resources as such into valued human capabilities."[34] For instance, the bicycle will not automatically extend the capability to travel freely unless the rider is bodily-enabled and the road is well paved.

9.5   CULTIVATING GLOBALLY COMPETENT ENGINEERS

As suggested earlier, discrepancies in efforts to situate engineering ethics in the global context often derive from different understandings of what constitutes the *global*. The global ethical codes approach discusses the global from the perspective of professional societies. Scholars in this approach often tend to examine the similarities among codes of engineering ethics from different countries and try to establish a global code of ethics based on shared values and articles from different international codes of ethics. Their goal is to promote a *globalized engineering profession*. The functionalist theory approach tends to look at the nature of engineering. Scholars in this approach believe that *engineering itself is a kind of global culture*. The reason why engineers from different countries are all called engineers is that they share a common practical culture that allows them to understand each other's work. Hence, there are social and moral norms inherently embedded in engineering practice that are understandable and communicable to all engineers wherever they reside. The cultural studies approach examines the global from a more micro perspective and emphasizes the role of cultural differences in shaping the professional, practical, and sociopolitical contexts of engineering work. The cultural studies scholars view the global more in terms of *localized practices that are challenged by processes of globalization*. Finally, the global ethics and justice approach considers the global as *a set of universal ethical values shared by all cultures*. An important goal of engineering is to prevent these values from being hindered or diminished and to design technologies that help promote such values.

We recommend that educators, practitioners, and policymakers in global engineering consider all four approaches in a more systematic and integral way. If we imagine a typical global engineer who is working with engineers from other cultures, certainly the global engineer must have some "common language" to start to work with their colleagues since they are all *engineers*. Otherwise, engineering work would simply not happen if multiple fully incommensurable cultures are involved. However, we need to notice that they will likely encounter situations in which significant cultural differences of engineering work do exist. The global engineer needs to be aware of effective ways to navigate these cultural differences, which is crucial for achieving their common goals. We must also admit that not every global engineer is exclusively interested in pursuing economic benefits. Many of the global engineers might also be interested in how their design solutions can advance the well-being and agency of the users.

A limitation of current practices in global engineering is that people often pay attention to only one or two of the four approaches to engineering ethics and ignore or downplay the others. Here we would like to make some general and perhaps even oversimplified assumptions based on our preceding discussion. We acknowledge that these assumptions suggest the need for further empirical evidence to verify their reliability and validity. The assumptions are as follows:

a. in practice, professional engineers will likely pay more attention to global codes of ethics as they are quite familiar with the existing codes of ethics and have curiosity and imagination about whether these codes of ethics are applicable in other cultures.

b. so far as we know, only a few scholars (mainly philosophers) have claimed that there are some characteristics unique to the engineering profession that define engineering as a globalized profession. It is unclear if we can conclude that this phenomenon is due to the traditional mission of philosophy that is "seeking the truth" essential for defining an object.

c. some business leaders, often with extensive experience traveling to different countries and working with or studying people from different backgrounds, often focus more on the cultural differences of engineering practice. Another interesting assumption might be that business leaders or global engineers working are more pragmatic and assume that cultural sensitivity is critical to business success. As a result, they are more interested in the cultural studies approach to engineering ethics. Scholars trained in sociology, anthropology, cultural studies, communication studies, and other "interpretive" social sciences may also be more interested in exploring how cultural differences intersect with engineering ethics.

d. engineers working for non-profit organizations often tend to amplify the moral and political values that are lacking and need to be further enhanced in developing contexts. They view technologies as *instruments* for well-being rather than profits.

As engineering educators who are interested in preparing future engineers for the increasingly globalized future, we need to be careful about what kind(s) of "global engineers" we are training. Emphasizing one or two approaches to engineering ethics over others represents an incomplete approach that fails to project an appropriately comprehensive view of global engineering practice. Obviously, we are not training every student to become a professional engineer working in a multinational business company, nor do we expect that every engineering graduate will work for an international development or other NGO.

Instead, we propose that educators should strive to prepare students for a wide variety of personal and professional pathways, yet with the goal of enabling them to become truly global engineers capable of navigating ethical issues in diverse job roles and national/cultural contexts. Thus, engineering educators from the four different approaches to engineering ethics in the global context need more communication, collaboration, and coordination among themselves, as how to educate a globally professional and responsible engineer is a very real and daunting issue that has received much less attention than other topics in the field of engineering education.

DISCUSSION QUESTIONS

1. Do you think engineering ethics in the global context should be about discovering commonalities or differences of values in engineering practice? And briefly explain why.

2. How would you balance respect for diversity with commitments to respect for individual rights in the following two cases?

    a. You are a woman assigned to work in a Middle Eastern country that requires women to wear traditional clothing, but doing so conflicts with your religious faith; or, you are a man who is a member of a team whose members include women who are required to wear traditional clothing. If you decline the assignment, your career advancement might suffer.

    b. Your company is asked to design a more efficient weaving apparatus whose size is quickly adjustable to young children, and you are assigned to the project. You know that the primary market for the apparatus is countries that use child labor.

3. Since Nigeria became a member of OPEC (the Organization of Oil Producing Countries) in 1970, the country's oil boom has led to increased corruption, lower living standards for the poor, and much political instability. Foreign oil companies (among whom Shell Oil has received much notoriety) are accused of having disregarded the safety and livelihood of local people when drilling and laying pipelines. The Ogony people in particular have protested, but in vain. Acquaint yourself with the happenings (then and now), and describe what you feel the role of foreign oil companies should be in a country such as Nigeria.

4. The World Trade Organization (WTO) was established to oversee trade agreements, enforce trade rules, and settle disputes. Some troublesome issues have arisen when WTO has denied countries the right to impose environmental restrictions on imports from other countries. Thus, for example, the United States may not impose a ban on fish caught with nets that can endanger other sealife such as turtles or dolphins, while European countries and Japan will not be able to ban imports of beef from U.S. herds injected with antibiotics. Yet, other countries ban crops genetically modified to resist certain pests, or products made therefrom, unless labeled as such. Investigate the current disputes and discuss how such problems may be resolved, not overlooking the fact that now a multinational company covering countries A and B has an opportunity

to pressure A to relax environmental regulations under the guise of reduced export opportunities to country B, and vice versa regarding exports from B to A.

5. Some professional societies explicitly affirm their codes of ethics as applying internationally, for example, the IEEE code. In a 1996 case, which involved a payment of bribes in another country, the National Society of Professional Engineers (NSPE) ruled that its code, too, applies internationally.[35] Examine the NSPE code to see whether all of its parts are straightforwardly applicable internationally. (See appendix B.)

6. Corporations' codes of ethics also have to take into account international contexts. Compare and contrast the benefits and liabilities of the types of ethics programs (a and b) at Texas Instruments at two different times, described as follows:

   a. Texas Instruments (TI) always had a long-standing emphasis on trust and integrity, but during the 1980s it greatly intensified its efforts to make ethics central to the corporation.[36] In 1987, TI appointed a full-time ethics director, Carl Skooglund, who was then a vice president for the corporation. Skooglund reported to an ethics committee that in turn reported directly to the board of directors. His activities included raising employees' ethical awareness through discussion groups and workshops on ethics, making himself directly available to all employees through a confidential phone line, and—especially relevant here—addressing specific cases and concerns in weekly newsletters and detailed brochures called "Cornerstones."

   b. In 1995, TI's popular chairman died suddenly, prompting a rapid review of its policies.[37] In two years, it made 20 acquisitions and divestitures, including selling its defense-industry business, leaving it with more non-U.S. employees than U.S. employees. The new chairman called for rethinking its ethics programs to have both a greater international focus and more emphasis on a competitive and "winning" attitude. Before his retirement, Carl Skooglund scrapped the Cornerstone series, focused on specific issues and cases, and replaced it with three core values: integrity (honesty together with respect and concern for people), innovation, and commitment (take responsibility for one's conduct).

7. Imagine an American computing corporation has donated 1,000 computers to a developing country such as Haiti where children need these computers to access rich learning materials on the Internet. Please list at least two possible "conversion factors" that may potentially limit these computers from helping the children improve their capability to learn knowledge.

8. You are a U.S.-based automotive engineer. Your boss asked you to go to China on his behalf and review and sign off on the final tooling for the new model body trim at your company's factory in Chongqing. Your boss said that the trip would be quick, as the local engineering manager had informed him that the tools were all ready to go. But, when you arrive on site, you see critical components strewn around the tool room still incomplete, and the Chinese engineering manager—whom you had never previously met—trying to hide the obvious shortcomings of the tooling. From a technical standpoint, you know exactly what needs to be done to get the tooling pulled together. What would you do?[38]

   a. Meet with one of the Chinese manager's subordinates to discuss the problem, then suggest they send a summary of the issues to both you and their manager.

   b. Invite the Chinese manager out to lunch to discuss the problem and possible solutions.

c. Seek out the Chinese manager's immediate supervisor, tell him about the situation, and offer to help solve the problem.

d. Ask the group if there have been any issues with completing the assembly.

e. Refuse to sign off on the tooling and explain why.

f. Call your boss in the United States to explain the situation and ask for his advice.

You may want to consult with Hofstede's cultural dimensions theory to get a general sense of professional cultures in China: https://www.hofstede-insights.com/country-comparison/china/

9. Based on the brief comparison of the ways in which American and Asian engineers perceive contracts in their professional work, how would you evaluate the effectiveness of the following responses?

As a sales engineer, you led negotiation of a contract with a Chinese customer that commits them to purchasing a certain number of parts from your firm each year at a discounted price. The contract stipulated that if the minimum number of parts was not purchased each year, the price would go up substantially. Having bought only 10 percent of the product they originally committed to, the price was raised significantly after the first twelve months. Now you are sitting down with a senior executive and several of his associates in one of their conference rooms, and you can see the executive is very angry. Before you even have a chance to speak he starts berating you in Chinese, then throws an ashtray that shatters on the floor. What would you do?[39]

a. Try to leave the room immediately and contact your boss.

b. Tell the executive that his behavior is not acceptable and he needs to cool off.

c. Try to calm down the angry executive and explain why you need to stick to the written contract.

d. Open up a conversation with the group about potentially renegotiating the contract.

e. Give the executive time to express his frustration, then explain that you value their business and want to discuss how to strengthen your relationship.

f. Stand up, go face-to-face with the angry executive, and tell him the price increase is not negotiable.

10. Conduct an interview with a student or a professor from your own field about their experience in engineering education and/or engineering practice and compare if their experience has any commonalities with your learning and/or professional experience.

KEY CONCEPTS

—**Globalization:** increasing integration of nations through trade, investment, transfer of technology, and exchange of ideas and culture.

—**The global ethical codes approach:** an approach to engineering ethics in the global context that aims to build up engineering codes of ethics that are expected to be applicable across cultures.

—**The functionalist theory approach:** an approach to engineering ethics in the global context that considers engineering itself as a culture. There are some fundamental, shared characteristics internal to the engineering profession that apply globally and might prove foundational for creating ethical codes.

—**Cultural studies approach:** an approach to engineering ethics in the global context that emphasizes the importance of culture to professional norms, engineering practice, and the sociocultural context of engineering.

—**Global ethics and justice approach:** an approach to engineering ethics in the global context that views a major obligation of engineers is to employ their expertise to promote well-being of people in underserved communities.

—**Minimal moral realism:** the philosophical view that acknowledges the necessary minimum of common values, standards, and basic attitudes in the global world.

—**Human capabilities:** some fundamental human rights that a person needs to be able to satisfy in order to live in a reasonable quality of life.

—**Conversion factors:** personal, environmental, or social factors that may hinder the conversion of resources into valued human capabilities.

REFERENCES

1. R. J. M. Sutherland, "The Sequel to Ronan Point," *Proceedings, 42d Annual Convention, Structural Engineers Association of California* (October 4,5,6, 1973), p. 167, cited by James R. Chiles, *Inviting Disaster: Lessons from the Edge of Technology* (New York: HarperBusiness, 2002), p. xx.
2. Angus Kress Gillespie, *Twin Towers: The Life of New York City's World Trade Center* (New York: New American Library, 2002), pp. 117–18.
3. Building Performance Study Team, *World Trade Center Building Performance Study: Data Collection, Preliminary Observations, and Recommendations* (Washington, DC: U.S. General Printing Office, May 2002).
4. Benjamin R. Barber, "Democracy and Terror in the Era of Jihad vs. McWorld," in Ken Booth and Tim Dunne (eds.), *Worlds in Collision: Terror and the Future of Global Order* (New York: Palgrave Macmillan, 2002), p. 249.
5. Daniel Yergin and Joseph Stanislaw, *The Commanding Heights: The Battle for the World Economy* (New York: Touchstone, 2002), p. 383.
6. Qin Zhu, "Engineering Ethics Studies in China: Dialogue between Traditionalism and Modernism," *Engineering Studies* 2, no. 2 (2010): 85–107.
7. Saif alZahir and Laura Kombo, "Towards a Global Code of Ethics for Engineers," *Proceedings of the 2014 IEEE International Symposium on Ethics in Science, Technology and Engineering* (2014). Available at: https://ieeexplore.ieee.org/abstract/document/6893407
8. WFEO, "WFEO Model Code of Ethics" (2006). Available at: https://www.wfeo.org/wp-content/uploads/code_of_ethics/WFEO_MODEL_CODE_OF_ETHICS.pdf
9. WFEO, "Who We Are" (2008). Available at: https://www.wfeo.org/about/
10. Michael Davis, "'Global Engineering Ethics': Re-inventing the Wheel?" in Colleen Murphy, Paolo Gardoni, Hassan Bashir, Charles Harris, and Eyad Masad (eds.), *Engineering Ethics for a Globalized World* (Basel, Switzerland: Springer, 2015), pp. 69–78.
11. Charles Harris, Michael Pritchard, and Michael Rabins, *Engineering Ethics: Concepts and Cases* (4th ed.) (Belmont, CA: Wadsworth, 2009).
12. Arthur E. Schwartz, "Engineering Society for Codes of Ethics: A Bird's-Eye View," *The Bridge: Linking Engineering and Society* 47, no. 1 (2017): 21–26.
13. Ibid.
14. Michael Davis, "Defining Engineering from Chicago to Shantou," *The Monist* 92, no. 3 (2009): 325–38.
15. Heinz C. Luegenbiehl, "Ethical Principles for Engineers in a Global Environment," in Ibo van de Poel and David E. Goldberg (eds.), *Philosophy and Engineering* (Dordrecht, Netherlands: Springer, 2017), pp. 147–59.
16. Ibid.

17. Sven Ove Hansson, "Theories and Methods for the Ethics of Technology," in Sven Ove Hasson (ed.), *The Ethics of Technology: Methods and Approaches* (London, UK: Rowman & Littlefield International, 2017), pp. 1–14.

18. Heinz C. Luegenbiehl, "Ethical Principles for Engineers in a Global Environment."

19. K. Danner Clouser and Bernard Gert, "A Critique of Principlism," *Journal of Medicine & Philosophy* 15, no. 2 (1990): 219–36.

20. Heinz C. Luegenbiehl, "Ethical Autonomy and Engineering in a Cross-Cultural Context," *Techne* 8, no. 1 (2004): 57–78.

21. Charles Harris, Michael Pritchard, and Michael Rabins, *Engineering Ethics.*

22. George Wang and Russel G. Thompson, "Incorporating Global Components into Ethics Education," *Science and Engineering Ethics* 19, no. 1 (2013): 287–98.

23. Ibid.

24. Qin Zhu and Brent K. Jesiek, "A Pragmatic Approach to Ethical Decision-Making in Engineering Practice: Characteristics, Evaluation Criteria, and Implications for Instruction and Assessment," *Science and Engineering Ethics* 23, no. 3 (2017): 663–79.

25. Gary Downey, Juan Lucena, and Carl Mitcham, "Engineering Ethics and Identity: Emerging Initiatives in Comparative Perspective," *Science and Engineering Ethics* 13, no. 4 (2007): 463–87.

26. Hans Küng, "Global Ethic and Human Responsibility," paper presented at the *High-Level Group Meeting on Human Rights and Human Responsibilities in the Age of Terrorism*. Available at: http://www.oneworlduv.com/wp-content/uploads/2011/06/hkung_santaclara_univ_global_ethic_human_resp_2005.pdf

27. Charles Harris, Michael Pritchard, and Michael Rabins, *Engineering Ethics.*

28. Ibid.

29. Ibid.

30. Ilse Oosterlaken, David J. Grimshaw, and Pim Janssen, "Marrying the Capability Approach, Appropriate Technology and STS: The Case of Podcasting Devices in Zimbabwe," in I. Oosterlaken and Jeroen van den Hoven (eds.), *The Capability Approach, Technology and Design* (Dordrecht, Netherlands: Springer, 2012), pp. 113–33.

31. Ibid.

32. Gary Downey, "The Local Engineer: Normative Holism in Engineering Formation," in Steen H. Christensen, Carl Mitcham, Li Bocong, and Yanming An (eds.), *Engineering, Development and Philosophy: American, Chinese and European Perspectives* (Dordrecht, Netherlands: Springer, 2012), pp. 233–51.

33. Jeroen van den Hoven, "Human Capabilities and Technology," in I. Oosterlaken and Jeroen van den Hoven (eds.), *The Capability Approach, Technology and Design* (Dordrecht, Netherlands: Springer, 2012), pp. 27–38.

34. Ibid.

35. National Society of Professional Engineers, *Opinions of the Board of Ethical Review*, case 96–5.

36. Francis J. Aguilar, *Managing Corporate Ethics* (New York: Oxford University Press, 1994), pp. 120–35, 140–43.

37. Dawn-Marie Driscoll and W. Michael Hoffman, *Ethics Matters: How to Implement Values-Driven Management* (Waltham, MA: Center for Business Ethics at Bentley College, 2000).

38. Brent K. Jesiek and Sang E. Woo (eds.), "GEC Scenario #3: Chongqing Tooling" (2018). Available at: https://geec.info/gec03

39. Brent K. Jesiek and Sang E. Woo (eds.), "GEC Scenario #20: Contract Conditions" (2018). Available at: https://geec.info/gec20

CHAPTER
10

TECHNOLOGY AND ENGINEERING LEADERSHIP IN FUTURE SOCIETIES

William A. Wulf, president of the National Academy of Engineering, calls for greater attention to broader social issues in the study of engineering ethics.[1] In addition to studying the micro issues concerning decisions made by individuals and corporations, we must also consider macro issues about technology, society, and groups within society, including engineering professional societies and the engineering profession in its entirety. In support of his view, Wulf cites philosopher John Ladd, who much earlier argued that an overly narrow focus on codes of ethics neglects broader issues about "technology, its development and expansion, and the distribution of the costs (e.g., disposition of toxic wastes) as well as the benefits of technology. What is the significance of professionalism from the moral point of view for democracy, social equality, liberty and justice?"[2]

In tune with the suggestions of Wulf and Ladd, we have linked micro and macro issues throughout this book, especially in developing the model of engineering as social experimentation and in discussing environmental and global issues. This chapter links our themes more explicitly to broader studies of technology in the interdisciplinary field called STS—an acronym for Science, Technology, and Society and for Science and Technology Studies—and also in the branch of philosophy called Philosophy of Technology.[3] It also underscores the importance of leadership by engineers in addressing broader issues about technological progress and about other areas of engineering. We will contextualize the discussions on the social impacts of technology and engineering leadership in the context of future societies that are enabled and shaped by emerging technologies such as artificial intelligence (AI), data science, and robotics.

10.1 CAUTIOUS OPTIMISM

We begin by discussing general attitudes toward technology. We then shift to more focused, although still general, points in thinking about technological progress. These grounds concern the prospects for human freedom and wisdom. Does technology control society? Is technology value-neutral or value-laden? Given the uncertainty surrounding technological development, are there grounds for hope in looking to the future? Are these attitudes toward technology being challenged or shaped by AI-enabled technologies?

10.1.1 Optimism, Pessimism, Realism

Both values and facts are involved in assessing when technological *change* constitutes technological *progress*. Progress means advancement toward valuable goals, hopefully using permissible means. Typically, debates between optimists and pessimists turn on more than disagreements about the facts and estimates of risks. They involve differing judgments about moral values, especially values of social justice, human fulfillment, and respect for the environment.

Often we think of technological progress narrowly, as enabling the improved performance of specific tasks. If the task is creating warmth in dwellings, we see a straight-line progression from wood fires in caves, to wood-burning fireplaces, to coal-burning furnaces, to gas furnaces. A fully honest reckoning, however, will take into account the sum total of benefits and losses provided by a technology, including impacts on the environment and social structures. Given the enormous complexity and variety of technology, can we make overall assessments of technology in its entirety?

In fact, most of us do develop global attitudes about the major aspects of our lives, for example, love, money, health, nature—and technology. Scholars typically group these attitudes into three categories: optimism, pessimism, and a third category sometimes called realism (being realistic about power) or contextualism (paying close attention to variations within specific contexts) that emphasizes the moral ambiguities of technology. Thus, Ian Barbour establishes such a threefold distinction: technology as liberator, technology as threat, and technology as a morally ambiguous instrument of power.[4] As authors, we are "cautious optimists" whose views straddle Barbour's first and third categories—optimism combined with realism. But let us briefly outline each of his three categories, citing representative thinkers.

General optimism about technology as liberator emerged with and helped fuel the emergence of modern science and industry. Early spokespersons for this emergence were understandably enthusiastic and even utopian in their vision of the steady development of techno-science: science as an unlimited wellspring for new technology, and technology in turn as advancing science. Around 1600, Francis Bacon proclaimed that "knowledge is power," and around 1800 Auguste Comte envisioned a technocracy (as it is now called) in which technologists govern society for the good of all. Following the technology-involved horrors in the twentieth century, especially world wars, a more nuanced optimism emerged that celebrates technology while calling for greater wisdom in its application.

Wulf voices strong optimism of this sort. In support of his view, he cites the list of top 20 engineering achievements of the twentieth century identified by the National Academy of Engineering.[5] That list, cited in chapter 1, bears repeating: electrification, automobiles, airplanes, water supply and distribution, electronics, radio and television, agricultural mechanization, computers, telephones, air-conditioning and refrigeration, highways, spacecrafts, Internet, imaging technologies in medicine and elsewhere, household appliances, health technologies, petrochemical technologies, laser and fiber optics, nuclear technologies, and high-performance materials. Wulf cautions that technologies have become so complex, and interactive in their complexity, that some negative impacts are literally unforeseeable. Nevertheless, it is precisely this new awareness of complexity that provides hope that humanity—and engineering—will act responsibly.

Similar optimism is expressed by Emmanuel G. Mesthene, former director of the Harvard Program on Technology and Society. Mesthene acknowledges that in solving some problems, technology generates new challenges. For example, automobiles solved transportation problems and provided greatly expanded mobility, but it led to smog, congested freeways, and the death of tens of thousands of people each year in the United States alone. Nevertheless, Mesthene sees strong grounds for overall optimism about technology's power to overcome poverty and create wealth, fight starvation and disease, raise the quality of life by steadily increasing opportunities, and above all to open up new possibilities. Hope is especially justified because humanity has become self-conscious about the unintended side effects of technology, and hence the need for wisdom in dealing with them.

> Technology, in short, has come of age, not merely as technical capability, but as a social phenomenon. We have the power to create new possibilities, and the will to do so. By creating new possibilities, we give ourselves more choices. With more choices, we have more opportunities. With more opportunities, we can have more freedom, and with more freedom we can be more human.[6]

As one more spokesperson for technological optimism, we cite Alvin M. Weinberg, a pioneer in the development of atomic energy and the person who coined the expression "quick technological fix," or "quick fix." Technological progress offers a more effective and less coercive remedy to intractable social problems. A quick fix for reducing deaths on the highways would be to design safer highways and cars. To solve water shortages in western states, the engineer would develop ways to generate more water at a cheaper price, perhaps by designing nuclear desalination plants.

In the age of AI, some engineers and policy scholars have expressed optimistic attitudes toward future societies. Robots will be developed to complete tasks that humans do not want to do or cannot do well, especially those jobs that are dull, dirty, and dangerous. The deployment of self-driving cars will significantly reduce the number of deaths in accidents. Other scholars argue that the automation of human jobs will liberate humans from work overburden and allow them to focus on activities that they find "intrinsically rewarding" (e.g., creating art, learning, playing games, raising children, or spending time with friends).[7]

Recently, designers, engineers, and ethicists in the United States have explored various ways to integrate values into technologies that are expected to generate positive social changes in the use context. Value-sensitive design (VSD) was developed by Batya Friedman and her colleagues at the University of Washington to emphasize the ethical values of stakeholders in fields of information systems design and human–computer interaction.[8] In Europe, there has been a movement "design for values (DFV)" led by some major engineering universities in the Netherlands such as Delft University of Technology. There are at least three goals that DFV attempts to achieve in design: (1) mitigating the mismatch between values that technologies embody and the values users hold; (2) incorporating moral and social values into engineering design; and (3) generating or stimulating certain values in the use of technologies.[9] Experts at the 4TU Centre for Ethics and Technology are now applying the DFV method to create technologies that respond to the emerging ethical, social, and political challenges engendered by the COVID-19 pandemic. For instance, researchers at Delft studied the opaque nature of algorithmic governance for the enforcement of quarantine measures and how people are affected unjustly by these measures.

In contrast, technological pessimists see a predominance of bad over good in major technological trends. Pessimists emphasize that technologies can disrupt communities, cause massive layoffs, alienate workers who are reduced to menial tasks, and create a sense of lost control as large organizations come to dominate our lives. Although few engineers embrace such pessimism, it is an attitude they must understand and confront in others.

Much pessimism about technology flows from how it threatens cherished values. Sometimes the values are moral, religious, and aesthetic values that are shuffled aside amid the distractions of technology-driven consumerism. Ralph Waldo Emerson, for example, complained that technology tends to narrow the human personality: "Look up the inventors. Each has his own knack; his genius is in veins and spots. But the great, equal, symmetrical brain, fed from a great heart, you shall not find."[10] Such a view seems implausible today. Not only do many engineers achieve breadth of understanding, but narrowness about technology is equally commonplace among moralists, religious thinkers, literary people, and even some scientists.[11]

Much pessimism is based on uncovering general patterns of technological thinking and dominant technological trends that subvert traditional values such as freedom and community. Ominous visions of technology emerged in the post–World War II era. Lewis Mumford, for example, began his career optimistic about technological progress, but he came to depict a world of impersonal, technology-driven bureaucracies to which the individual had to conform.

Technology moves in the direction of concentrating power in ways that erode democratic freedoms:

> the dominant minority will create a uniform, all-enveloping, super-planetary structure, designed for automatic operation. Instead of functioning actively as an autonomous personality, man will become a passive, purposeless, machine-conditioned animal whose proper functions, as technicians now interpret man's role, will either be fed into the machine or strictly limited and controlled for the benefit of depersonalized, collective organizations.[12]

The French thinker Jacques Ellul went even further in characterizing technology as "autonomous," literally beyond the control of human beings. Conceiving of technology as "technique"—that is, the modes of thinking and types of organizational structures driving the development of machines—Ellul wrote that "technique has become the new and specific milieu in which man is required to exist. . . . It is artificial, autonomous, self-determining, and independent of all human intervention."[13]

Albert Borgmann, an influential philosopher whose response to contemporary technology veers in a pessimistic direction, uses the example of the traditional fireplace to illustrate how easily we overlook the changes brought about by technologies. The hearth was part of what centered and unified a family: "It was a *focus,* a hearth, a place that gathered the work and leisure of a family and gave the house a center. Its coldness marked the morning, and the spreading of its warmth the beginning of the day."[14] Valuable technology, in Borgmann's view, promotes and sustains the values of family and community, and it also engages individuals' skills and caring—as in cutting firewood or starting the morning fire. Much contemporary technology does not do that. Instead, it consists of "devices": artifacts that serve a specific purpose, but whose inner workings we have no grasp of and which we view as mass-produced, disposable items. In addition, much of it erodes valuable relationships and activities. Television, for example, seems to liberate us by opening us to a wider world, but it also reduces occasions for family activities and absorbs time for reading.

Philosophers Shannon Vallor and Pak-hang Wong have recently concerned that robot caregivers may potentially reduce and remove the opportunities for nurses to develop their moral skills and virtues of care.[15] These moral skills and virtues of care are often acquired through nurses' specific practices and experience with their patients (e.g., conversation, body contact, and empathizing with patients). The moral skills nurses have developed cultivate practical wisdom and moral habituation that constitute the true virtues of care. If robot caregivers are as effective as human caregivers, they may devalue human care and therefore make care know-how useless or unnecessary. More broadly, a widespread concern in the United States regarding automation nowadays is that the deployment of robotics will replace a large number of human jobs particularly truck drivers. American public often hold a much higher expectation for safety standards of self-driving cars than those of traditional cars. Therefore, such a public attitude toward self-driving cars often leads to a dilemma: on the one hand, most people feel worried about the safety issues of self-driving cars and they do not feel comfortable with participating in safety tests. On the other hand, without sufficient people participating in these tests, it will be extremely difficult for engineers to further improve the reliability of self-driving cars.[16]

Many people waver between technological optimism and pessimism according to the technologies of greatest concern or interest to them at a given time and whether what we value is threatened or promoted by them. Such general attitudes play practical roles in influencing our responses to choosing careers (most engineers are optimistic), leaving careers (some engineers become disillusioned), voting as citizens, and risking our money as investors. We believe, however, that it is better to explore the more specific issues that underlie the disputes between optimists and pessimists, guided by Barbour's third attitude toward technology—with an overall optimistic accent.

That third attitude is that technology is "an ambiguous instrument of power whose consequences depend on its social context."[17] Power can be used for good or evil, and for greater or lesser good. All technology involves trade-offs, and the trade-offs can be made wisely or selfishly. Again, technologies can be used for good or bad purposes: a knife can cut bread or kill an innocent person. Often values become embedded in technological products and approaches in ways that create unfair power imbalances. And major technologies carry a momentum that cannot be fully controlled by individuals, although they remain under the control of larger groups. These interwoven themes, discussed in what follows, can be affirmed while retaining a strong sense of optimism and hope that is so essential in engineering. The optimism and hope, however, are selectively targeted toward those specific technologies that are reasonably foreseen to produce genuine benefits to humanity, and even then they are tempered with an awareness of the risks involved. This is the import of our first theme of cautious optimism about moral agency and decency, about engineering professionalism as having a moral core, and about responsible technology as integral to human progress.

10.1.2   Technology: Value-Neutral or Value-Laden?

One issue underlying the clash of optimism and pessimism about technology concerns how values are related to technology: Is technology value-neutral or value-laden? Many people think of technology as value-neutral, and hence their optimism or pessimism is actually about humanity's capacity for wisdom in guiding technology. Thus, optimists about human capacities for wisdom envision the steady advance of technology as generating new instruments that can be used to solve problems and make steady progress. Pessimists emphasize that in advancing human powers, technology tends to multiply the scale of stupidity in making choices—witness the dangers of nuclear war, of environmental destruction, of a crassly materialistic society preoccupied with pleasure.

However, most scholars believe that things are more complex, and as authors we agree. Technology, properly understood, is not altogether value-neutral. It already embeds values, and optimism or pessimism are better focused on what those values are.

Clearly, this debate turns on how we define technology in the first place. According to the value-neutral view, technology consists of *artifacts* or devices—machines, tools, structures—perhaps together with *knowledge* about how to make and maintain devices. As such, it is neither good nor bad, but merely a means that can be used for good or bad purposes. A screwdriver can be put to many uses, including building homes or killing persons, but by itself it has no intrinsic value or even tendency toward desirable or undesirable ends. This view of technology is often dubbed *instrumentalism:* technology consists of devices and knowledge that are mere instruments, with no single connection to any particular values or ends.[18]

In opposition to the instrumentalist view, those who view technology as value-laden insist that it consists of the *organizations* and general *approaches* that make technological development possible. Organizations and approaches are guided by values. Hence, in the context in which they are developed and used, artifacts and knowledge embody the dominant values of those who make and use them. Thus, an artificial heart emerges from the value to extend and improve the quality of human lives, and we could not understand what it is a technological object without grasping those values.

Mary Tiles and Hans Oberdiek state the point clearly:

> values become embodied in technologies. Just as artists naturally express their artistic values in their art, so do the makers of technologies. If, for instance, price is more important than safety in the minds of manufacturers, their products will undoubtedly embody that trade-off.[19]

Tiles and Oberdiek quickly add that the values embedded in a device or process are fluid rather than fixed.[20] As an example, they note that lighthouses were designed to protect ships against dangerous shoals; one could not understand what lighthouses are, as a technology, without grasping their function and that value. Today, however, when electronic devices have replaced the original function of lighthouses, their primary value is historic and aesthetic, as picturesque reminders of another era. These symbolic values are important, as land developers, nonprofit preservationist organizations, and large segments of the general public will attest.

Another view against the value-neutral thesis in the philosophy of technology is the mediation theory. The mediation theory argues that artifacts created by engineers actively mediate the ways in which humans experience and engage the world.[21] Artifacts are not value-neutral objects but mediators that shape human perceptions and behaviors. In his seminal essay "Do Artifacts Have Politics," Langdon Winner argues that there are two ways artifacts can have political implications: (1) the invention, design, or arrangement of a particular technical artifact creates a particular political order or power relationship in a community (e.g., the low bridges over the parkways on Long Island, New York, designed by Robert Moses limit access of racial minorities and low-income groups to Jones Beach);

and (2) certain technical systems require or are strongly compatible with particular kinds of political relationships (e.g., effective operation of the nuclear power system requires an authoritarian regime).[22]

Peter-Paul Verbeek's technological mediation theory argues that the mediating role of technologies often affects human perceptions and behaviors and shapes the moral relations and interactions between humans and others.[23] One classic example is the design of speed bumps. A speed bump mediates the behavior of the driver. The driver has to slow down the car and be aware of people living in the community when approaching a speed bump. It creates a moral relationship between the driver and people (especially children) living nearby. More recently, studies in human–robot interaction (HRI) have shown that robots are able to influence, persuade, or coerce humans in different ways. For instance, humans may forego a previously desired action if a robot protests against it.[24]

The narrow instrumentalist definition of technology as value-neutral artifacts misleads by leaving out essential aspects of technological change. Worse, it provides a ready-made basis for engineers and other participants in technological development to deny responsibility.[25] "I am merely making things; responsibility for them lies entirely with the user," thinks the engineer seeking to abandon moral responsibility. As we have also emphasized, however, engineers share responsibility with many others, which brings us to the next point.

10.1.3   The Co-shaping of Technology and Society

Technological determinism is the thesis "that technology somehow causes all other aspects of society and culture, and hence that changes in technology dictate changes in society."[26] In a strong version, the thesis of technological determinism denies human choice: We are victims of technology rather than in control of it. Technological determinism undermines shared responsibility for technological projects. For, responsibility presupposes freedom, and engineers are not responsible for changes that are caused by the "inner logic" of the technical system which is independent of human control.

Is technological determinism true? It has some intuitive appeal, for each of us has at times felt pushed or pulled by technology. On the one hand, can we genuinely choose not to use a telephone, ride in cars, or rely on a computer? To be sure, we are usually happy to have such technologies available, and hence their attraction (pull) strikes us as expanding rather than limiting freedom. Nevertheless, the impact of such technology in shaping our lives is pronounced and pervasive. On the other hand, at every turn our lives are shaped by large, technology-driven organizations and structures over which we have no control: traffic lights, the telephone company, the Internal Revenue Service, and increasingly sophisticated terrorists. When we become victims of identity theft, impersonal health maintenance organizations, or layoffs because of shifts in the global economy, we experience how limited freedom is in an increasingly complex technological society. And as we witness large-scale human events, such as wars, genocide, and mass starvation, the presence or absence of technology seems to be the primary causal factor.

The thesis of technological determinism, however, is not directly about us as individuals. None of us controls every aspect of our lives. An appreciation of our vulnerability, as individuals, to economic and political forces is part of humility and intelligence. Technological determinism is the view that the primary structures of human society are determined by technology, rather than human beings (as a group) controlling technology. A few optimists hold this view, confident that technology overall is beneficial for humanity. But technological determinism is most vigorously supported by pessimists like Ellul, who believe that "technology is autonomous"—driving us in ways that tend to subvert human freedom and values.

How does Ellul defend such a sweeping thesis? He identifies each of the main groups and features of moral choice that are usually assumed to govern technology and then tries to show how each source of control is illusory. Managers of corporations are driven to develop and apply technologies as dictated by profit rather than moral values. Scientists and even engineers tend to be naïve about unintended side effects of technologies; consider, for example, Einstein's initial support for nuclear weapons followed by his opposition to them. Politicians are either crassly self-interested or ideologically directed. Consumers are uninformed and duped by advertising, and citizens are easily manipulated. As science progresses, new technological possibilities become irresistible—if something can be done, it will be done. And the entire process is driven by increasing concentrations of wealth, large corporations, and government support.

In fact, a very large body of careful interdisciplinary study has refuted any strong version of technological determinism.[27] Human choices matter! Technology is not a juggernaut with a will of its own that renders all of humanity its victims. To be sure, once major technological trends become entrenched, they tend to carry a momentum of their own. But those trends emerge as a combination of human freedom exercised within constraints from past technologies and other factors. There is always a two-way interaction between human choice and technological momentum.

A telling example is the automobile. In the United States, according to early 1990 estimates, there were 1.7 automobiles for every U.S. citizen; one in seven jobs were in car-related industries; one-fifth of retail dollars centered on cars; 10 percent of arable land went to the car infrastructure, and in Los Angeles two-thirds of the land space is used for cars.[28] The automobile's rise to dominance seems inevitable, once the basic technology of internal combustion engines merged with Henry Ford's assembly line production to make available a financially accessible product. So do the effects of its dominance, which include the depletion of world oil supplies, pollution, and tens of thousands of deaths each year. In fact, despite this seeming inevitability, the emergence of the automobile is clearly the cumulative product of decisions by corporations, consumers, and government. If technology dramatically influences us, we also shape the directions of technology.

In STS studies, this two-way interaction often goes under the heading of social constructionism. Social constructivists highlight the importance of human perceptions and interpretations, emphasizing how different groups can see a technological change in very different ways. As Wiebe E. Bijker illustrates,

> a nuclear reactor may exemplify to a group of union leaders an almost perfectly safe working environment with very little chance of on-the-job accidents compared to urban building sites or harbors. To a group of international relations analysts, the reactor may, however, represent a threat through enhancing the possibilities for nuclear proliferation, while for the neighboring village the chances for radioactive emissions and the (indirect) employment effects may strive for prominence.[29]

Critics of social constructionism, however, see it as neglecting the full possibility of moral reasoning about the values that ought to govern the assessment of technological change. When done carefully, such assessments focus on particular technologies and leave room for the kind of shared responsibility for implementing justified values that we have emphasized in this book.

Langdon Winner holds such a view. As one of many examples, Winner discusses the development of the mechanical tomato harvester, which plucks and sorts tomatoes with a single pass through fields. The cost of harvesting tomatoes was reduced significantly, although tougher (and less flavorful) varieties of tomatoes had to be developed to withstand the machinery. Yet, tens of thousands of jobs were permanently lost, and thousands of small growers were forced out of business by the high costs of the machines they could not afford. Funding for developing the new technology came from California taxpayers, thereby supporting the financial interests of large agribusiness at the expense of less powerful constituencies. Winner's point is that democratic values require public understanding and debate of such changes, and too often that does not occur.

Winner calls upon engineers to develop greater "political savvy" about power relationships within their corporations and within the economic system in which they work. Equally important, engineers need to develop "political imagination"—an understanding of how their work affects public life.

> As part of mastering the fundamentals in their fields, engineers and other technical professionals ought to be encouraged to ask: Can we imagine technologies that enhance democratic participation and social equality? Can we innovate in ways that help enlarge human freedom rather than curtail it? How can planning for technological change include a concern for the public good as distinct from narrowly defined economic interests?[30]

Winner develops these suggestions with discussions of the interplay of engineering, politics, and free enterprise that we find illuminating. Indeed, his insights resonate with our theme of shared responsibility among engineers, managers, and the public for technological ventures in pursuing social experiments.

10.1.4 Uncertainty, Ambiguity, and Social Experimentation

Uncertainty about general trends in technological change, as well as about specific technologies, lies at the heart of debates about technological optimism. Although the pace of scientific and engineering advancement is breathtaking, there is a lag in moral, social, and political understanding. The contemporary world leaves ample room for disagreements about which risks and benefits, in what degrees of each, surround new technologies and the cumulative effects of older ones. The model of engineering as social experimentation highlights this dimension of engineering, whatever one's specific beliefs about the relevant facts concerning technological change.

The social experimentation model underscores how more is at stake than straightforward disagreements about the hard facts. Perceptions of risk and benefit turn in part on how facts are presented to individuals. Statistics are easily manipulated in the direction one favors. Environmental impact statements can be phrased in language that foregrounds, backgrounds, or selectively omits detailed information. Distinct from purely factual disagreements, there are different responses to risks that center on values. Some differences pertain to individual psychology: some people are more risk-averse than others, either in general or with regard to specific activities such as flying on an airplane. Other differences pertain to the core values endorsed by individuals. Individuals' environmental ethics, for example, will be reflected in their responses to clear-cut logging and strip mining. And safety is acceptable risk—acceptable in light of one's settled value principles and knowing the pertinent facts. In all these ways, values need to be applied contextually and with nuance, rather than globally with regard to all technology.

DISCUSSION QUESTIONS

1. Do you agree or disagree with the following passage from Alvin M. Weinberg? In defending your view, discuss how values shape what counts as a quick fix, as distinct from an unsuccessful "fix."

> Edward Teller [inventor of the H-bomb] may have supplied the nearest thing to a Quick Technological Fix to the problem of war. The hydrogen bomb greatly increases the provocation that would precipitate large-scale war—and not because men's motivations have been changed, not because men have become more tolerant and understanding, but rather because the appeal to the primitive instinct of self-preservation has been intensified far beyond anything we could have imagined before the H-bomb was invented.[31]

2. The distinguished British philosopher Bernard Williams (1929–2003) once wrote that "Nuclear weapons are neither moral nor immoral—they are just piles of chemicals, metals and junk."[32] Identify and assess the instrumentalist concept of technology that Williams seems to be using. Can we comprehend weapons of mass destruction without grasping the aims with which they are developed and their intended functions? As an additional example, discuss the notion of technology reflected in a definition of

handguns as "structured metal and bullets," and the claim that "guns don't kill people; people do."

3. Each of the following claims, concisely stated by Merritt Roe Smith and Leo Marx, has been explored in science and technology studies (STS). With regard to each claim, (a) clarify what is being claimed, (b) identify the element of truth (if any) in the claim, and (c) identify relevant truths neglected in the claim.

> The automobile created suburbia. The atomic bomb divested Congress of its power to declare war. The mechanical cotton-picker set off the migration of southern Black farm workers to northern cities. The robots put the riveters out of work. The Pill produced a sexual revolution.[33]

4. Tiles and Oberdiek point out that the values embedded in technology are fluid rather than fixed. Think of, and discuss, two examples illustrating this theme, in addition to their example of lighthouses. For example, research and discuss the controversial drug RU-486, which was developed as an abortion agent and later was found to have promise in treating various diseases.

5. Robert Moses was given unprecedented power to shape the landscape of New York City and surrounding areas.[34] In exercising that power, he used several ways to block minorities and low-income people, who depended on public transportation, from having access to the state parks he developed, including blocking proposals to extend railway access to them. His most ingenious way, however, was to order that key overpasses and bridges be build a few feet lower than normal in order to block buses from using convenient access roads. This is not an isolated case of how a distorted conception of social justice has in the past distorted engineering ethics, even though most distortions are less conscious and deliberate. Reflecting on neighborhoods you are familiar with, can you think of an example of where the interests of a dominant economic or racial group shaped an engineering project? Are such distortions less likely to occur today, and if so, why—and owing to which shared values?

6. One of the most complex and also most studied urban transformations is Boston's recently completed Central Artery/Tunnel Project. Research that project and discuss how the five themes discussed in this chapter apply to it—both the themes of this book and their analogs in STS and Philosophy of Technology studies. As references, good starting points include: Clive L. Dym and Patrick Little, *Engineering Design: A Project-Based Introduction* (New York: John Wiley & Sons, 2000), pp. 233–63; and Thomas P. Hughes, *Rescuing Prometheus: Four Monumental Projects that Changed the Modern World* (New York: Vintage, 1998), pp. 197–254.

7. The term *Luddite* is used to denote reactionary opposition to technological development. In fact, the term derives from Ned Ludlum, a stocking maker who destroyed his stocking frames in response to a reprimand from his employer. The term came to refer to violent outbursts among textile workers in early nineteenth century England against their employers. In most instances the rebellion was not against all technology, but only against specific innovations that were perceived as threatening jobs and in some cases reducing the quality of products. The rebellion was primarily due to extremely low wages and poor working conditions at a time of general economic depression. Hence, the Luddite movement illustrates how pessimism about technological trends is typically rooted in wider economic and political conditions.[35] Research the issue and write a paper linking the Luddite movement and a related contemporary topic concerning technology of your choosing.

10.2 MORAL LEADERSHIP

As managers, business entrepreneurs, corporate consultants, academics, and government officials, engineers provide many forms of leadership in developing and implementing technology—where "technology" is understood broadly to include artifacts, knowledge, organizations, and approaches. In this concluding section we focus on engineers as moral leaders within their professions and communities who contribute to technological progress. We will sample a few current activities that illustrate leadership within the profession, and we will take note of ongoing challenges that will require continuing moral leadership.

10.2.1 Morally Creative Leaders

Because it takes so many forms, leadership is difficult to define. Leadership is an achievement word: It indicates success in moving a group toward goals. But suppose the goals are evil? Tyrants like Hitler and Stalin were leaders only in a neutral, instrumental sense that places no moral restrictions on the goals achieved. They were not leaders in the honorific sense that implies praise. When a leader's goals are not only permissible but also morally valuable, we will speak of moral leadership.

Moral leaders, then, are individuals who direct, motivate, organize, creatively manage, or in other ways move groups toward morally valuable goals. Leaders might be in positions of authority within a corporation, or they might not be. Leadership can be shown by individuals participating at all levels of organizations. Accordingly, leadership should not be confused with "headship," that is, with being the head of a group.[36]

In speaking of the moral leadership provided by engineers and engineering societies, we set aside any notion of engineers as a group leading society. Most emerging professions have at times had dreams of governing, if not dominating, society. A dream for technologists in this regard was presented early in the twentieth century as a technocracy in which engineers and scientists were best qualified to govern technology-driven societies. Frederick Taylor, the inventor of "scientific management," argued that technologists were best qualified to govern because of their technical expertise, as well as their logical, practical, and unprejudiced minds. As Edwin Layton pointed out in criticism, however, "Taylor was convinced that the scientific laws he had discovered were moral as well as material in character. By identifying the good with mechanical efficiency, he blurred the distinction between 'is' and 'ought.' "[37]

Today most of us believe that no single profession holds the key to the moral governance of society. Indeed, leadership typically requires moving above any narrow professional interest in grappling with increasing social diversity and cross-disciplinary complexity. Certainly moral leadership within democracies is not a matter of the imposition of values by a governing elite. Nevertheless, engineers have their share of moral leadership to contribute—to their professional societies, to their profession as a whole, and to their communities.

If moral leadership does not refer to dominance by an elite group, what does it mean? It means employing morally permissible means to stimulate groups to move toward morally desirable ends. Precisely what means are most effective depends on the situation. Sometimes political savvy is most important; at other times, it is a largely nonpolitical commitment to moral ideals. Again, sometimes conflict resolution is most important to forging unity amidst diversity; at other times the key ability is to stir things up to provide the stimulus for change.

Moral leaders are morally creative. That does not mean they discover or improvise new moral values from scratch. Moral values are the product of centuries and millenia of gradual development, not instantaneous invention. Moral creativity consists in identifying the most important values that apply in a particular situation, bringing them into focus through effective communication within groups, and forming workable commitments to implement them. As with other forms of creativity, moral creativity means achieving valuable newness, in this case morally valuable newness. But the newness consists in identifying new possibilities for applying, extending, and putting values into practice, rather than inventing values (whatever that might mean). That may require fresh moral insight, but even more it requires deep commitments grounded in integrity.

10.2.2 Participation in Professional Societies

Not surprisingly, moral leadership within engineering is often manifested in leadership within professional societies. Professional societies do more than promote continuing education for their members. They also serve to unify a profession, and to speak and act on behalf of it (or a large segment of it). Professional societies provide a forum for communicating, organizing, and mobilizing change within and by large groups. That change has a moral dimension.

Many of the current tensions in professional societies exist because of uncertainties about their involvement in moral issues. This was illustrated in the Bay Area Rapid Transit (BART) case. One chapter of the California Society of Professional Engineers felt it should play a role in supporting the efforts of the three engineers who sought to act outside normal organizational channels in serving the public. Another chapter felt it was inappropriate for the society to do so. In this and other controversial cases, professional societies have often been reluctant to become involved.

It is unlikely that existing professional societies will, and it is perhaps undesirable that they should, take any univocal pro-employee or pro-management stand. Their memberships, after all, are typically a mixture of engineers in management, supervision, and nonmanagement. Yet professional societies can, should, and are playing a role in conflicts involving moral issues, although rank-and-file engineers remain skeptical because they still consider the societies to be management-dominated.[38] Through membership participation on committees, these societies provide a sympathetic and informed forum for hearing opposing viewpoints and making recommendations. Through their guidelines for employment practice and conflict resolution they can help forestall debilitating disputes within corporations. On a national level, they can lobby for earlier vesting and portability of pensions. Details of the extent and form of such activities deserve ongoing discussion within engineering ethics. Clearly there is an ongoing need for individual engineers to provide moral leadership.

Just as moral responsibilities are shared, moral creativity in the professions is a shared phenomenon. Nevertheless, individuals can make a dramatic difference. To cite just one example, Stephen H. Unger is largely responsible for persuading the Institute of Electrical and Electronics Engineers (IEEE) to move beyond the traditional focus of most societies on punishing wrongdoers toward supporting responsible engineers.[39] After investigating the activities of the BART engineers, he succeeded in getting IEEE to present the three with awards for outstanding professional service. He also helped organize and lead the Committee on Social Implications of Technology, which later became the IEEE Society on Social Implications of Technology (SSIT). For several decades this group has institutionalized an ongoing concern for moral issues. Steve Unger in turn credits the late Victor Paschkis of Columbia University for awakening in him an awareness of the engineer's social responsibility. And it was also Victor Paschkis with his pioneering Society of Social Responsibility in Science that greatly influenced the career path of the engineer coauthor of this book.

Many other individuals have spurred professional societies to foster the study of engineering ethics. They have helped to sponsor ethics workshops, conduct surveys on matters of ethical concern, inform their members of developments related to ethics, and encourage schools of engineering to support regular and continuing education courses in engineering ethics.

To mention just one important development, in 1988 the National Society of Professional Engineers created The National Institute for Engineering Ethics, giving it the mission to promote ethics within the engineering profession. That organization, which in 1995 was restructured to involve many other professional societies and to operate independently (it is currently based at Purdue University), has developed educational videos, computer disks, and newsletters. The focus has been on education, rather than propaganda for any narrow perspective.

Making general appraisals of the role of professional societies ultimately entails examining a profession's "macroeconomics": that is, examining how they do and should function as a group within contemporary society. For example, to what extent is it desirable for the engineering profession to set standards in such areas as the disposal of toxic wastes?[40]

Or, to take another kind of topic, is the trend toward increasing rule-making on behalf of professionalism within engineering in the public interest? Here many issues are involved, at least given the model of professionalism derived from developments in medicine and law:

1. Should the engineering profession be allowed to have the authority to decide which students and how many students will be admitted to schools of engineering? Should laypersons representing the public have a say?

2. Should licensing of all engineers in industrial practice be mandatory, as it is for doctors and lawyers? There would be potential benefits: for example, greater assurance that minimal standards of training and skill would be met by all engineers. But there would also be drawbacks, if only that bureaucratic red tape would increase.

3. Should continuing education be mandatory for all engineers?

Ultimately these "macro" issues return us to the "micro" issues of individual responsibility. For it is individuals involved in their professional societies who are the ultimate loci of action and hence of leadership.

This leads us to engineers' obligations to their profession. The code of ethics of ABET (incorporated as the Accreditation Board for Engineering and Technology, Inc.) suggests that engineers should obey the code in order to "uphold and advance the integrity, honor and dignity of the engineering profession." Should such statements be dismissed as remnants of the natural esprit de corps of an emerging major profession? Or are there special professional obligations to the engineering profession engineers should recognize?

Surely something can be said in defense of a duty to respect and defend the honor of the profession. Effective professional activity, whether in engineering or any other profession, requires a substantial degree of trust from clients and the public. Total absence of such trust would undermine the possibility of making contracts, engaging in cooperative work, exercising professional autonomy free of excessive regulation, and working under humane conditions. Building and sustaining that trust is an important responsibility shared by all engineers. It is also an area where moral leadership within professional societies is especially important.

We might add that there is always the danger that the idea of an obligation to one's profession can become perverted into a narrow, self-interested concern. Such would be the case, for example, if a profession deliberately limited the number of its practitioners to create a greater demand for them and hence manipulate their salaries or fees upward. For this reason it may be preferable to understand talk about obligations to the profession as shorthand for certain obligations to the public. Engineers as individuals and as a group owe it to the public to sustain a professional climate conducive to meeting their other obligations to the public.

10.2.3 Leadership in Communities

Do engineers have special responsibilities as citizens that go beyond those of nonengineers? For example, should they provide greater leadership than others in social debates about industrial pollution, automobile safety, and invasion of privacy in AI systems?

Answering this question would require a clarification of the obligations citizens have in public policy issues. But even here there is considerable disagreement. One view holds that no one is strictly obligated to participate in public decision making. Instead, such participation is a moral ideal for citizens to embrace and pursue as their time allows. A contrasting view holds that all citizens are obligated to devote some of their time and energies to public policy matters. Minimal requirements for everyone are to stay informed about issues that can be voted on, while stronger obligations arise for those who by professional background are well grounded in specific issues as well as for those who have the time to train themselves as public advocates. But whether or not there is strict obligation here, certainly there is a need for moral leadership in identifying and expanding the areas of possible good that can be achieved.

For example, engineers are not as well represented on many legislative and advisory bodies as they might be in the United States. Perhaps they are too modest about offering their services, or maybe they see a number of complications arising from service of this kind.

For engineers in private practice, the notion of public service can be particularly troubling. For example, there is the matter of advertising: While an engineer who is employed by a company will bring recognition and honor to the company through volunteer activities, any such efforts on the part of a self-employed engineer could be interpreted as self-serving attempts to gain publicity and perhaps even to secure valuable inside information. But most voluntary service involves mixtures of altruism and self-interest, and as long as elements of self-interest do not distort the motives of helping, they should not be seen as self-serving ruses.[41]

Moral leadership of engineers in private practice does not limit to their participation in public service. It can also be exemplified in the proactive role of engineers in criticizing and disclosing the potential risks and harms brought to the public by corporations and their products. Timnit Gebru, former co-lead of the Ethical Artificial Intelligence Team at Google, has been studying the justice and equity issues in AI and promoting diversity and inclusion in professional communities of AI. In December 2020, her employment was terminated by Google after she refused to either withdraw a paper on risks of very large language models or remove the names of all Google employees on that paper. Gebru's advocacy for justice in AI in this regard exemplifies some possible moral leadership roles that engineers in private practice can assume and cultivate.

10.2.4   Ideals of Voluntary Service

Should the engineering profession encourage the pro bono, voluntary giving of engineering services without fee or at reduced fees to especially needy groups? Is this an ideal that is desirable for engineering professional societies to embrace and foster among individuals and corporations? Certainly it can be a vital avenue for providing moral leadership within communities.

Voluntarism (or philanthropy) of this sort has long been encouraged in medicine, law, and education. By sharp contrast, engineering codes of ethics have either been silent on this question or taken stands that discourage voluntarism. For example, the ABET (formerly ECPD) code was revised during the 1960s to state: "Engineers shall not undertake nor agree to perform any engineering service on a free basis." Most other codes also insisted that engineers are obligated to require adequate compensation for their work—meaning compensation at the present fee scale. Such statements are now being revised in light of Supreme Court rulings suggesting they restrain free trade. Nevertheless, there continues to be a sentiment against encouraging engineers to donate their services without full compensation.

Robert Baum has challenged this sentiment.[42] He acknowledges that engineers have fewer opportunities to donate their services as individuals than do doctors and lawyers. This is because engineering services tend to require shared efforts and to demand the resources of the corporations for which most engineers work. But this merely shows that engineers might best help the needy through group efforts. (It is also true that increasing numbers of doctors and lawyers work for corporations.)

Baum argues, for example, that Native Americans often lack the resources for the engineering studies needed for negotiating with the Bureau of Land Management, which has authority to grant leases on Native American land. There is money for lawyers, but no money for costly environmental impact studies required for, say, challenging a proposed government project that is harmful in the view of a Native American group. There are similar problems with the health issues of polluted water and soil on reservations. In addition, there are financially disadvantaged groups, especially the elderly and some minorities living in both urban and rural areas, whose minimal needs are at present not met: needs for running water, sewage systems, electrical power, and inexpensive transportation. This could be remedied if access to engineering services were made available at lower-than-normal costs.

There are many options that the profession of engineering might explore. These include encouraging engineers to serve in government programs like VISTA, urging government to expand the services of the Army Corps of Engineers, encouraging engineering students to focus their senior projects on service for disadvantaged groups, and encouraging corporations to offer 5 percent or 10 percent of their services free or at reduced rates for charitable purposes.

Are professional societies obligated to foster voluntarism among engineers in providing engineering services at reduced fees? Baum leaves the question open. His main concern was to argue that needy groups ought to have access to engineering services, but not to resolve the question of who should provide them (groups of engineers, corporations, local government, federal government, etc.). He suggests, however, that engineers do have one important duty in serving the needy: "to participate in dialogues concerning the needs of specific individuals and groups and the possible ways in which these needs might be met."[43] Baum feels that through discussion between engineers and disadvantaged groups, solutions may be found.

We would add that a morally concerned engineering profession should recognize the rights of corporations and individual engineers to voluntarily engage in philanthropic engineering service. Furthermore, it would be desirable for professional societies to endorse the voluntary exercise of this right as being a desirable ideal, an ideal of generosity that goes beyond the call of duty. While good deeds beyond the scope of one's primary work cannot compensate for unethical conduct inside it, a profession fully dedicated to the public good should recommend participation by its practitioners in all aspects of community life. Many individual engineers and some engineering societies are already engaged in such volunteer services. They range from tutoring disadvantaged students in mathematics and physics, to "urban technology" interest groups and senior engineering students who advise local governments on their engineering problems.

To conclude, there is an ongoing need for moral leadership in engineering, as in other professions. A primary forum for that leadership is substantial involvement in professional societies which, in addition to furthering technical knowledge and representing engineers collectively, help establish high standards of moral integrity within the profession. Another forum for moral leadership is in community service. Moral leadership does not consist of moral elitism and dominance, but instead moral creativity in helping to guide, organize, and stimulate groups toward morally desirable goals.

Human excellence and good lives—including morally good lives—take many different forms.[44] How technology enters into those lives will vary considerably. The collective exercise of the profession of engineering and of engineering professional societies is reasonably limited to the widely shared values of professionalism developed within a framework of the core values of democracy, human safety and health, and sustainable development. In addition, as both citizens and professionals serving as volunteers, individual engineers contribute their vision of good lives, and their outlooks will legitimately shape their careers and the kinds of work they find meaningful.

DISCUSSION QUESTIONS

1. The NSPE code lists as a "professional obligation" that "Engineers shall seek opportunities to participate in civic affairs; career guidance for youths; and work for the advancement of the safety, health, and well-being of their community." (III, 2a) Do you agree that this is an obligation, or is it instead a desirable but morally optional ideal?

2. Critics argue that "overconsumption" by U.S. consumers is having a devastating impact on the environment through pollution, destruction of ecosystems, and depletion of limited natural resources. E. J. Woodhouse argues that it is unfair to ask engineers to risk their jobs by taking stands against overconsumption as part of their paid work. Nevertheless, it is desirable for engineers, individually and in professional groups such as Computer Professionals for Social Responsibility and the Union of Concerned Scientists, to apply their expertise through voluntary service in the following directions: "1) Stimulate thought and discussion about overconsumption, both within engineering and more generally; 2) Focus on optional [i.e., morally permissible and desirable] ethical behaviors, not just on mandated tasks; 3) Take a collective rather than individual approach, expecting schools of engineering and professional organizations to take the lead; and 4) Rather than thinking of professional responsibilities as occurring only at work, envision ways that engineers can act responsibly as citizens and consumers."[45] Do you agree or disagree with Woodhouse, and why?

3. Most states do not require engineers employed by industrial corporations to be registered or licensed by the state (based on meeting certain minimal requirements of education, knowledge, and experience), although they require registration of engineers in charge of activities that affect the public health and safety. This "industrial exemption" has come under increasing criticism. Do you agree or disagree with the following reasons for abolishing the industrial exemption and for requiring all engineers to be licensed? Are they all good moral reasons for requiring registration?

   a. Registration assures the company that an engineer has met the prescribed statutory requirements of the law enacted to protect the public health, safety, and welfare.

   b. An engineering staff composed of registered professional engineers enhances the prestige and public relations potential of the firm.

   c. Registration improves the morale of the engineer by attesting to his or her qualifications, competence, and professional attitude. It also encourages engineers to take full responsibility for their work.

   d. Registration improves company-client relations by attesting to engineering staff competence and satisfies the legal requirements of many states and municipalities that require projects to be under the control of a registered engineer.

   e. Registration promotes high standards of professional conduct, ethical practice, integrity, and top-quality job performance.[46]

4. Currently only state, not national, registration is possible for engineers in the United States. Assess the following views of requiring national registration of all engineers. Are there other reasons for and against national registration?

   a. "The method of obtaining national registration as well as the approach would act as a vehicle for elevating the engineering profession to standards that doctors and lawyers now enjoy."

   b. "I feel that a national registration law administered by a Federal agency would be very detrimental to the engineering profession because it would transfer our local responsibilities and authority to a Federal power. . . . [In] practice I think it would become a dictatorial agency, administering the law for its own purposes, with very little regard for the engineering profession."[47]

5. Identify and discuss any moral duties, rights, and ideals pertinent to the following example:

   An engineer who had also had experience as a carpenter-contractor was asked by his church to assist in the construction of a new building. He finally served as the general contractor, the engineer-inspector, and the construction foreperson. He used labor donated by members of the church, including many carpenters, painters, cement finishers, and so on, even though they were not members of the skilled trades. No salaries were paid to any church members, but credit for labor was allowed against a pledge made by each member. . . . Before the engineer accepted the multiple responsibilities, bids from general contractors had been taken, and all were out of reach of the church group.[48]

6. Defend your view as to whether engineers have special obligations beyond those of nonengineers to enter into public debates over technological development. If you think they do not have special obligations, is it nevertheless especially desirable (as a moral ideal) for them to contribute to these debates? Should professional societies and moral leaders within the profession encourage such participation?

7. In order to make their voices heard more clearly, engineers and scientists have formed societies to discuss and promote social responsibility in their professions. Examples are the Federation of American Scientists (founded after World War II by participants in the Manhattan Project), the Computer Professionals for Social Responsibility, and the IEEE Society on Social Implications of Technology. Select one of these groups and research the success and obstacles it has confronted.

KEY CONCEPTS

—**Competing attitudes toward technological promise and perils:** (1) optimism, (2) pessimism, (3) "realism" (suggesting being realistic about power) or "contextualism" (suggesting close attention to variations within specific contexts), and (4) cautious optimism (combining 1 and 3).

—**Quick (technological) fix:** using technology to solve otherwise intractable social problems. A contrast is with **social engineering:** inducing change in the motivation and habits of individuals, as by using powerful social institutions with the authority to control human conduct.

—**Opposing definitions of technology:** (1) value-neutral (or "instrumentalism") definitions say that technology consists of artifacts or devices—machines, tools, structures—perhaps together with knowledge about how to make and maintain devices; (2) value-laden definitions say that technology consists of value-guided organizations and general approaches, in addition to artifacts and knowledge.

—**Technological determinism:** technology is largely autonomous and causes and dictates all other aspects of society, such that we are victims of technology rather than in control of it.

—**Social constructionism:** approaches in science and technology studies that emphasize two-way causal interactions between technology and society, and that highlight the importance of human perceptions and interpretations, emphasizing how different groups can see a technological change in very different ways.

—**Moral leadership:** success in moving a group toward morally desirable goals using morally desirable procedures. **Moral leaders** are individuals who direct, motivate, organize, creatively manage, or in other ways move groups toward morally valuable goals.

—**Voluntarism** or **philanthropy:** Pro bono (free, or reduced charge), voluntary giving of engineering services without fee or at reduced fees to especially needy groups.

REFERENCES

1. William A. Wulf, "Great Engineering Achievements of the 20th Century and Challenges for the 21st," talk given at Stony Brook. We thank Mike Rabins for making an outline of the talk available to us.
2. John Ladd, "The Quest for a Code of Professional Ethics: An Intellectual and Moral Confusion," in Deborah G. Johnson (ed.), *Ethical Issues in Engineering* (Englewood Cliffs, NJ: Prentice Hall, 1991), p. 135. First published in Rosemary Chalk, Mark S. Frankel, and Sallie B. Chafer (eds.), *AAAS Professional Ethics Project: Professional Ethics Activities in the Scientific and Engineering Societies* (Washington, DC: AAAS, 1980), pp. 154–59.
3. Stephen H. Cutcliffe and Carl Mitcham, Introduction to *Visions of STS: Counterpoints in Science, Technology, and Society Studies* (Albany, NY: State University of New York Press, 2001), p. 2.
4. Ian G. Barbour, *Ethics in an Age of Technology* (New York: HarperCollins, 1993), p. 3.

5. William A. Wulf, "Great Engineering Achievements of the 20th Century and Challenges for the 21st," talk given at Stony Brook.

6. Emmanuel G. Mesthene, "Technology and Wisdom," in Robert C. Scharff and Val Dusek (eds.), *Philosophy of Technology* (Malden, MA: Blackwell, 2003), p. 619.

7. Miles Brundage, "Scaling Up Humanity: The Case for Conditional Optimism about Artificial Intelligence," in European Parliament (ed.), *Should We Fear Artificial Intelligence* (Brussels: European Union, 2018), pp. 13–18. Available at: https://www.europarl.europa.eu/RegData/etudes/IDAN/2018/614547/EPRS_IDA(2018)614547_EN.pdf

8. Batya Friedman and David G. Hendry, *Value Sensitive Design: Shaping Technology with Moral Imagination* (Cambridge, MA: The MIT Press, 2019).

9. TU Delft Design for Values Institute, "*Vision: Why Design for Values*," available at: https://www.delftdesignforvalues.nl/vision-why/

10. Ralph Waldo Emerson, "Works and Days," in Arthur O. Lewis, Jr. (ed.), *Of Men and Machines* (New York: E. P. Dutton, 1963), p. 69.

11. Cf. C. P. Snow, *The Two Cultures and A Second Look* (New York: Cambridge University Press, 1969), p. 32.

12. Lewis Mumford, quoted by Merritt Roe Smith, "Technological Determinism in American Culture," in Merritt Roe Smith and Leo Marx (eds.), *Does Technology Drive History?* (Cambridge, MA: MIT Press, 1994), p. 29.

13. Jacques Ellul, "The Technological Order," *Technology and Culture* 3 (Fall 1962): 10.

14. Albert Borgmann, *Technology and the Character of Contemporary Life* (Chicago: University of Chicago Press, 1984), pp. 41–42.

15. Shannon Vallor, "Moral Deskilling and Upskilling in a New Machine Age: Reflections on the Ambiguous Future of Character," *Philosophy and Technology* 28, (2015): 107–24.

16. Pak-Hang Wong, "Rituals and Machines: A Confucian Response to Technology-Driven Moral Deskilling," *Philosophies* 4, no. 4 (2019). Available at: https://doi.org/10.3390/philosophies4040059

17. Ian G. Barbour, *Ethics in an Age of Technology,* p. 15.

18. This sense of instrumentalism is different from John Dewey's use of the term to label his form of pragmatism (chapter 2).

19. Mary Tiles and Hans Oberdiek, *Living in a Technological Culture: Human Tools and Human Values* (New York: Routledge, 1995), p. 46.

20. Ibid., p. 60.

21. Peter-Paul Verbeek, *Moralizing Technology: Understanding and Designing the Morality of Things* (Chicago, IL: University of Chicago Press, 2011).

22. Langdon Winner, "Do Artifacts Have Politics?" *Daedalus* 109, no. 1 (1980): 121–36.

23. Peter-Paul Verbeek, *Moralizing Technology.*

24. Qin Zhu, Tom Williams, Blake Jackson, and Ruchen Wen, "Blame-Laden Moral Rebukes and the Morally Competent Robot: A Confucian Ethical Perspective," *Science and Engineering Ethics* 26, no. 5 (2020): 2511–2526.

25. Samuel Florman comes close to portraying engineers as nonresponsible and blameless, beyond meeting state-of-the-art engineering. *Blaming Technology* (New York: St. Martin's Press, 1981), p. 190.

26. Robert C. Scharff and Val Dusek, Introduction to "Is Technology Autonomous?" *Philosophy of Technology,* p. 384.

27. Stephen H. Cutcliffe and Carl Mitcham, Introduction to *Visions of STS* (Albany, NY: State University of New York Press, 2001), p. 4; Robert Pool, *Beyond Engineering: How Society Shapes Technology* (New York: Oxford University Press, 1997).

28. Mary Tiles and Hans Oberdiek, *Living in a Technological Culture: Human Tools and Human Values,* pp. 130–38; Peter E. S. Freund and George Martin, *The Ecology of the Automobile* (New York: Black Rose Books, 1994).

29. Wiebe E. Bijker, "Understanding Technological Culture Through a Constructivist View of Science, Technology, and Society," in Stephen H. Cutcliffe and Carl Mitcham (eds.), *Visions of STS,* p. 26.

30. Langdon Winner, "Engineering Ethics and Political Imagination," in Deborah G. Johnson (ed.), *Ethical Issues in Engineering* (Englewood Cliffs, NJ: Prentice Hall, 1991), p. 381. Earlier published in Paul Durbin (ed.), *Philosophy and Technology,* vol. 7 (Dordrecht: Kluwer, 1990).
31. Alvin M. Weinberg, "Can Technology Replace Social Engineering?" p. 38.
32. Bernard Williams, "Morality, Scepticism and the Nuclear Arms Race," in Nigel Blake and Kay Pole (eds.), *Objections to Nuclear Defense* (London: Routledge & Kegan Paul, 1984), p. 100.
33. Merritt Roe Smith and Leo Marx, Introduction to *Does Technology Drive History?* p. xi.
34. Robert A. Caro, *The Power Broker: Robert Moses and the Fall of New York* (New York: Random House, 1974), pp. 318, 951–58. Discussed by Langdon Winner in *The Whale and the Reactor,* pp. 22–23.
35. Rudi Volti, *Society and Technological Change,* 4th ed. (New York: Worth Publishers, 2001), pp. 21–23; Malcom I. Thomis, *The Luddites* (New York: Schocken Books, 1972).
36. Charles R. Holloman, "Leadership and Headship: There Is a Difference," in William E. Rosenbach and Robert L. Taylor (eds.), *Contemporary Issues in Leadership,* 2nd ed. (Boulder, CO: Westview Press, 1989), p. 109.
37. Edwin T. Layton, Jr., *The Revolt of the Engineers: Social Responsibility and the American Engineering Profession* (Baltimore: Johns Hopkins University Press, 1986), p. 41.
38. Albert Flores (ed.), *Designing for Safety: Engineering Ethics in Organizational Contexts, Workshop on Engineering Ethics: Designing for Safety* (Troy, New York: Rensselaer Polytechnic Institute, 1982), p. 80.
39. Edwin T. Layton, Jr., *The Revolt of the Engineers: Social Responsibility and the American Engineering Profession,* pp. xi–xii.
40. John Ladd, "The Quest for a Code of Professional Ethics," in Chalk, Frankel, and Chafer (eds.), *AAAS Professional Ethics Project: Professional Ethics Activities in the Scientific and Engineering Societies* (Washington, DC: American Association for the Advancement of Science, 1980), p. 158.
41. Mike W. Martin, *Virtuous Giving: Philanthropy, Voluntary Service, and Caring* (Bloomington, IN: Indiana University Press, 1994).
42. Robert J. Baum, "Engineering Services," *Business and Professional Ethics Journal* 4 (1985): 117–35.
43. Ibid., p. 133.
44. Lawrence C. Becker, "Good Lives: Prolegomena," *Social Philosophy and Policy* 9, no. 2 (1992): 15–37.
45. E. J. Woodhouse, "Curbing Overconsumption: Challenge for Ethically Responsible Engineering," *IEEE Technology and Society Magazine* 20, no. 3 (Fall 2001): 29.
46. G. J. Kettler, "Against the Industry Exemption," in James H. Schaub and Karl Pavlovic (eds.), *Engineering Professionalism and Ethics* (New York: Wiley, 1983), p. 534.
47. Comments from Professional Engineers, National Registration, *The American Engineer* (July 1964): 25–30.
48. Philip L. Alger, N. A. Christensen, and Sterling P. Olmsted, *Ethical Problems in Engineering* (New York: John Wiley & Sons, 1965), p. 236. Quotation in text used with permission of the publisher.

GENERAL RESOURCES ON ENGINEERING ETHICS

The following is a sampling of journals, websites, books, and videotapes in engineering ethics.

JOURNALS

1. *Science and Engineering Ethics*
2. *Business & Professional Ethics Journal*
3. *Teaching Ethics*
4. *Ethics and Information Technology*
5. *Nanoethics*
6. *Journal of Responsible Innovation*
7. *IEEE Technology and Society Magazine*
8. *IEEE Transactions on Technology and Society*
9. *AI and Ethics*

WEBSITES

Most of these websites allow direct transfer to each other and to additional helpful sites.

1. National Institute for Engineering Ethics
   http://www.niee.org
2. National Society of Professional Engineers
   http://www.nspe.org
3. The Online Ethics Center for Engineering and Science
   http://www.onlineethics.org/
4. The Ethics Codes Collection
   http://ethicscodescollection.org/
5. Illinois Institute of Technology Ethics Education Library
   http://ethicscodescollection.org/
6. National Center for Professional & Research Ethics
   https://ethicscenter.csl.illinois.edu/
7. Ethics Unwrapped
   https://ethicsunwrapped.utexas.edu/
8. International Dimensions of Ethics Education in Science and Engineering
   https://www.umass.edu/sts/ethics/

GENERAL BOOKS

These books, listed chronologically by date of first editions, provide general coverage of engineering ethics.

Murray I. Mantell. *Ethics and Professionalism in Engineering.* New York: Macmillan, 1964.

Philip L. Alger, N. A. Christensen, and Sterling P. Olmsted. *Ethical Problems in Engineering.* New York: John Wiley & Sons, 1965.

Edwin T. Layton, Jr. *The Revolt of the Engineers: Social Responsibility and the American Engineering Profession.* Baltimore: Johns Hopkins University Press, 1986. Earlier edition in 1971.

Robert J. Baum and Albert Flores, eds. *Ethical Problems in Engineering.* 2nd ed. Troy, NY: Center for the Study of the Human Dimensions of Science and Technology, Rensselaer Polytechnic Institute, 1980. Vol. 1: *Readings,* edited by Albert Flores. Vol. 2: *Cases,* edited by Robert J. Baum. First edition published in 1978.

Stephen H. Unger. *Controlling Technology.* 2nd ed. New York: John Wiley & Sons, 1982, 1994.

Mike W. Martin and Roland Schinzinger. *Ethics in Engineering.* 4th ed. Boston: McGraw-Hill, 1983, 1989, 1996, 2004. Shorter version published as *Introduction to Engineering Ethics.* Boston: McGraw-Hill, 2000.

James H. Schaub and Karl Pavlovic, eds. *Engineering Professionalism and Ethics*. New York: John Wiley & Sons, 1983.

Deborah G. Johnson, ed. *Ethical Issues in Engineering*. Englewood Cliffs, NJ: Prentice Hall, 1991.

Eugene Schlossberger. *The Ethical Engineer*. Philadelphia: Temple University Press, 1993.

Charles E. Harris, Jr., Michael S. Pritchard, and Michael J. Rabins. *Engineering Ethics: Concepts and Cases*. 6th ed. Belmont, CA: Wadsworth, 2009, 2014, 2019.

Michael Davis. *Thinking Like an Engineer*. New York: Oxford University Press, 1998.

Caroline Whitbeck. *Ethics in Engineering Practice and Research*. New York: Cambridge University Press, 1998.

Kenneth K. Humphreys. *What Every Engineer Should Know About Ethics*. New York: Marcel Dekker, 1999.

Joseph R. Herkert, ed. *Social, Ethical, and Policy Implications of Engineering*. New York: Institute of Electrical and Electronics Engineers, 2000.

Carl Mitcham and R. Shannon Duval. *Engineering Ethics*. Upper Saddle River, NJ: Prentice Hall, 2000.

Edmund G. Seebauer and Robert L. Barry. *Fundamentals of Ethics for Scientists and Engineers*. New York: Oxford University Press, 2001.

Daniel A. Vallero. *Biomedical Ethics for Engineers: Ethics and Decision Making in Biomedical and Biosystem Engineering*. Cambridge, MA: Academic Press, 2007.

W. Richard Bowen. *Engineering Ethics: Outline of an Aspirational Approach*. London, UK: Springer-Verlag, 2009.

P. Aarne Vesilind and Alastair S. Gunn. *Hold Paramount: The Engineer's Responsibility to Society*. 3rd ed. Boston, MA: Cengage, 2011, 2016.

Ibo van de Poel and Lamber Royakkers. *Ethics, Technology, and Engineering: An Introduction*. Malden, MA: Wiley-Blackwell, 2011.

Colleen Murphy, Paolo Gardoni, Hassan Bashir, Charles E. Harris, and Eyad Masad, eds. *Engineering Ethics for a Globalized World*. Cham, Switzerland: Springer International.

Wade L. Robison. *Ethics within Engineering: An Introduction*. New York, NY: Bloomsbury Academic, 2016.

Heinz Luegenbiehl and Rockwell Clancy. *Global Engineering Ethics*. Oxford, UK: Butterworth-Heinemann, 2017.

Rania Milleron and Nicholas Sakellariou, eds. *Ethics, Politics, and Whistleblowing in Engineering*. Boca Raton, FL: CRC Press, 2018.

Ali E. Abbas, ed. *Next-Generation Ethics: Engineering a Better Society*. Cambridge, MA: Cambridge University Press, 2019.

Martin Peterson. *Ethics for Engineers*. New York, NY: Oxford University Press, 2020.

Deborah Johnson. *Engineering Ethics: Contemporary & Enduring Debates*. New Heaven, CT: Yale University Press, 2020.

Behnam Taebi. *Ethics and Engineering: An Introduction.* Cambridge, UK: Cambridge University Press, 2021.

CASE STUDY BOOKS

These books, listed chronologically, are examples of books that use cases to explore broad issues in engineering ethics.

Robert M. Anderson, Robert Perrucci, Dan E. Schendel, and Leon E. Trachtman. *Divided Loyalties: Whistle-Blowing at BART.* West Lafayette, IN: Purdue University Press, 1980.

John H. Fielder and Douglas Birsch, eds. *The DC-10 Case.* Albany, NY: State University of New York Press, 1992.

Douglas Birsch and John H. Fielder, eds. *The Ford Pinto Case.* Albany, NY: State University of New York Press, 1994.

Diane Vaughan. *The Challenger Launch Decision: Risky Technology, Culture, and Deviance at NASA.* Chicago: University of Chicago Press, 1996.

Rosa Lynn B. Pinkus, Larry J. Shuman, Norman P. Hummon, and Harvey Wolfe. *Engineering Ethics: Balancing Cost, Schedule, and Risk—Lessons Learned from the Space Shuttle.* New York: Cambridge University Press, 1997.

John R. Wilcox and Louis Theodore. *Engineering and Environmentalism: A Case Studies Approach.* New York: John Wiley & Sons, 1998.

Michael E. Gorman, Matthew M. Mehalik, and Patricia H. Werhane. *Ethical and Environmental Challenges to Engineering.* Upper Saddle River, NJ: Prentice Hall, 2000.

William M. Evan and Mark Minion. *Minding the Machines: Preventing Technological Disasters.* Upper Saddle River, NJ: Prentice-Hall, 2003.

Sarah K. A. Pfatteicher. *Lessons amid the Rubble: An Introduction to Post-Disaster Engineering and Ethics.* Baltimore, MD: Johns Hopkins University Press, 2010.

Steven K. Starrett, Amy L. Lara, and Carlos Bertha. *Engineering Ethics: Real World Case Studies.* Reston, VA: American Society of Civil Engineers, 2017.

Robert McGinn. *The Ethical Engineer: Contemporary Concepts and Cases.* Princeton, NJ: Princeton University Press, 2018.

VIDEOTAPES

"Gilbane Gold." NSPE, P.O. Box 1020, Sewickley, PA 15143. Also contact The National Institute for Engineering Ethics, http://www.niee.org, which is developing a new instructional video.

"The Story Behind the Space Shuttle Challenger Disaster." Mark Maier, Organizational Leadership, Chapman University, One University Drive, 92866.

"Truesteel Affair." Fainlight Productions, 47 Halifax St., Boston, MA 02130.

"The 59 Story Crisis: A Lesson in Professional Behavior." The Online Ethics Center for Engineering and Science, http://www.onlineethics.org/

"The Insider." Available in video stores. Although about the chemist, Jeffrey Wigand, who challenged the tobacco industry, the video has relevance to whistleblowing in engineering.

"A Civil Action." A 115 min Touchstone movie and video based on the book by Jonathan Harr (Random House, 1995). Screenplay by S. Zaillian. Featured actors: John Travolta and Robert Duvall.

"Erin Brockovich." A 130 min Universal Pictures movie and video (2000). Screenplay by Susannah Grant, features actors Julia Roberts and Albert Finney.

"Testing Water . . . And Ethics." Institute for Professional Practice, 13 Lanning Road Verona, NJ 07044-2511.

"Incident at Morales." National Institute for Engineering Ethics (Purdue University), http://www.niee.org

"Ethicana." American Society for Civil Engineers (ASCE), available at ASCE's online store.

"Henry's Daughter." National Institute for Engineering Ethics (Purdue University), http://www.niee.org

APPENDIX
B

SAMPLE
CODES OF ETHICS
AND GUIDELINES

NSPE: National Society of Professional Engineers Code of Ethics
   https://www.nspe.org/resources/ethics/code-ethics
IEEE: The Institute of Electrical and Electronics Engineers
   https://www.ieee.org/about/corporate/governance/p7-8.html
AIChE: American Institute of Chemical Engineers
   https://www.aiche.org/about/governance/policies/code-ethics
ASCE: American Society of Civil Engineers
   https://www.asce.org/code-of-ethics/
ASME: American Society of Mechanical Engineers
   https://www.asme.org/wwwasmeorg/media/resourcefiles/aboutasme/get%20
   involved/advocacy/policy-publications/p-15-7-ethics.pdf
ACM/IEEE/CS: Joint Task Force on Software Engineering Ethics and
   Professional Practices
   https://ethics.acm.org/code-of-ethics/software-engineering-code/
SPE: Society for Petroleum Engineering
   https://www.spe.org/en/about/professional-code-of-conduct/
WFEO: World Federation of Engineering Organizations
   https://www.wfeo.org/wp-content/uploads/code_of_ethics/WFEO_MODEL_
   CODE_OF_ETHICS.pdf
More codes of ethics of professional organizations and corporations worldwide
   can be found at the Ethics Codes Collection at the Illinois Institute of
   Technology: http://www.ethicscodescollection.org/

AUTHOR INDEX

A

Abbas, Ali E., 250n.15
Adams, Donald D., 249n.8
Adler, Matthew D., 149n.21
Agnew, Spiro T., 196
Aguilar, Francis J., 268n.36
Ahearne, John F., 144, 150n.30
Alderman, Frank E., 226n.35
Alger, Philip L., 193n.28, 226n.38, 291n.48
alZahir, Saif, 254, 267n.7
Ames, Roger T., 89n.32, 90n.33, 90n.35, 90n.37
An, Yanming, 268n.32
Anderson, Robert M., 186, 195n.60, 195n.62
Andrew G., 49, 54n.18
Angel, Stephen, 90n.34
Ansari, Ali, 231, 250n.9
Applegate, Dan, 180, 185
Aristotle, 69, 71–73, 75, 85, 89n.26
Armstrong, Neil, 1
Arnould, Richard J., 148n.8

B

Babbage, Charles, 204, 225n.18
Bacon, Francis, 270
Baggins, Frodo, 29
Baltimore, David, 209–210
Baram, Michael S., 193n.18
Barber, Benjamin R., 267n.4
Barbour, Ian G., 246, 251n.41, 270, 274, 289n.4, 290n.17
Baron, Marcia, 192n.6
Bashir, Hassan, 267n.10
Basl, John, 250n.15
Battin, Margaret P., 91n.71
Baum, Robert J., 91n.64, 286, 291n.42
Bayles, Michael, 32n.30
Beardsley, Tim, 226n.33

Beauchamp, T., 193n.22
Becker, Lawrence C., 251n.38, 291n.44
Bell, Daniel A., 90n.41
Bell, Trudy E., 120n.25
Benjamin, Martin, 91n.74
Bennett, Scott, 169
Bergel, Kurt, 251n.35
Bertha, C., 53n.3
Bijker, Wiebe E., 278, 290n.29
Birsch, Douglas, 194n.44
Blackstone, William T., 242, 251n.30
Blake, Nigel, 291n.32
Blankenzee, Max, 187–188, 189
Bocong, Li, 268n.32
Boisjoly, Roger, 111, 114, 117, 120n.19, 120n.28, 196
Bok, Derek, 206, 225n.24
Bok, Sissela, 197, 198, 224n.3, 224n.5
Borgmann, Albert, 273, 290n.14
Borrelli, Peter, 119n.5
Bosso, Christopher, 250n.15
Bowie, N., 193n.22
Bradley, F. H., 80, 90n.55
Bradsher, Keith, 7, 31n.14
Brandt, Richard B., 58, 60, 88n.5
Broome, Taft H., Jr., 98, 119n.7
Brown, Marvin T, 192n.14
Bruder, Robert, 187–188, 189
Brummer, James J., 32n.36, 33n.49
Brundage, Miles, 290n.7
Bruno, Andrea, 193n.17, 194n.41
Bryant, Diane, 160, 178
Burke, R. J., 192n.12
Bustillo, Miguel, 250n.20

C

Cade, Ebb, 208
Cahn, Steven M., 193n.39

298

SUBJECT INDEX

Italicized page numbers indicate real or hypothetical cases in the text or discussion questions.